Rheinwerk
Computing

Have you been to our website?

For code downloads, print and e-book bundles, extensive samples from all books, special deals, and our blog, please visit us at:

www.rheinwerk-computing.com

Rheinwerk Computing

The Rheinwerk Computing series offers new and established professionals comprehensive guidance to enrich their skillsets and enhance their career prospects. Our publications are written by the leading experts in their fields. Each book is detailed and hands-on to help readers develop essential, practical skills that they can apply to their daily work.

Explore more of the Rheinwerk Computing library!

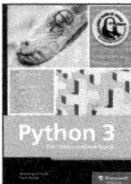

www.rheinwerk-computing.com

Jason Hodson

Applied Machine Learning

Rheinwerk
Computing

Editors Megan Fuerst
Acquisitions Editor Hareem Shafi
Copyeditor Rachel Gibson
Cover Design Silke Braun
Photo Credits iStockphoto: 1219460734/© andresr; Shutterstock: 1704315859/© shuttersv
Layout Design Vera Brauner
Production Kelly O'Callaghan
Typesetting III-satz, Germany
Printed and bound in the United States of America, on paper from sustainable sources

ISBN 978-1-4932-2758-7
1st edition 2026

© 2026 by:
Rheinwerk Publishing, Inc.
2 Heritage Drive, Suite 305
Quincy, MA 02171
USA
info@rheinwerk-publishing.com
+1.781.228.5070

Represented in the E.U. by:
Rheinwerk Verlag GmbH
Rheinwerkallee 4
53227 Bonn
Germany
service@rheinwerk-verlag.de
+49 (0) 228 42150-0

Library of Congress Cataloging-in-Publication Control Number: 2025053188

Contents at a Glance

Contents

4   Starting with the Data   55

7   Implementing, Monitoring, and Measuring the Model

8   Closing Thoughts

Preface

As I started my learning journey in the analytics space, I found a notable gap in the available resources. Introductory content is often too narrow in focus, and much of the existing material is not targeted toward early-stage learners. An important component of learning about analytics—and machine learning in particular—is developing a mental model of the connections among topics and themes. Therefore, this book is designed to help you understand which capabilities you should learn *and* how they fit together.

My goal with the following chapters is to give you a condensed master's-level class in machine learning that you can apply immediately. I've intentionally incorporated storytelling components to make the content engaging. Technical material can be dry, so my hope is that these stories help you work through and retain the information!

Who Is This Book For?

If you're already an expert in machine learning with experience using machine learning algorithms on the job, then this isn't the book for you. That said, this book is also not a comprehensive guide to machine learning. It focuses on explaining the practical applications of machine learning, creating a foundation for the end-to-end machine learning process.

The content in this book assumes that you have some basic knowledge of Python, which is the only main prerequisite. For example, understanding the general syntax of pandas will be helpful. If you don't have any experience with Python, I'd recommend consulting resources that best fit your learning style. A great introductory option is W3 Schools (*https://www.w3schools.com/*), where you can find high-level, Python-specific learning content for free.

This book was written with three primary personas in mind:

- Analyst with some coding experience
- Undergraduate student looking for applied experience
- Nontechnical leader of an analytics team

Analyst with Some Coding Experience

Across my various corporate roles at large companies like Walmart and Allstate, I've seen this persona in various departments, domains, and career stages. Data analytics careers tend to attract curious self-learners. However, the sheer amount of content on what's required for a successful data career can be incredibly overwhelming (speaking from personal experience). One recommendation to navigate this complexity is to find a technical mentor; see the upcoming text box.

Even with a technical mentor, I found it incredibly challenging to go from being a data analyst who knew how to write Python code to understanding how to properly leverage machine learning in my job. I was slowly building my knowledge with each capability in the process, but I didn't understand how everything fit together, which made it really challenging to understand the "why" behind what I was learning. It wasn't until my master's course in machine learning that all of the concepts began to fit together in my mental model.

I understand it's not always practical to invest the money and time needed to get a master's degree. If you fit into this persona, view this book as a fast and significantly discounted mini-master's class in machine learning. As a successful data analyst, you already know how important understanding the business context is to effectively fulfill your role. The same is true when leveraging machine learning. Understanding the business problem and the data is more important than mastering the machine learning algorithms. You can build the best machine learning model known to man, but if it doesn't answer the business problem, it's useless!

> **Practical Advice: Find a Technical Mentor**
>
> I personally benefited from a number of technical mentors who were more than gracious with their time. If you're early in your technical career, I recommend identifying someone who can help you on your journey. Ideally, this would be an individual within your organization, as they'll be more familiar with your company's data and technology. If you can't find a mentor at your company, look for someone with similar experience on a platform like LinkedIn. I think you'll find that people are happy to help!

Undergraduate Student Looking for Applied Experience

A common critique of entry-level analysts is that they don't understand the business application of machine learning (more on this soon). The use cases in this book serve as practical experience you can add to your code portfolio, providing you with talking points about applying machine learning to realistic business cases that you can use when interviewing for jobs.

This also enables you to showcase an end-to-end machine learning example. The use cases in this book span from understanding the initial business problem to putting a model to use. This adds to your interview talking points and can help differentiate you from others interviewing for entry-level data science or data analyst roles.

Working with Entry-Level Data Scientists

A consistent observation I have when working with junior data scientists is their interest in the technical problem, not the business problem. (If you're one of these individuals, don't take this observation the wrong way!) If you're in a research role in the data science and artificial intelligence (AI) field, this may be more appropriate. However, if you're working in a corporate setting, you must be interested in the business problem because that's where your role adds value.

I once worked with a junior data scientist who was incredible with Python and thoroughly understood the principles of machine learning and predictive modeling. However, their ability to understand the business problem they were trying to solve was a persistent challenge and they needed to be assisted continuously by senior members of the team. While junior employees aren't expected to know everything off the bat, it's important to combine your technical skills with an understanding of the bigger picture if you want to be successful in your career.

Nontechnical Leader of an Analytics Team

Many companies leverage rotational programs to give their leaders a well-rounded perspective. If you're someone without a technical background finding yourself leading an analytics team, you can view this book as a crash course on the machine learning process.

The code samples are likely optional for you. If you don't have any previous coding experience, now is not the time to learn how to write Python code. If you do have some previous coding experience, running through the code may give you some additional street cred with your new team and can help you build trust with them.

Maureen Kalas, a former data science leader at Allstate, is a favorite leader of mine, and she fits this persona. She was often my inspiration for making this content accessible to other nontechnical leaders leading an analytics team. Here is a note from Maureen.

Advice for Team Leaders

Occasionally, business leaders take on *stretch assignments*, finding themselves leading teams with skills far different from their own. You may be one of these leaders, experienced in a particular business area with an appreciation for the power of analytics, which has now led you to your new role. Perhaps you'll even lead the analytics team charged with solving the very business problems you faced before. This can be a perfect match!

The learning journey for a nontechnical analytics leader looks different from the path most data science leaders follow. You already have the business experience you need, but now you must determine just how much technical knowledge is required for you to be a successful leader. No one will look to you to code or to train junior analysts. However, you have three key responsibilities to your business partners, all of which require some level of technical knowledge:

- Understand the business opportunity and whether machine learning can help solve a critical problem.
- Understand how the machine learning model will work in the real world.
- Build trust and confidence in the solution with the business.

Business partners tend to be rightfully skeptical of new products. The nontechnical leader has an advantage in speaking the language of the business partner and understanding business processes. However, as the analytics leader, you'll be challenged on every aspect of the technical solution you propose—and if you can't sell your product, you'll lose your business partner's commitment.

So, what's the right level of technical knowledge to acquire? Your job will be to ask good questions—both of your analysts, to guide development of the model and help them understand the business problem they are trying to solve, and of your business partners, to help them understand the benefits and risks of the technical solution. Your goals should include understanding the data, learning how the model works at a high level, and, most importantly, knowing what will trouble business leaders most about the solution.

Where can you gain this knowledge? I recommend starting with a resource that brings together the complex topics of analytics and machine learning, identifies common themes, and illustrates how it all fits together. This book masterfully guides you through the most important machine learning concepts and can serve as a springboard to go deeper when you're ready for the next level. Take advantage of free resources to supplement what you learn here (including a refresher in statistics!), and of course, work with your team to better understand your specific applications.

The Structure of This Book

This book is divided into eight chapters. The first three contain introductory material to get you acquainted with machine learning and the use cases discussed throughout the book. Then, in Chapter 4 through Chapter 7, we dive into the technical content. The final chapter concludes the material and offers key takeaways.

Here is an overview of each chapter:

- **Chapter 1: Introduction**
  In this introductory chapter, we explain the value of applied machine learning and cover relevant topics such as learning how to troubleshoot code and understanding the impact of generative AI.

- **Chapter 2: Getting Started**

  In this chapter, we walk you through setting up the accounts you'll need to follow along with the book's content. This includes a free account with Anaconda, a cloud-based platform for writing and running code.

- **Chapter 3: Introduction to Our Use Cases**

  This chapter is, well, exactly what the title states. To help you understand the fundamentals of machine learning, the book employs three fictional use cases, which we introduce in this chapter. Each use case aligns to a specific dataset.

- **Chapter 4: Starting with the Data**

  This chapter is where we begin the focus of this book. We cover the most important concepts for preparing data for a machine learning model and apply them to the three use cases and their datasets.

- **Chapter 5: Picking Your Model**

  This chapter contains the core machine learning content of the book. We explore some of the most common models and work through examples and related considerations for each algorithm. After this chapter, you'll have a hands-on understanding of how these algorithms work. We don't dive headfirst into the statistics and math behind these models; instead, we focus on the practical components you should be aware of.

- **Chapter 6: Evaluating the Model and Iterating**

  This chapter focuses on error metrics and ways to interpret a machine learning model's output. The error metrics are our way to keep a model in line with what we're trying to get it to do. While this chapter falls more on the math side of this book's content, it's a critical component for building effective models. We'll also cover how to interpret a machine learning model, which is incredibly useful in practice and also a fun peek behind the curtain of the model.

- **Chapter 7: Implementing, Monitoring, and Measuring the Model**

  This chapter focuses on generating predictions from your model, as well as steps to consider after your model is built. This includes monitoring your model as well as measuring its impact from a business perspective.

- **Chapter 8: Closing Thoughts**

  This chapter wraps the content in a bow and highlights key takeaways from the book.

Downloadable Supplements

You can find downloadable supplements for this book at *www.rheinwerk-computing.com/6170*. This includes the code examples (also available at *https://github.com/jason-hodson/applied_machine_learning/tree/main*) and, for readers of the print book, full-color versions of any key visualizations.

Conclusion

I hope these first few pages have given you a taste of my writing style. While the book will focus on practical advice, I also try to sprinkle in some humor (and the occasional sarcastic comment) to keep things light.

Much of this book is written through my perspective—one of someone who lives in a hybrid world between data science and business. I've found this space to be a niche that allows me to quickly add value in an organization, so this book is my way of sharing how I've done this using machine learning techniques. Now, let's begin with Chapter 1!

Chapter 1
Introduction

Welcome to an introduction to applying machine learning! In this chapter, we'll begin our journey toward a foundational understanding of the application of machine learning. Everyone needs to start somewhere, so let's dive in.

Machine learning and artificial intelligence (AI) have had no shortage of hype over the past few years. Amidst all the buzzwords and headlines, it's easy to be overwhelmed by the potential of these technologies. In this book, we're focused on practical application—how you can begin to use machine learning *today*, as part of your daily work.

This chapter will set the stage for the rest of the book by addressing some key questions about terminology and how internet searches and generative AI fit into the bigger picture of machine learning. Let's get started!

1.1 Aligning on Nomenclature

If you've already started exploring free online resources about machine learning, you've probably noticed that the same topic can appear under multiple names. Here's some detail on why this is, for those who are curious.

The analytics community has strong ties with the scientific community, and the scientific community places great value on open-source material. This is an admirable quality: In their view, making information readily available increases innovation of and progress on important issues. The downside to this is that as innovation occurs, there isn't necessarily one individual or company driving the innovation, which creates deviations in naming conventions.

To compound this issue, companies often take open-source technology, build on top of it, and then sell it. This creates additional naming conventions for the same or similar concepts. American capitalism is alive and well.

> **Example: OpenAI**
> OpenAI is a great example to demonstrate the complexities of open-source software. Their product ChatGPT became a sensation and brought the concept of large language models (LLMs) to the general public. The interesting nuance with OpenAI is they've become an organization with two distinct entities.

The first is a nonprofit with the sole focus of research, which is the entity contributing open-source material. This is the behind-the-scenes research advancing the field of AI, not the ChatGPT product you're likely familiar with. The second is their commercial entity, which is focused on using their open-source research to build commercialized products for businesses and consumers. This part of the business delivers ChatGPT with the goal of making money for the company.

While OpenAI sells ChatGPT as a product, other companies can use OpenAI's open-source material to build their own products on the same technology. Microsoft has done this with their Copilot product. Walmart has done this with their internal LLM for associates. Each product subsequently develops its own nomenclature for similar features.

While analytics terms are continuously evolving, we'll focus on a handful of key data terms throughout the book, as listed in Table 1.1. My hypothesis is that almost no one reads a book's appendix, so I've included Table 1.1 right at the beginning of the book to make sure you read it.

Terminology	Definition	Synonyms
Row	One line of data in the table or spreadsheet	Observation, record
Variable	Set of data that represents information about the same topic or item	Column, attribute, field, variable
Machine learning	Algorithm that generates a prediction	AI, analytics, predictive modeling
Analytics	General terminology referring to the use of data to make business decisions	Evidence-based decision making, business analytics, data analytics
LLM	Model that takes a prompt as an input, then outputs an answer	ChatGPT, Copilot, Gemini
Library	An aggregated set of functions that perform specific tasks, making the use of Python easier than writing all the raw code yourself	Package, functions, code
Dummy coding	Converting a categorical column into a numeric column by creating a new column for each value of that categorical column, with either a 1 or 0 to indicate the value	One-hot encoding

Table 1.1 Common Terms in the Machine Learning Space with Many Names

1.2   Learning to Google (or Prompt)

An underlying skillset in leveraging machine learning is learning how to search for the right answer to your problem. Traditionally, this meant learning how to Google—knowing which websites were the best references and how to navigate their respective interfaces. With the emergence of LLMs as a search tool, you now have more options, as we'll discuss in this section.

1.2.1   What Can You Find with Google?

There are a lot of resources online, so knowing which resources are probabilistically the best is important. The Google algorithm may do some of this for you, but knowing which websites to focus on will help expedite your research and troubleshooting process. For example, some resources help you gain a general understanding of the issue while others are better at providing answers for specific use cases or questions. The resources that we'll discuss are by no means extensive, but you'll likely see them come up on a recurring basis.

We'll use a simple search of "how to build random forest model sklearn" as an example. When you search for this on Google, you get the results shown in Figure 1.1. We'll break down these results in the following sections.

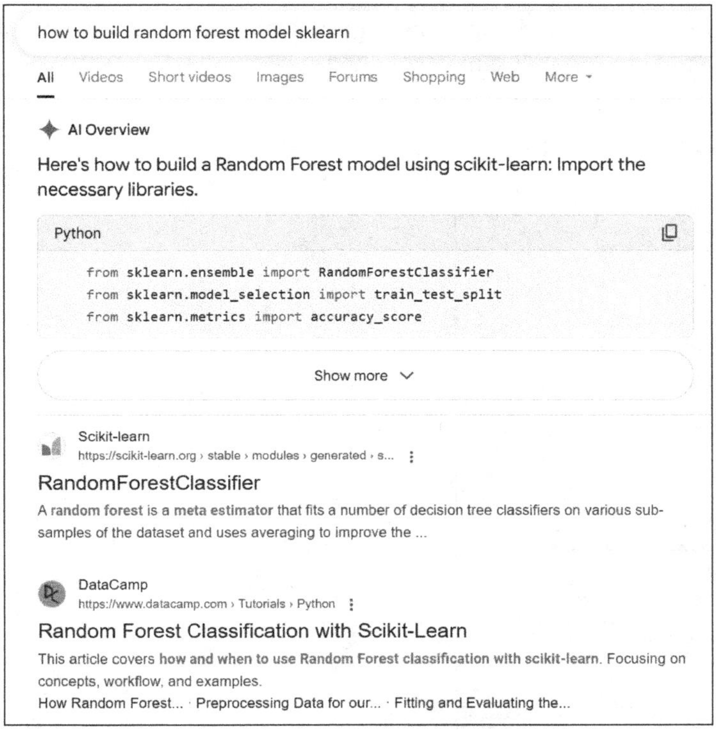

Figure 1.1 Google Search for Random Forest Model

> **Scikit-learn: Python's Machine Learning Swiss Army Knife**
>
> If you aren't familiar with scikit-learn (also known as sklearn), that will change. sklearn is the most common machine learning library in Python. It has a vast array of tools and models, making it a great tool for learning the different applications of machine learning. The models we'll explore in this book are all from sklearn, which has some significant benefits that you'll see later in the book.

Generative AI Result

Even as of writing this book in 2025, the generative AI results are impressive. They give you all the necessary code to build a random forest model. However, a word of caution here: Generative AI is not always right. This is why understanding how the code works is important. Getting to a place where AI can supplement and speed up your coding process will be much more valuable than solely relying on AI to write the code.

After expanding to view the full result (see Figure 1.2 and Figure 1.3), you may be thinking, "Why did I pay for this book if I can just Google the results?" It's a fair question, but here are a few things to consider:

- These results have no idea about your data (see our discussion of prompting in Section 1.2.2).
- If you don't already have the underlying knowledge of a random forest model and how it works, your ability to correctly edit it is limited at best.

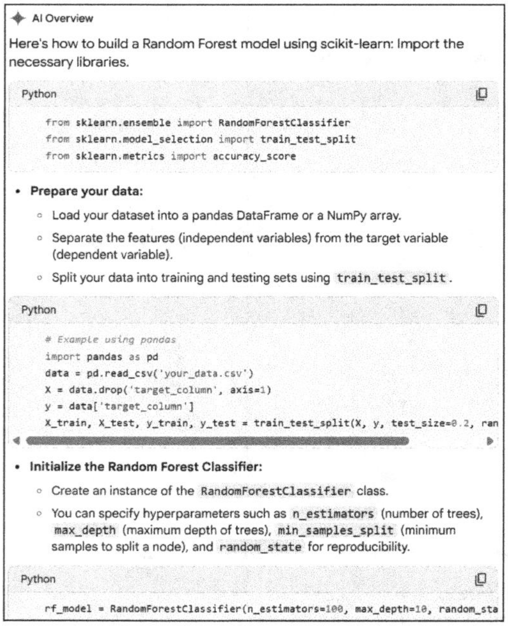

Figure 1.2 Google Gemini AI Results for Building a Random Forest Model (Part 1)

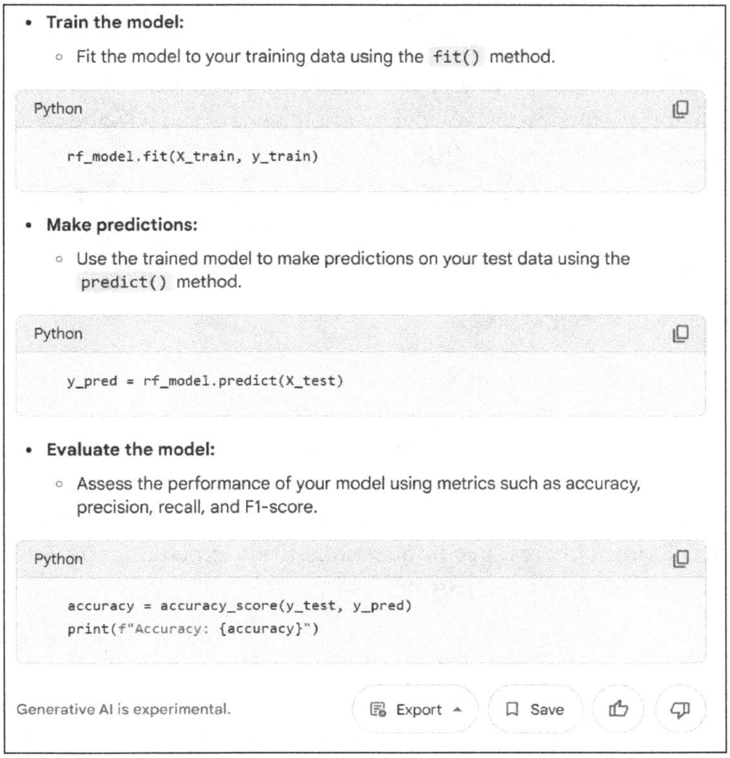

- **Train the model:**
  - Fit the model to your training data using the `fit()` method.

Python

```
rf_model.fit(X_train, y_train)
```

- **Make predictions:**
  - Use the trained model to make predictions on your test data using the `predict()` method.

Python

```
y_pred = rf_model.predict(X_test)
```

- **Evaluate the model:**
  - Assess the performance of your model using metrics such as accuracy, precision, recall, and F1-score.

Python

```
accuracy = accuracy_score(y_test, y_pred)
print(f"Accuracy: {accuracy}")
```

Generative AI is experimental.   Export ▲   Save   👍   👎

Figure 1.3 Google Gemini AI Results for Building a Random Forest Model (Part 2)

So, while Google's AI result may seem extremely convenient, it should be considered with caution. As a rule of thumb, avoid blindly copying and pasting this code, as it almost always requires edits.

Python Library Documentation

You'll often see the actual Python library documentation in your search results (the first non-AI summary result in Figure 1.1). The quality of documentation can vary, and it can be dense at times, but the libraries leveraged in this book all contain robust and high-quality documentation. The packages that don't have the best documentation are often the ones with less usage, which usually means they have niche applications. For example, network analysis in Python doesn't have an established, go-to library like data manipulation has with pandas.

The developers and maintainers of the library usually provide simple examples of how to use each function. These no-frills examples are especially helpful when you're using new functions or packages. As you become more familiar with each library's documentation, you'll find that many of the other sites that appear in your Google search simply use the library's documentation examples. This is why we recommend starting with the library's documentation.

> **What Is a Python Library?**
>
> The concept of a Python library may seem very abstract. Python is considered a *high-level* programming language, which means you don't need to understand how the core Python language works to be able to code with it.
>
> Python libraries are what make Python accessible to individuals who aren't computer programmers. The libraries are built by writing Python code and packaging it for others to use. It cannot be overstated how much time this saves you when building a machine learning model.

Stack Overflow

You'll notice that Stack Overflow, a community-based platform where users can post questions and answers, isn't in our search view—it was six additional sites lower in the results. Stack Overflow seems to rank higher for queries with more specialized use cases. The nature of this platform invites specific questions, likely explaining why Stack Overflow isn't showing up in our top search results.

GeeksforGeeks

GeeksforGeeks, a personal favorite, has been one of the most helpful resources. The platform is article-based, so it's a middle ground between more specific questions (Stack Overflow) and really general information (library documentation). The biggest limitation is their concise, blog-post-like format, so you get more polish and clarity at the expense of content.

Reddit

Reddit is a place to find really niche answers. However, my success rate at finding the answers I'm looking for in the machine learning space is relatively low. If you're a general Reddit user, I promise this isn't a dig on the entire platform. (I'm aware of the passion of the Reddit user base!)

Paid Sources

There are a number of paid sources that you'll likely come across. One example you see in the search results in Figure 1.1 is DataCamp. They have some free resources available at the time of writing, but the bulk of their content is behind a paywall. I personally love DataCamp and recommend it to people who are getting started in programming. However, their content leans into the category of course learning versus ad hoc research and a source for questions.

Towards Data Science and Medium are blogs with great content behind a paywall. In my experience, though, there isn't a noticeable difference between GeeksforGeeks' content and these two paid sources.

Specificity of Python Resources to Answer Questions

It can be helpful to consider these resources based on where they fall on a scale from highly generalized (1) to highly specific (5):

1. Library documentation
2. AI summary
3. GeeksforGeeks
4. Stack Overflow
5. Reddit

1.2.2   Prompting

The speed at which generative AI progresses is enough to make your head spin, so this section of the book will probably age poorly very quickly. I wrote the content for this book over the course of 2025. At the beginning of my writing process, many companies were still working through the security implications of enabling their workforces with generative AI, and the conversational interface was the main way to interact with ChatGPT. By the time I was revising my final draft, OpenAI had released their own web browser, and most companies were well on their way to integrating AI into their processes. The impact of LLMs and their ability to assist you in writing your code has a lot of potential. Throughout the book, you'll see examples of how I use generative AI when writing code and working through machine learning problems. Here are some considerations for leveraging generative AI tools when working on machine learning projects:

- **Use them a research partner**
  While LLMs don't always have the most up-to-date information, they can act as your research partner when starting a project. Using LLMs to help you with initial research on your machine learning project can be helpful, though you need to be aware of your company's data privacy before doing so. A prompt such as "I'm being asked to build a machine learning model to predict the retail sales for my company. What type of data should I consider including in the model?" can help you jump-start the data acquisition for your machine learning project.

- **Use them as a starting place for code**
  It's helpful to use an LLM as a starting place for your code. Pretty much any machine learning project will require you to write custom code. However, as you build more models, you'll find that modeling is repetitive. The custom code is almost always specific to preparing your data.

1.3   Predictions for Generative AI's Impact on Machine Learning

Before we move forward, let's take a moment to think about the future impact of generative AI and how it will shape the landscape of machine learning. If the following predictions are wildly off base by the time you're reading this, you're welcome to send me a LinkedIn message to remind me of how poorly I predicted the future:

- **Generative AI will have a market correction.**
  Ever since OpenAI launched ChatGPT to the public in November 2022, companies have gone all in on generative AI. There will most likely be a market correction on the use of generative AI. Eventually, corporations may realize that they cannot use machine learning to reduce their headcounts as much as they thought (or perhaps hoped).

- **Generative AI doesn't have all the necessary inputs to fully automate roles.**
  The basis of all machine learning is inputting the necessary data to create a useful model. The complexity of enabling an LLM with every possible data point to automate every employee's role is simply not cost-effective. This is especially true for small and medium-sized companies. The cost of purchasing licenses to all the necessary data and putting it into a clean data structure is significant.

  Another layer to this is the companies that host platforms may make it more expensive and challenging to extract data from those platforms. Take LinkedIn as an example—Microsoft makes it incredibly difficult for companies to directly connect to LinkedIn's data with an application programming interface (API). It's reasonable to expect an increased frequency and intensity of these examples as companies seek to profit from the data they're generating.

- **A generative AI correction highlights the need for human problem-solving.**
  Corporations are realizing that they still need employees, as people can handle problem-solving tasks that most company data infrastructures cannot. This highlights the importance of being a curious and critical thinker. In a machine learning space, you should be highly skilled at working with business leaders to understand their current challenges. Being an expert coder will likely have lower and lower returns.

1.4   Summary

This chapter outlined common terms and definitions, ideas for troubleshooting questions and problems, and predictions on generative AI. With these building blocks in place, it's time to get acquainted with the key tools that you'll use for our applied machine learning scenarios.

Chapter 2
Getting Started

This chapter walks you through setting up the accounts you'll need to follow along with the use cases we'll discuss throughout the book. This is your first step toward hands-on experience with building a machine learning model.

This book aims to be practical and provide hands-on examples you can leverage in job interviews. The majority of readers can use both GitHub and Anaconda to follow along with the examples. By committing your changes to GitHub, you can include the repositories on your resume and LinkedIn profile. Using Anaconda allows you to follow along with the code samples while limiting the frustrations that can come with downloading and maintaining software on your computer.

This chapter will show you how to sign up for both GitHub and Anaconda. We'll go into additional detail about why these tools are used in the book, the value of learning them, as well as the practical uses of each solution.

2.1   GitHub

GitHub is arguably one of the most used platforms in the programming world. If you haven't used it before, think about it as a centralized location to store your code.

If you haven't done much programming, GitHub may seem arbitrary and unnecessary. If you're working on a small project, then yes, it may be a bit arbitrary. However, once a business begins using your model on a large scale, what happens when you get feedback about a new data source that needs to be added? How do you test these changes while keeping the existing model working? If you're like me before I used GitHub, my answer was "file_name_v2" or "file_name_testing." This can become quite a challenge to manage, and it's also an automatic sign to a hiring manager that you don't have much real-world experience.

Looking for some more motivation to use GitHub? Here are a few practical uses of it in the real world:

- **Version control**
  Version control is a very important concept in any machine learning context. Imagine a situation where you've updated and tested a model and believe it's working

perfectly. However, a few days later, your stakeholders find a significant issue with the update to the model. If you're using version control, you can simply roll back your changes to the last working version of the model.

- **Reference for production code**
  Some platforms that allow you to write, run, and save your code are glorified file systems. When you're writing code, you should have the most recent version running in production. GitHub enables this process by preventing code you're still developing from impacting any code that's currently being used. As your code gets more complicated, the value of this increases. In a machine learning context, this means the code creating your predictions is stored in GitHub, not in your file system.

  This nuance may seem arbitrary, but again, as you're building on larger scales, this type of differentiation is critical to reducing the risk of human error. For example, say you have a model predicting sales and your stakeholder wants to add a new data source to the model that they believe will improve its accuracy. You don't know how long it'll take to make this change, but you know the current model needs to keep running. With GitHub, you can develop the new code on either your computer or the platform you're using to code. These changes have no impact on the code in GitHub until you specifically want it to. (This is called committing and pushing your code.) If the code running the model your stakeholders are using references what's on GitHub, then you're able to fully develop and test new improvements without impacting the current process.

- **Avoid skepticism in job interviews**
  Having base-level knowledge of GitHub won't get you a job, but not having any working knowledge of it could result in you not being selected for a role. In the machine learning space, GitHub is so common that not having used it can be seen as a red flag.

- **Showcase your work**
  Giving a hiring team insight into your work and how you write code is a great way to differentiate yourself. Not only does it reduce uncertainty for the hiring team, but it also shows you took initiative beyond your day job in pursuing the craft. A definite win-win!

Hopefully, you're now convinced of the benefits of GitHub. Let's get your account set up and check out how we'll use GitHub to support your learning throughout the book.

2.1.1   Creating an Account

To prevent these instructions from becoming outdated, we'll keep our discussion of account setup general. Here's hoping the user experience (UX) for GitHub's account signup doesn't degrade!

Follow these steps to set up your account:

1. Search for "create GitHub account" on your search engine (or navigate to *https:// github.com/*).

2. Find the link that directs you to creating the account, as shown in Figure 2.1.

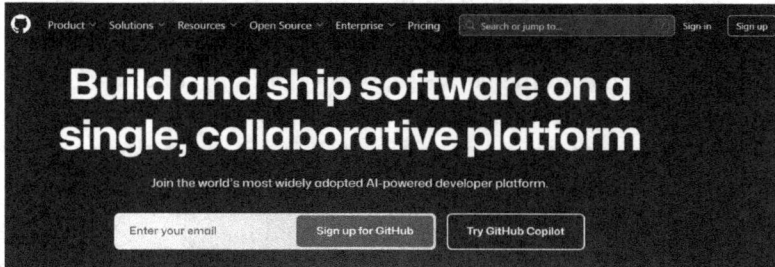

Figure 2.1 GitHub Home Page with Sign Up Button

3. Input your email, desired password, and desired username on the sign up screen shown in Figure 2.2. You may have to try a few different combinations since these cannot be duplicated across users.

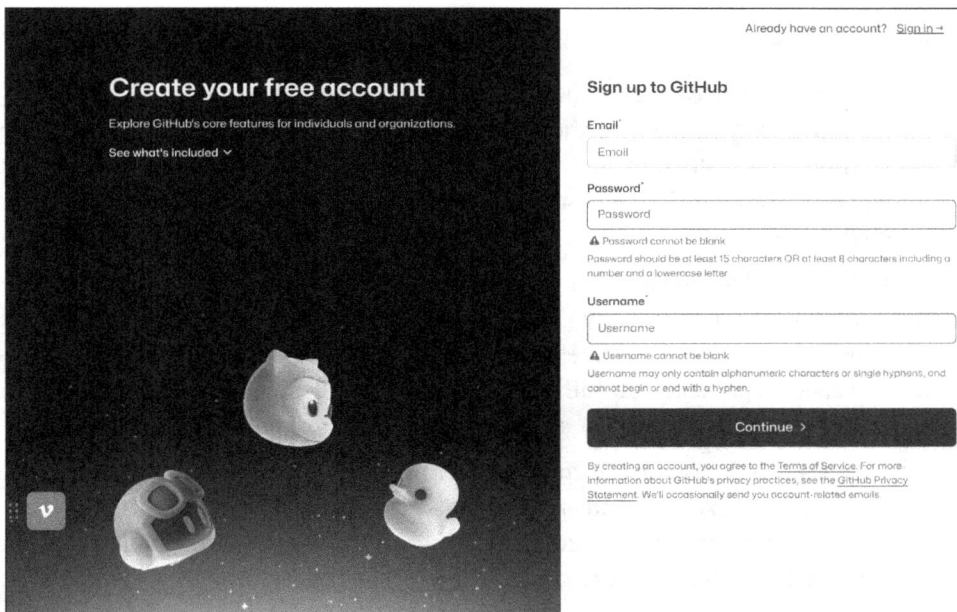

Figure 2.2 GitHub Account Creation Page

4. There will be some degree of multiple authentication required. As of the time of writing, a code is sent to the email you provide to authenticate the account.

5. You now have a GitHub account!

2.1.2   GitHub in This Book

Simply creating the account may have felt arbitrary (because it was). The goal of this book is not to make you an expert in using GitHub. If you're looking to learn more, GitHub has a number of free resources on their homepage that will provide you with more detail.

The code in this book is available in the book's repository (*https://github.com/jason-hodson/applied_machine_learning/tree/main*). You can create your own repository where you'll upload the code you write using this book. This can also become practice to create future repositories that you can share as part of future interview processes.

I've repeated this a few times, but research suggests that something needs to be said five to seven times to ensure it's retained (*http://s-prs.co/v617000*). Skipping the GitHub portion of these chapters may seem like an easy way to save time, but these use cases can become content you can bring to interviews to highlight what you've learned. It's much more impactful to show a hiring team a repository of code you've created than just saying you've read a book about machine learning!

2.2   Anaconda

Learning how to write code and build machine learning models is hard enough. However, making sure you have all the right libraries and dependencies makes it even harder. When I started learning to write R and Python code, issues with my libraries not aligning correctly were the biggest culprits behind my urge to throw my computer through the window.

All libraries have dependencies, or another library (often many libraries) that they're dependent on to work correctly. This gets really complicated when libraries have updates with new features, bug fixes, and so on. There's often a range of compatible versions of a specific package that fulfill the package dependency. Across a large number of libraries, this gets *very* complex and challenging to manage.

Anaconda is a cloud-based platform data scientists use to develop and run code. The magic of a platform like Anaconda is that it will manage these dependencies for you—a behind-the-scenes feature that's so incredibly useful. Without this magic, libraries are just folders and files in a directory, and it's all on you to understand the dependencies.

Library Version Hell

Early in my exposure to Python code, I worked with a team of data scientists. They had built a predictive model to score resumes, which was a complicated process with many moving parts. There were a number of libraries required to complete all of components of the model.

A required update to one of the libraries eventually created a trickle-down effect of broken library dependencies. Unfortunately, their environment was managed outside of Anaconda. The result of this package dependency breaking was four entire working days dedicated to fixing the environment issue.

When managing these dependencies manually, you have to run your code to identify whether or not the code will work. With a highly complex model such as this one, there was a lot of code to run and iterate through until the dependencies were all worked out.

Since this book is not focused on Python and library management, you can stick to Anaconda's free light cloud version. You won't need to install any large files to have access to the software and its features (the Windows download for Anaconda is almost 1 GB).

If you already have Python and an integrated development environment (IDE) on your computer or your company has a machine learning operations (MLOps) platform, you should be relatively safe to proceed with without Anaconda. The only section you may run into version issues is in Chapter 6, Section 6.5 with a library called Shapley Additive exPlanations (SHAP). For everyone else, we'll get started with Anaconda and discuss how we'll use it in this book in the following sections.

IDE, MLOps, and SHAP

If you read the last paragraph and immediately went to Google, you're not alone. For now, know that SHAP is a Python package. It will be discussed at great length in Chapter 6, which covers machine learning interpretability.

An IDE is where you write and run your code. Examples of this include Jupyter Notebooks and Microsoft's Visual Studio. An IDE is critical to effectively writing your code. Without one, you'll be running your code from a command line, and *no one* wants that. If you don't know what a command line is, pull up your Windows search bar and search for "cmd." The black box that shows up is the command line—not exactly an ideal user experience for programming.

MLOps platforms contain at least one IDE (though they often have multiple), but they also have other integrated capabilities such as scheduling. In practice, these are almost necessary to effectively bring a model into full use. Examples of MLOps platforms include Domino, Databricks, Azure Machine Learning, and Amazon SageMaker.

2.2.1 Creating an Account

Anaconda is not quite as universally known as GitHub. For that reason, we'll be a bit more explicit and ensure you're able to correctly and confidently identify what you're downloading.

Start by Googling "Anaconda" (see Figure 2.3).

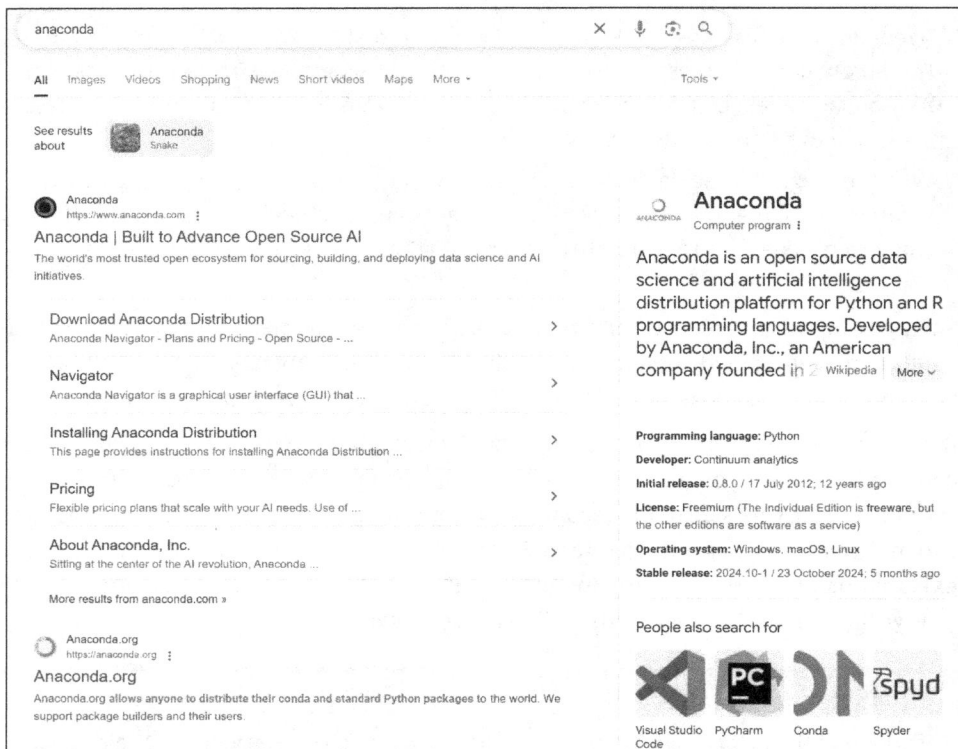

Figure 2.3 Google Search for Anaconda

Click on the **Anaconda** link and you'll arrive at their home page that prompts you to create an account, as shown in Figure 2.4.

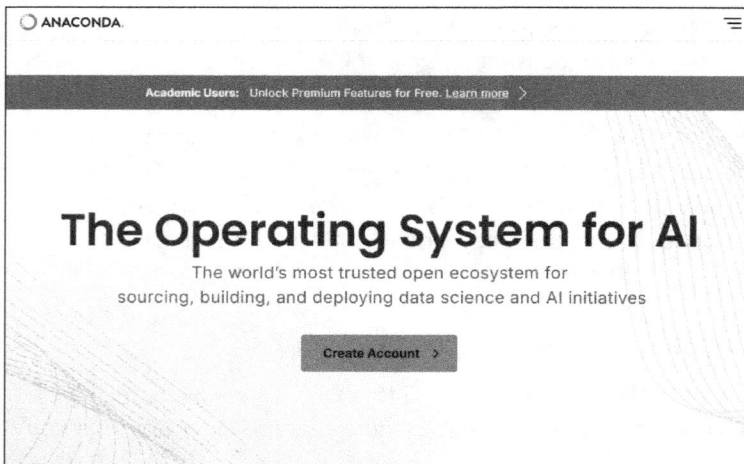

Figure 2.4 Anaconda Create Account Page

After creating your account, navigate to the **Notebooks** tile, as shown in Figure 2.5.

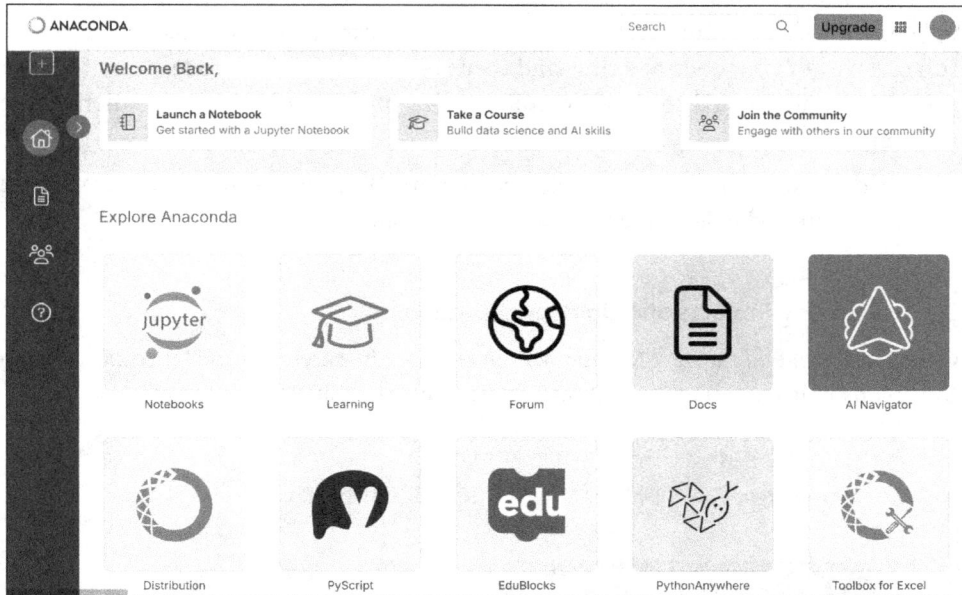

Figure 2.5 Anaconda Cloud Page

The **Notebooks** tile will open up a Jupyter Notebook for you to start writing your code, as shown in Figure 2.6.

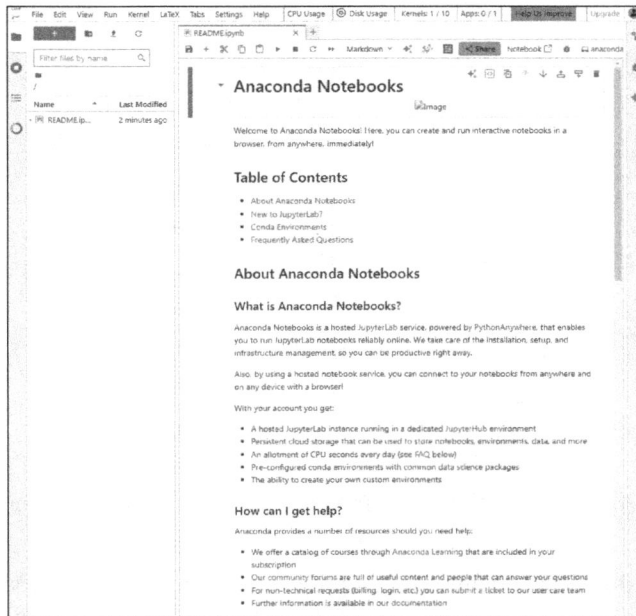

Figure 2.6 Anaconda Notebook

This is where you'll follow along with the rest of the book! Jupyter Notebook is one of the standard IDEs used by people writing Python code. As you'll see when you start writing code, it's broken into *code chunks*, which are blocks of code that allow you to only run one part of your code at a time and then see the result of the code. Even if it's not clear to you now, you'll see how helpful this way of writing code can be when you're learning.

At the time of writing, Anaconda includes a helpful README file that you can reference if you're interested in learning more about Anaconda notebooks.

2.2.2   Creating Projects and Uploading Data

Once you've opened up the **Notebooks** tile and see the screen from Figure 2.6, click on the green circle button on the left and select **Manage Projects** (see Figure 2.7).

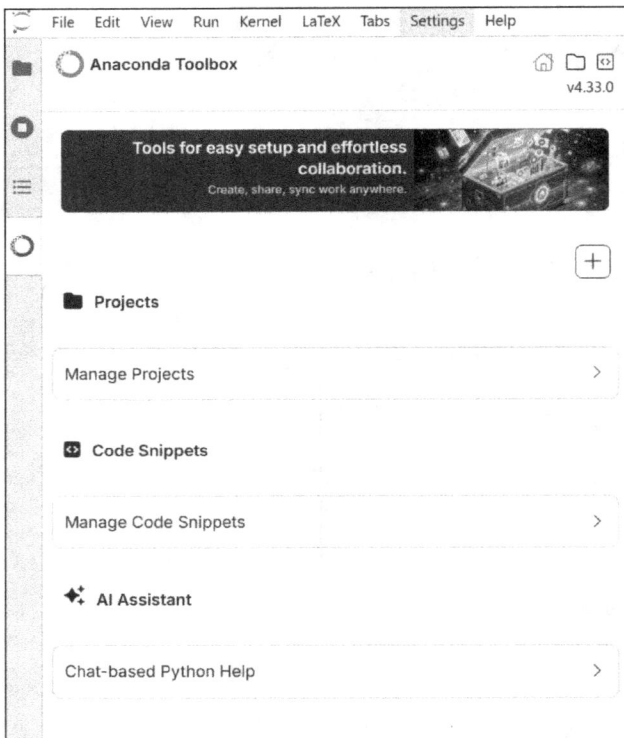

Figure 2.7 Open Anaconda Project Pane

Once on this screen, click the **+** button and then **New Project**. Name your project as you wish.

Once you create the project, you'll add two items to the project (see Figure 2.9). The first is a notebook; click the icon with three dots and choose **New Notebook**. You can add notebooks as you work through the use cases and examples throughout the book.

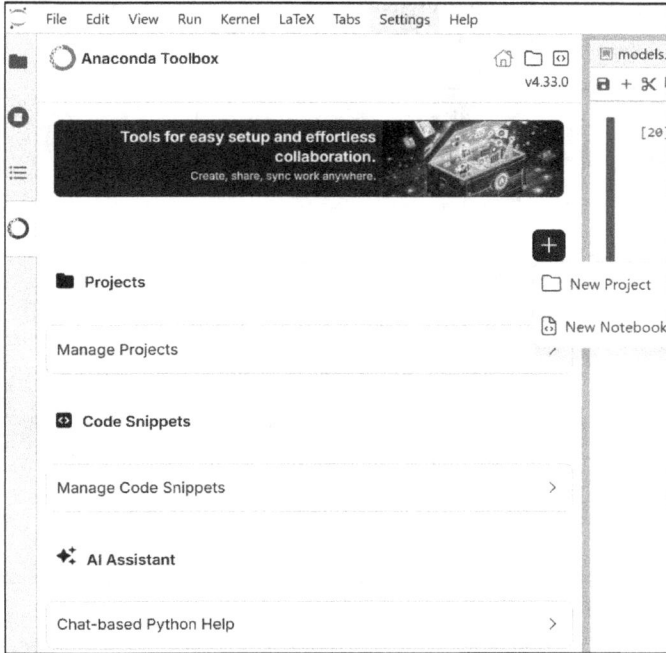

Figure 2.8 Create New Project

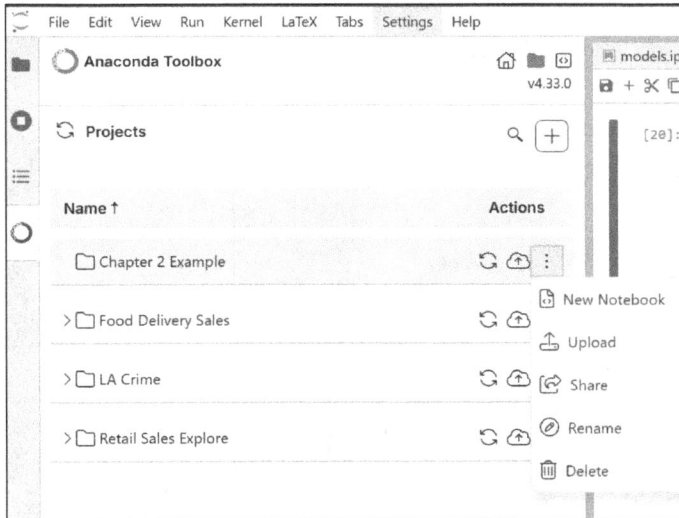

Figure 2.9 Adding Items to Project

The second item to add is the data for our use cases. These are stored in the GitHub repository for the book. We'll walk through one example of doing this. The first step is to download the file from GitHub, as shown in Figure 2.10. Navigate to the GitHub repository for this book (*https://github.com/jason-hodson/applied_machine_learning*).

From there, navigate to the data files. Under each use case folder, there is a folder called *data* that contains all of the data for each use case.

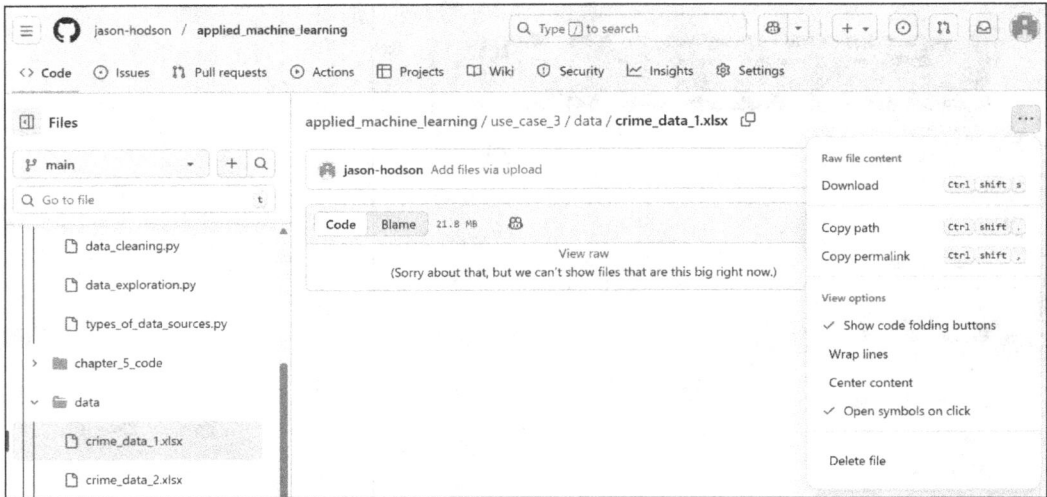

Figure 2.10 Download File from GitHub

Once you've downloaded the file from GitHub, you can then select the **Upload** option (refer back to Figure 2.9) to upload the data to Anaconda, which will take you to the screen in Figure 2.11.

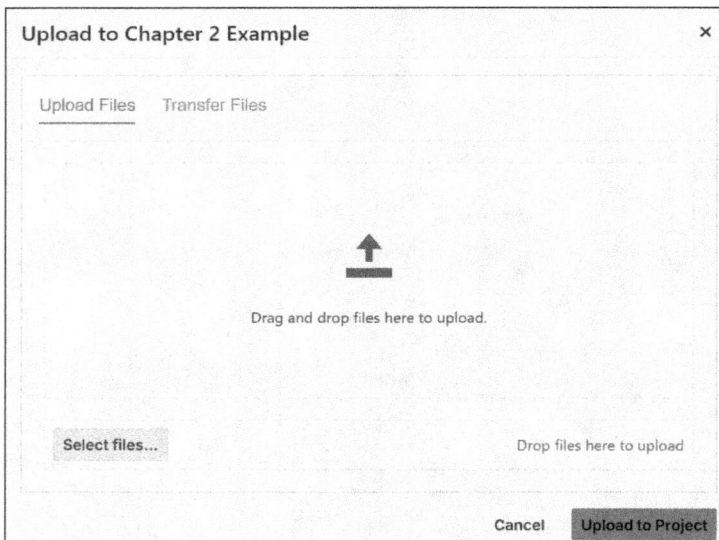

Figure 2.11 Upload File to Anaconda

You'll need to do this for each data source for each use case. You can choose to either do this right away or wait until you start working on the use cases in Chapter 4.

2.2.3   Anaconda in This Book

As mentioned previously in this section, using Anaconda's cloud notebook to run Jupyter Notebooks simplifies your setup process. Once you have an Anaconda account, all the code and examples will be run through notebooks on Anaconda's cloud service. As a reminder, if you already have another means of writing and running your Python code, that will also work! For those who don't, Anaconda keeps your focus on the machine learning content rather than on downloading software.

However, free services do have drawbacks, as we'll discuss in this section. These limitations are something you need to be aware of when we start getting into the code in Chapter 4.

Should You Use the Paid Version?

All of the code from this book can be run with the free version of Anaconda. However, it may be worth considering the paid version while you work through the book. The cost at the time of writing is $15 per month. What you'll get from the paid version is more storage (less of a concern) and more memory (more of a concern). As you'll see in our discussion, memory usage needs to be front of mind with the free version. This can be a distraction when it comes to learning. While either version will work, know the paid version is an option (and no, I'm not getting any kickback by including this).

Storage

At the time of writing, Anaconda provides you with 5 GB of storage for free. For the purposes of the use cases in this book—and most coding practice projects—this shouldn't be a limitation you frequently need to account for. When you get into the real world, though, 5 GB of data is quite small. Most enterprise cloud solutions have hundreds of GB of data storage available by default, and even personal computers have more than 200 GB of storage. At the time of writing, Microsoft's 365 personal account has 1 TB of storage. The point is that storage is rarely your limitation, regardless of the platform, given how cheap it is to store information.

A warning based on personal experience: Avoid saving copy after copy of your data. You might be tempted to do this if you're testing a lot and want to save your data as a comma-separated values (CSV) or Excel file and then export it to view in Excel. However, if you're creating multiple iterations and copying all your data, this level of duplication can eat up storage quickly.

Everyone's brains work differently, so you don't necessarily need to stop doing this. The flexibility of reviewing data in Excel is a great approach. However, my advice is to stop hoarding the files and delete them after you're done with them. It's quite unlikely that you'll need all those files in the future (you likely have no idea what filters or adjustments were made to create them), so just delete them occasionally. Even if the available storage isn't a concern, having a ton of files to look through and not knowing which one is the right one can be frustrating for anyone you're working with—including yourself.

Memory

The available memory in the free version of Anaconda is something you'll almost certainly need to keep in mind. Memory is more expensive than storage, so the limitations are generally easier to reach. If you're not familiar with memory, it's the processing power of a computer or server. The more memory, the more complex calculations can be performed.

On the other hand, the computational power of enterprise-grade cloud computing can sometimes mask inefficiencies and bad practices. A beefed-up environment can handle more computationally heavy tasks, so the only feedback is the longer time it takes for your code to run. In an Anaconda environment, there is a much lower ceiling; testing for the use cases in this book crashed the environment multiple times.

Crashing in this case means your memory limit has been reached, and the kernel (the computational engine executing the code) needs to restart. All of your code remains the same, but any code you've run will need to be rerun. Check the last code you ran to avoid continuously needing to restart the kernel.

Knowing what actions are computationally expensive can take some time to figure out. Generally, less computationally expensive tasks include the following:

- Filtering a DataFrame
- Concatenating two DataFrames
- Aggregating or grouping data

Actions that tend to be more computationally expensive include:

- Merging or joining data on large datasets
- Performing transforms that require the majority of the dataset to be considered (e.g., dummy coding)
- Saving data to a CSV or Excel file

The use cases in this book will highlight many computationally expensive tasks and how to avoid them, along with their alternatives.

2.3   Summary

You now have access to the tools we'll be leveraging throughout the book. These will be important both for learning the content but also for creating examples and talking points for yourself in interviews. Now, we'll get into the last nontechnical chapter and introduce you to the three use cases for the book.

Chapter 3
Introduction to Our Use Cases

Many learning resources use simple examples in perfect scenarios, making the learning curve of applying machine learning more challenging, and in the worst case, misleading. In this chapter, we'll focus on concrete use cases that demonstrate the realities of using machine learning in practice.

The following three use cases not only introduce you to machine learning but also help you understand the realities you'll encounter when you start applying it on the job.

If you're already working in an analyst role, you'll likely already be familiar with some of the personalities in each use case. If you're a student, these examples aren't meant to scare you! They should give you an idea of what you'll encounter once you're applying what you've learned in the real world.

Throughout the chapter, there are also sections of questions and answers. Some of these questions are subjective, so I've also enlisted some friends and colleagues to provide additional perspectives for these particular scenarios. My hope is that this makes the content more engaging, entertaining, and informative!

The following sections introduce each use case and provide important context for each application of machine learning. Additionally, there will be real stories from myself and others to contextualize and emphasize the importance of these topics.

3.1 Importance of Understanding the Business Problem

Mastering machine learning is as much about understanding the business case as it is about the models and techniques. The following sections present anonymized real-world stories that demonstrate the importance of understanding the business problem (adapted from personal lessons learned over my career). These are not exclusively machine learning examples, but they all focus on a misalignment in communication that makes it impossible for the analytics and data tasks to be successful.

3.1.1 Business Reviews

Let's consider Peter, who was new in his role as an analytics consultant and working with a leader that many of his peers struggled to support. For the sake of the example, we'll call this leader Tom.

Tom was also relatively new to his role. In all fairness, he was put in a difficult position with his stakeholders. They placed a number of expectations on Tom with little support for him to meet those expectations. This required Peter to expand beyond simply working with the data and become an internal management consultant who helped Tom identify and solve data, process, and communication problems.

Tom would openly acknowledge that he can be difficult to work with. Much of what he was looking to deliver was in the form of PowerPoint decks. The scope of these decks was always changing, and Peter kept spinning his wheels. For the better part of two years, each quarter they'd go through and spend an absurd amount of time creating the decks, editing them, and repulling data.

Towards the end of the two years, things finally started clicking: Peter needed to ask better questions earlier in the process. It wasn't enough to simply take a directive and do the work. The missing piece was his own understanding of the exact purpose of the decks and how they were being used—in essence, the business problem. While there may have been some tension around asking for more specific details at the beginning of the project, Tom would have certainly appreciated the better output and time savings.

3.1.2 Definition of Success

Now, let's consider Michael, who was supporting a university recruiting team within talent acquisition. One of their main focuses was recruiting for, as well as managing, the internship program. As Michael learned more about their team and the internship program specifically, his skepticism about the return on investment (ROI) for the program grew. It required a significant amount of money to recruit and build relationships with the organization, and the time investment for regular employees at the company was significant, as they needed to take time away from their day jobs to coach and develop the interns.

Michael was never asked to measure the ROI for the university recruiting teams, but he was curious about it. While curiosity can be a great attribute in the data space, it can also get you into trouble if you don't take the proper steps.

He decided to do an analysis and then present it to the head of the university recruiting team. He dug into the data and created a novel analysis that hadn't been done before. The data said the company was just as well-off hiring entry level talent instead of recruiting, hosting, and converting interns into full-time employees. He had been proactive, done the proper analysis, and effectively communicated the results. So, why was the head of the university recruiting team livid after the presentation and sending IMs to Michael's boss's boss's boss? Michael didn't properly define and understand the use case.

Michael should have first talked with the university recruiting team, built a relationship with them, and asked them questions about how they define success for their team. Had he done that, he would have found they consider the impact of branding and goodwill to be part of their team's ROI. Michael hadn't factored this into his analysis at all. Even if he did it retroactively, he had already lost credibility with the leader, making the situation harder for himself.

3.2   Use Case 1: The Retail Tyrant

With the motivation for understanding the business problem in place, let's move on to the business problems that will be the driving force for this book. For our first use case, meet our hypothetical stakeholder, Chris, who is a challenging leader to say the least. You've been tasked with helping Chris build a predictive model that predicts expected sales for the next three months.

3.2.1   Details of the Request

The reason for the request is rooted in your company's overall strategy. Online sales are expected to be a significant area of growth for your company, but the sales numbers are currently small in comparison to the company's core brick-and-mortar retail business. Chris likes to look good with management, so he's already putting pressure on your team to generate a "favorable" forecast for the next three months of sales. The sales history data is what we uploaded into Anaconda in Chapter 2.

Chris has recently become the head of your company's online retail division. Because you're part of your company's enterprise analytics team, you've heard from your colleagues about how challenging it is to work with Chris. This means you need to put your best foot forward to impress Chris, which will in turn impress your boss.

You have a dedicated team in the analytics department that supports time series forecasting, which oftentimes is where this type of request would go. However, that team's capacity is fully dedicated for the foreseeable future with a request from your CEO. In typical Chris fashion, this makes him feel like he's getting "B" team support, further complicating your dynamic with Chris.

On the bright side, you've been provided with the proper dataset to build the model from, so not everything is doom and gloom!

> **Use Case Note**
>
> The content for this use case in Chapter 4 through Chapter 7 will reflect the unique experience of working with Chris. The intent is to replicate the challenges associated with a difficult stakeholder. Don't worry, you'll still learn all of the practical techniques you'll need to know—just with a bit more nuance to work through.

3.2.2   History of the Request

The online retail division of your company's business has always been small relatively in comparison to your core brick-and-mortar business. The original head of the department, Jesus, was a scrappy leader who tried to make things work; however, he wasn't set up for success.

There is an institutional defensiveness of the brick-and-mortar retail business, so other leaders didn't exactly do Jesus any favors. This created challenges for Jesus that slowed down work and limited his ability to be successful in moving projects forward. The lack of movement in Jesus's work led to him being asked to leave the company.

After Jesus left, Chris was placed in the role as the head of the online retail division. Chris came from a competing business with a very different culture. In his time so far at your company, there's been a lot of turnover and movement of leaders out of Chris's team and into other areas of the business. Chris seemed to put a trance over the C-suite at the company right away. While all of Chris's peers and team found it challenging to work with and for him, the C-suite thought Chris was going to be the one to turn around their retail division—until recently.

Over the past few months, the CEO of your company has started expecting more results from Chris, wanting to see movement on the goals that Jesus didn't achieve. This led to Chris becoming more and more critical and challenging to work with. This is also why you have received the request to build a predictive model. The CEO has tasked Chris with short-term projections for the next quarter of sales. Chris doesn't have predictive modeling experience, so he needs to leverage your expertise in machine learning.

Chris is notorious for changing requirements and moving the goal posts. Members of your team who have supported Chris in the past say this has compounded the challenges with supporting him. Being clear with Chris about the requirements of your work will be critical in successfully executing this project.

3.2.3   Relationship with the Stakeholder

In person and during your meetings so far, Chris has been cordial with you. However, when you've asked for more clarity, Chris forwards your responses directly to the head of the analytics department (your boss's boss), stating he shouldn't need to make these types of decisions.

Forwarding Emails

Believe it or not, there are some stakeholders like Chris in the real world who do forward your emails directly to your boss or the head of your department. These types of leaders can be the most challenging to work with. Understanding the root cause is often the most important factor in effectively managing the situation. In my experience, there are two primary reasons a stakeholder will seem to escalate all or most communications you send.

Most often, it's rooted in a lack of trust. Whether or not it's warranted, this is the most common cause. The stakeholder is sending your communication to someone they believe will create the intended outcome they want. Building trust can take time, but the single biggest way you can build trust with a challenging stakeholder like this is to show you understand their needs and start making movement towards those needs.

The second reason is the stakeholder's perception that they are too important to deal with someone who doesn't have a big enough title. This is a much more toxic situation, one you often have to navigate with help from your boss and/or the leader to whom the stakeholder is defaulting their escalations. Make sure you document your communications and clearly communicate with your leadership about the progress of a project so they can help you navigate the situation in the best possible way.

3.2.4   Use Case Questions

Understanding how to dissect a business case and manage your stakeholders is important. This section is set up in a question-and-answer format to encourage you to test your understanding of the topics. The following section contains the answer(s) to each question. While these topics are not technical, it's important to spend some time thinking about these types of questions. Sure, Chris may seem a bit "extra," but these types of leaders do exist at companies!

1. What approach would you take to collecting requirements for a machine learning project with Chris?

   A. Discuss requirements exclusively during meetings/phone calls

   B. Discuss requirements exclusively via email

   C. Discuss requirements mostly via meetings/phone calls, but sometimes via email

   D. Discuss requirements mostly via email, but sometimes via meetings/phone calls

   E. Discuss requirements equally across both email and meetings/phone calls

2. Even if you're relatively independent and don't require support from your manager or leadership with other stakeholders, should you consider bringing in your manager or leadership to help manage Chris?

   A. Yes

   B. No

3. How frequently should you discuss technical specifics with Chris?

   A. Never

   B. Sometimes

   C. Often

3.2.5   Use Case Answers

Some of these questions are nuanced, so we'll explore multiple perspectives from different individuals on these important topics. Let's walk through the answers:

1. **What approach would you take to collecting requirements for a machine learning project with Chris?**
   With Chris's personality, **D** and **E** are likely the two best answers. This is probably the most subjective question, so we'll give you multiple perspectives and factors to consider to help you create your own decision-making process for managing similar stakeholders.

   The first consideration is how Chris likes to change requirements. If the definition of success isn't clear and keeps moving, you're set up for failure. This is why option **A**, discussing requirements exclusively via meetings/phone calls, isn't an optimal approach. If nothing is in writing, Chris will be able to cast his perspective on each conversation with no documentation or record for you to lean back on.

   The next consideration is Chris and his oversized ego. If you only communicate with Chris via email, Chris may view this project as not being important enough for his time. Meeting with a leader like Chris can be intimidating, so it's understandable to want to avoid it. However, face time in meetings can help humanize you to even the most difficult of leaders. While meetings may give Chris more opportunities to do what he does, no meetings will likely make the dynamics more challenging for you. This is why we don't recommend option **B**.

   Though it may feel like a contradiction to our last point about needing to meet with Chris to cater to his ego, leaders like Chris also detest their time being wasted. So, option **C** (discuss requirements mostly via meetings/phone calls, but sometimes via email) is not optimal because too many meetings run the risk of making Chris feel like you're wasting his time. More meetings also make it more challenging to gather the requirements of the project.

   This ultimately varies based on the specific leader, but try to find a balanced approach that favors written communication for leaders like Chris.

2. **Even if you're relatively independent and don't require support from your manager or leadership with other stakeholders, should you consider bringing in your manager or leadership to help manage Chris?**
   The answer is **A**, yes. Personally, I've had to learn when to set my ego aside and ask for help before a situation gets out of hand. This is especially true when working with

a leader like Chris, whose ego prefers to work with people who have the most important job title. While this is wrong and it can be frustrating that leaders like Chris are allowed to operate this way, it's the reality of the situation.

Perspective of Analytics Leader

I met with a leader who has led various analytics team across a number of large companies. When working with him, I always felt the ability to handle difficult stakeholders was a notable differentiator for him as a leader. I sat down with him to get his perspective on the topic of managing difficult stakeholders. Here is the summary of that conversation:

Q: What is your initial approach when working with a stakeholder you know to be difficult?

A: My first question is, what is this type of person? Challenging leaders come in many forms. Sometimes they're compensating for something like a lack of knowledge or confidence in a space, sometimes they're very blunt, which comes across as challenging, but sometimes they are just a jerk. Knowing the type of challenging leader you're working with is important because you have to tailor your approach to that specific style.

Q: What tips and tricks do you have for someone working with a difficult stakeholder?

A: My best piece of advice is to separate yourself from the emotions of the situation. When you're emotionally invested in the situation, you're going to be more likely to provide an emotional response. Especially in live conversations (whether it's face-to-face or on a call), an emotional response can be detrimental to the success of managing a stakeholder. No matter your familiarity with the stakeholder, this is something you can always practice.

The next tip is to match the stakeholder's sense of urgency. While you may not be able to drop everything in the moment, if the stakeholder views the topic as urgent and a high priority, ensure you're not downplaying that in the conversation. That can lead to a challenging individual becoming more challenging.

The last piece of advice is the most commonly discussed. Building a relationship with the stakeholder will go a long way in ensuring they're able to trust you. More times than not, they'll also go easier on you during challenging situations.

Q: Do you have a story you could share that illustrates this advice?

A: I sure do! While it's not specific to machine learning, I think the principles translate well to predictive modeling use cases, too. I was working with a group of directors on building out a scorecard for our company's talent acquisition organization. My team and I worked on building the dashboard for this and the directors loved it. It consolidated their efforts into one place instead of needing to go into multiple different dashboards to get the information they needed. However, their boss, the vice principle (VP) of talent acquisition, hated the dashboard when she was shown it.

Historically, she has been known as a difficult stakeholder to work with. She often has an opinion on the topics being discussed but has a hard time articulating those opinions.

During the conversation, when she came to show her displeasure for the dashboard, I separated myself from the work. She wasn't attacking me or my team; she was attacking the work. I also matched her sense of urgency and tone. She indicated data being a priority but wasn't articulating it well. I immediately set up time with just her to understand her needs. Through that conversation and asking a lot of questions, I found she actually liked the scorecard but felt there needed to be a dashboard that was more flexible for her daily needs. I found the variety of questions she was getting from the head of HR and other executives around the company necessitated a dashboard that could be filtered and sliced very easily.

I worked with my team to build a new dashboard for this purpose, and the head of talent acquisition could not have been happier. This helped build the relationship and has since made future asks and working with her smoother and more pleasant for everyone involved!

3. **How frequently should you discuss technical specifics with Chris?**

   You've probably been told not to get into any technical details with any nontechnical leader. However, consider a perspective on this that may be more novel than simply saying option **A**, never.

   Especially with a leader like Chris, you must handle their ego alongside their desire not to have their time wasted. This is a tough balancing act. When it comes to discussing technical content, option **C**, often, is not advised because there is a high probability that Chris views hearing all the technical detail as a waste of his time. Option **B**, sometimes, is best because you can use your knowledge of machine learning as a subtle reminder to leaders like Chris that you do add value and know things they don't know. Here are the considerations for using this tactic:

   - Timing: If you start using this tactic right away with a leader like Chris, they're more likely to see it as you wasting their time and not being able to effectively communicate at the appropriate level. Leveraging this tactic later on at a critical moment in the project, after you've already built some trust, will have the highest probability of success.

   - Frequency: If you find yourself defaulting to this tactic often, you're probably not doing it right. This will begin to threaten Chris's ego and will become counterproductive. Use this tactic situationally in critical moments. For example, if Chris is questioning your approach or methodology, that can be an effective time to throw in a bit more technical detail than you otherwise normally would.

   - Depth: If you start going into the core math behind machine learning and explaining exactly how entropy is calculated and used in your algorithm, you've probably taken the example too far. Simply throwing out the term a like "entropy" in enough context for a leader like Chris to infer the definition will suffice.

Entropy

Entropy is the measure of randomness or uncertainty in your data. In this example, there are a number of words you could replace entropy with that you know a stakeholder would explicitly understand. But, if you're trying to make a point, it's a fancy-sounding word that someone like Chris probably doesn't know.

3.3   Use Case 2: Customer Retention

Compared to Chris, Mary is an incredibly understanding and easy stakeholder to support. Mary leads the customer experience and retention division of the meal delivery company you work for, which operates in India. In our next hypothetical example, Mary has tasked you with building a model to predict whether or not a specific customer will make another order within the next seven days.

3.3.1   Details of the Request

Your company has had proven success in building a brand that provides meal delivery services to customers. As your company wants to continue to grow, there's a focus on growth through customer retention. Customer acquisition costs through various marketing channels are high, so the CEO has asked Mary to better understand which factors impact customer retention and create a mechanism to predict it. These predictions can then be used by the marketing team to adjust promotional offers to customers based on the probability of the customer ordering again in one week.

For example, if a customer has a high probability of ordering again in one week, they won't be given an aggressive promotion. If a customer has a low probability of ordering again in one week, a more aggressive promotional offer will be provided. The company currently doesn't make any adjustments to their promotional offers to customers, meaning the existing data doesn't need to be adjusted to account for previous promotions offered by the company.

For this use case, we'll need to clean and prepare our data for the model while using a classification model to predict probabilities, which we'll explore in future chapters.

Use Case Note

Contrary to Chris, Mary is an easy stakeholder, and the content delivery will reflect that.

3.3.2   History of the Request

To date, your company hasn't had to focus much on customer retention because it's been in hyper-growth mode. Therefore, you're the first individual to look into this data and build a machine learning model to predict how likely a customer is to place another order within the next week. It's always a fun opportunity to explore something that hasn't been explored before!

While the company has been around for a few years, recent changes within the organization require you to focus on only the past few months of data. While it'd be preferable to have multiple years of data to build the model on, your manager has guided you to focus on this smaller dataset (at least to start with).

3.3.3   Relationship with the Stakeholder

This use case may be new, but you've worked with Mary before over the past few years. Mary is a fantastic leader that her direct team and other partners (like your team) enjoy working with and for. She has an energetic personality, and the running joke is that she has so much energy because she drinks multiple Diet Pepsis per day. Mary values you and the previous work you've done for her. She treats you as a partner, which enables you to do your best work. Her expectations are high, but she creates an environment for your success.

Across the company, Mary is known for valuing relationships and acknowledging her technical knowledge limitations. She is a classic "people person" with a healthy balance of self-awareness.

Mary has become a friend and mentor over the time you've supported her. As much as she values you, you equally value her. While Mary does not have any technical expertise, she's helping you develop a better understanding of the business and how to support leaders at her level.

3.3.4   Use Case Questions

This section is set up in a question-and-answer format. The following section contains the answer(s) to each question. While these topics are not technical, it's important to spend some time thinking about these types of questions. Understanding how to dissect a business case and manage your stakeholders is important.

1. What approach would you take to collect requirements for a machine learning project with Mary?

    A. Discuss requirements exclusively during meetings/phone calls

    B. Discuss requirements exclusively via email

    C. Discuss requirements mostly via meetings/phone calls, but sometimes via email

    D. Discuss requirements mostly via email, but sometimes via meetings/phone calls

    E. Discuss requirements equally across both email and meetings/phone calls

2. Even if you're relatively independent and don't require support from your manager or leadership with other stakeholders, should you consider bringing in your manager or leadership to help manage Mary?

    A. Yes

    B. No

3. How frequently should you discuss technical specifics with Mary?

    A. Never

    B. Sometimes

    C. Often

3.3.5   Use Case Answers

Some of these questions are nuanced, so we'll explore multiple perspectives from different individuals to give you insight into these important topics.

1. **What approach would you take to collect requirements for a machine learning project with Mary?**
   When working with the type of leader who treats you like a partner, the requirements gathering process is still extremely important. However, the level of caution required compared to a leader like Chris is more tempered. Leaders like Mary come in different forms, so the best advice is to ensure you're catering to their preferences. They tend to prefer option **E**, discussing requirements equally across email and meetings/phone calls. They appreciate the dialogue and face time with you, but they also like to have written documentation to reference back for themselves.

2. **Even if you're relatively independent and don't require support from your manager or leadership with other stakeholders, should you consider bringing in your manager or leadership to help manage Mary?**
   Yes (**A**), but with some nuance to keep in mind. Leaders like Mary values your partnership, so the decision-making criteria for when to bring in support is different. A leader like Chris prefers to work with individuals with important titles, but a leader like Mary may view getting your manager and leader more involved as you deferring responsibility. In general, it's best to let scenarios play out a bit longer before involving leaders.

   For example, if a project isn't going well but you still have a few options to explore that may resolve your problems, start looking into those options first. If they work out, fantastic! If they don't, then it makes sense to then bring in additional support. Mary also has high expectations and wants to deliver for her business, so you don't want to wait too long to ask for help and spin your wheels for weeks with no progress.

Working with Fantastic Stakeholders

More often than not, working with a great stakeholder like Mary means smooth sailing. I've had a few Mary-like stakeholders and their ability to make a positive impact on data projects is significant.

While working in the data space is often about managing communication and expectations, when you have a stakeholder who proactively communicates their needs, includes you in the right conversations, and trusts you are the expert in your space, the outcomes you're able to produce are elevated.

When working with one of my recent Mary-like stakeholders, I often found myself wondering if I was dedicating enough time to their team's work. After reflecting, I realized my output for them was much greater than that of my work for other stakeholders. The difference was how much smoother the communication and expectations were, allowing me to focus on the right work for their group.

As the data person supporting this type of stakeholder, I've also found my general mood and attitude to be so much more positive. I feel valued and get to focus on the work instead of the politics often required with stakeholders.

---

3. **How frequently should you discuss technical specifics with Mary?**

   For a leader like Mary, the answer is **A**, basically never. Even if you're just starting to work with a leader like Mary, she tends to think logically and defaults to trust. If Mary is a nontechnical leader, this creates an environment where she doesn't need to be reminded that you're the expert in your domain.

   The rare example is when something seems illogical. You may encounter situations where, having described a situation to a stakeholder, you realize how ridiculous it sounds to someone without the proper context. These are the situations where you may want to dive into some of the technical details to make your explanation sound less ridiculous.

   A quick anecdote from the data visualization and dashboard space: I was explaining to a leader like Mary how a Power BI graph couldn't have a dual axis in the way she wanted. As I was giving the explanation, I could see the "Jason, I know this can be done in Excel" look on her face. I then took the step back and explained the more technical nuance, and the leader was satisfied after hearing the additional context.

3.4   Use Case 3: Crime Predictions

The stakeholder for our final hypothetical use case is Jen. Compared to Chris and Mary, Jen is a more technical stakeholder. Jen is a leader responsible for community safety. She's worked with a law enforcement agency in the past, but her current initiative is to get insight into which areas of the city are most likely to have crime in the next month.

She's been given a significant federal grant to dedicate to crime prevention resources, so she wants a model that will help her decide which areas of the city to dedicate her funds to.

Note

Like Mary, Jen is a relatively easy stakeholder. However, Jen has the ability to do some analysis on her own, so you may have to go back and validate some of her assumptions. You'll see this nuance show up throughout the book, as some ideas introduced early on are corrected or clarified.

3.4.1   Details of the Request

Jen's tendency to roll up her sleeves and do the work herself is something you've always admired. When working with Jen in the past, you've seen how her ability to do some analytics work herself prompts her to generate better questions and more insightful feedback. Jen doesn't have a background in machine learning, which is why she's coming to you with the request. She's done some statistical analysis on the data and identified some patterns in the data, which she'd like you to validate.

3.4.2   History of the Request

Jen's role has significant overlap with the law enforcement agency, so she's become quite skilled at navigating the politics that are abundant in her role. She's voiced some displeasure with the law enforcement agency's team not being able to provide more detailed information about the location of crimes, but she has confidence in your ability to help her deliver what she needs.

The impact of a request like this is significant, as the goal is to ensure resources are deployed appropriately to reduce crime. It's important to avoid letting any personal preconceived notions about the work impact your approach. This can be a polarizing topic, so you want to lead with the data.

3.4.3   Relationship with the Stakeholder

You and Jen get along quite well. At times, Jen has acted as your mentor while you build your technical expertise. You value her perspective and respect her opinions. The conversations you have are an open dialogue, and Jen is not defensive when you challenge her perspectives on a topic.

With this type of positive relationship, you should feel comfortable recommending alternative approaches and methods of looking at the data. A stakeholder like Jen should be viewed as a partner and asset to help with the project, so collaborating with this type of stakeholder is key!

3.4.4   Use Case Questions

This section is set up in a question-and-answer format. The following section contains the answer(s) to each question. While these topics are not technical, it's important to spend some time thinking about these types of questions. Understanding how to dissect a business case and manage your stakeholders is important.

1. What is the recommended approach to collecting requirements for a machine learning project with Jen?

   A. Discuss requirements exclusively during meetings/phone calls

   B. Discuss requirements exclusively via email

   C. Discuss requirements mostly via meetings/phone calls, but sometimes via email

   D. Discuss requirements mostly via email, but sometimes via meetings/phone calls

   E. Discuss requirements equally across both email and meetings/phone calls

2. Even if you're relatively independent and don't require support from your manager or leadership with other stakeholders, should you consider bringing in your manager or leadership to help manage Jen?

   A. Yes

   B. No

3. How frequently should you discuss technical specifics with Jen?

   A. Never

   B. Sometimes

   C. Often

3.4.5   Use Case Answers

Some of these questions are nuanced, so we'll explore multiple perspectives from different individuals to give insight into these important topics.

1. **What is the recommended approach to collecting requirements for a machine learning project with Jen?**

   Options **C** (discuss requirements mostly via meetings/phone calls, but sometimes via email) and **E** (discuss requirements equally across both email and meetings/phone calls) are likely the best fit for a leader like Jen. It's easier to collaborate and brainstorm solutions with a leader like Jen during live conversations. However, leaders like Jen often tend to be busy; they also value written communication to help them be efficient with their time.

   The key differentiator for working with this type of leader is their desire to be an active participant in the project. This isn't a situation where you're catering to an ego. There is legitimate value in directly involving Jen in the conversation. She's likely already considered the challenges that you'll need to think through for the project, and her closeness to the underlying use case is invaluable. Having live conversations lets your collaboration flourish and creates a better outcome in a shorter timeline.

In my experience, leaders like Jen tend to be very busy individuals because they're highly capable and valued in their organizations. Capacity and availability constraints may lead to a more balanced approach of online and offline communication.

2. **Even if you're relatively independent and don't require support from your manager or leadership with other stakeholders, should you consider bringing in your manager or leadership to help manage Jen?**
   The answer to this is similar to Mary's: Yes (**A**), but with a caveat. Jen values your perspective, so she may view involving your manager too early as a sign that you're deferring your responsibility. However, leaders like Jen also have high expectations, and they'll be frustrated if you let a project stagnate for weeks at a time without getting help from your higher-ups.

   The guiding principle should be if you're making progress, you likely don't need to bring in your manager or other leaders. If you see the project stalling out, engage your leadership for help!

3. **How frequently should you discuss technical specifics with Jen?**
   The answer is usually often (**C**) with leaders like Jen. The level of detail may vary by the specific leader, but leaders like Jen love learning and understanding how things work. If you skim over the specifics too much, they may begin to question your approach and become more difficult to manage as a stakeholder. They want to be let into the specifics, so let them in!

   A good trick is to use them as an opportunity to practice simplifying technical concepts for other leaders. Rarely are they going to understand all the nuances of how something as complex as machine learning truly works, but their ability to connect concepts is a good entry point into explaining a technical concept for nontechnical audiences. Use their questions to inform how you'll communicate the same concept to the next level of leadership, who'll require a much simpler explanation of the model.

3.5 Summary

We introduced the use cases that we'll explore through the remaining chapters in this book. Each use case has an accompanying dataset you'll be able to access on the book's GitHub page at *https://github.com/jason-hodson/applied_machine_learning/tree/main*.

Now that you understand the use cases, we'll move to the technical section of the book, where you'll start working with the data.

Chapter 4
Starting with the Data

We're now starting our journey into the technical content. This chapter will prepare you for the realities of readying your data for a machine learning model.

Now that we've discussed our use cases, we can get into the data! Data in its raw form is almost never ready for a machine learning model, so before you begin building, you need to first understand and prepare your data. As you do more and more machine learning work, you'll find this is the most time-consuming step. In general, you should expect to dedicate about 80% of your time on any given project to preparing your data, which we'll cover in this chapter. There are common steps required for most datasets, including the dummy coding that's required to transform your categorical columns into something your model can consume. However, much of the data preparation work will vary based on your dataset and your project's objective.

In this chapter, we'll go through the different types of data sources, best practices for exploring your data to identify outliers and hidden errors, steps for dummy coding your categorical columns, and the execution of dimensionality reduction. Along the way, we'll integrate our use cases to demonstrate examples for each process in a real-world setting.

The content in this chapter (as well as Chapter 5, Chapter 6, and Chapter 7) often has a lot of grey area between where one topic starts and another ends, which can make learning the topics and themes more challenging. This book is organized with learning in mind. We introduce a topic and then go through how that topic applies to each of our use cases.

4.1 Types of Data Sources

The quantity of different data sources you could encounter is mind numbing. While this isn't a book about file extensions, databases, or application programming interfaces (APIs), the following sections will cover some basic information about the most common file types you're likely to encounter when working on machine learning projects. We'll explore some examples of both manual and automated data sources.

Here's a helpful way to think about data sources: Whether your source is manual or automated, the key difference is *how* you get it. Manual files still use code to import your data for cleaning, but it's slightly different code than automated sources. In the manual file approach, you need to manually retrieve the data from somewhere (such as a file system or website), save it to your computer, and then point your code to the file.

4.1.1   Manual

When you're starting out in the machine learning world, it's likely you'll need to be familiar with leveraging manual files rather than an automated source like a database. While automated data feeds are always preferred due to their stability, they're not a requirement for you to start adding value with machine learning techniques.

We'll explore the most useful manual data sources in machine learning in the following sections.

CSV and Excel

Comma-separated values (CSV, *.csv*) and Excel (*.xls* or *.xlsx*) files are the most common files people in the data space work with. The process for reading CSV and Excel files is similar. In practice, they're often interchangeable to stakeholders, which is why we're covering them together rather than separately.

If you're already an Excel user, you likely know some of the notable differences when performing an analysis or building out visuals in Excel. The most notorious example is creating multiple different tabs and forgetting to change from the *.csv* extension to the *.xlsx* extension, losing all but the first tab. Some of these same concepts that impact how you use Excel are important to know when writing Python code that prepares your data for the machine learning model process.

The most important consideration is ensuring you're using the proper functions. When you open your data in Excel, it may appear as if there's no difference compared to a CSV file—but there *is* a difference when you're importing your data! To demonstrate what happens when you use the wrong function, let's start by loading pandas and using the read_csv function on the data from our first use case (*Retail Sales File 1.xlsx*), as shown in Listing 4.1. The result is shown in Figure 4.1.

```
#import pandas package
import pandas as pd

#read in the specific file you're trying to read in
df = pd.read_csv("Retail Sales File 1.xlsx")
```

Listing 4.1 Read an Excel File Using the read_csv Function

> **The pandas Library**
>
> If you haven't worked with pandas yet, you'll soon become quite familiar with it. This is the main library for data manipulation within the Python ecosystem.

```
UnicodeDecodeError                    Traceback (most recent call last)
Cell In[2], line 1
----> 1 df = pd.read_csv("Retail Sales File 1.xlsx")

File /opt/conda/envs/anaconda-panel-2023.05-py310/lib/python3.11/site-packages/pandas/io/parsers/readers.py:912, in read_csv(filepath_or_buffer, sep, delimite
r, header, names, index_col, usecols, dtype, engine, converters, true_values, false_values, skipinitialspace, skiprows, skipfooter, nrows, na_values, keep_defa
ult_na, na_filter, verbose, skip_blank_lines, parse_dates, infer_datetime_format, keep_date_col, date_parser, date_format, dayfirst, cache_dates, iterator, chu
nksize, compression, thousands, decimal, lineterminator, quotechar, quoting, doublequote, escapechar, comment, encoding, encoding_errors, dialect, on_bad_line
s, delim_whitespace, low_memory, memory_map, float_precision, storage_options, dtype_backend)
    899 kwds_defaults = _refine_defaults_read(
    900     dialect,
    901     delimiter,
    (...)
    908     dtype_backend=dtype_backend,
    909 )
    910 kwds.update(kwds_defaults)
--> 912 return _read(filepath_or_buffer, kwds)

File /opt/conda/envs/anaconda-panel-2023.05-py310/lib/python3.11/site-packages/pandas/io/parsers/readers.py:577, in _read(filepath_or_buffer, kwds)
    574 _validate_names(kwds.get("names", None))
    576 # Create the parser.
--> 577 parser = TextFileReader(filepath_or_buffer, **kwds)
    579 if chunksize or iterator:
    580     return parser

File /opt/conda/envs/anaconda-panel-2023.05-py310/lib/python3.11/site-packages/pandas/io/parsers/readers.py:1407, in TextFileReader.__init__(self, f, engine, *
*kwds)
    1404     self.options["has_index_names"] = kwds["has_index_names"]
    1406 self.handles: IOHandles | None = None
--> 1407 self._engine = self._make_engine(f, self.engine)

File /opt/conda/envs/anaconda-panel-2023.05-py310/lib/python3.11/site-packages/pandas/io/parsers/readers.py:1679, in TextFileReader._make_engine(self, f, engin
e)
    1676     raise ValueError(msg)
    1678 try:
--> 1679     return mapping[engine](f, **self.options)
    1680 except Exception:
    1681     if self.handles is not None:

File /opt/conda/envs/anaconda-panel-2023.05-py310/lib/python3.11/site-packages/pandas/io/parsers/c_parser_wrapper.py:93, in CParserWrapper.__init__(self, src,
**kwds)
    90 if kwds["dtype_backend"] == "pyarrow":
    91     # Fail here loudly instead of in cython after reading
    92     import_optional_dependency("pyarrow")
---> 93 self._reader = parsers.TextReader(src, **kwds)
    95 self.unnamed_cols = self._reader.unnamed_cols
    97 # error: Cannot determine type of 'names'

File /opt/conda/envs/anaconda-panel-2023.05-py310/lib/python3.11/site-packages/pandas/_libs/parsers.pyx:550, in pandas._libs.parsers.TextReader.__cinit__()

File /opt/conda/envs/anaconda-panel-2023.05-py310/lib/python3.11/site-packages/pandas/_libs/parsers.pyx:639, in pandas._libs.parsers.TextReader._get_header()

File /opt/conda/envs/anaconda-panel-2023.05-py310/lib/python3.11/site-packages/pandas/_libs/parsers.pyx:850, in pandas._libs.parsers.TextReader._tokenize_rows
()

File /opt/conda/envs/anaconda-panel-2023.05-py310/lib/python3.11/site-packages/pandas/_libs/parsers.pyx:861, in pandas._libs.parsers.TextReader._check_tokenize
_status()

File /opt/conda/envs/anaconda-panel-2023.05-py310/lib/python3.11/site-packages/pandas/_libs/parsers.pyx:2021, in pandas._libs.parsers.raise_parser_error()

UnicodeDecodeError: 'utf-8' codec can't decode bytes in position 15-16: invalid continuation byte
```

Figure 4.1 Pandas Error Message

What a mess! This error message is telling us it can't understand the file format. Instead of troubleshooting the error message, pandas already has a function for *.xlsx* files. By swapping out the function to read_excel, as shown in Listing 4.2, we're now able to ingest the *.xlsx* file with no issues.

```
#read in excel file with read_excel function
df = pd.read_excel("Retail Sales File 1.xlsx")
```

Listing 4.2 Read in Excel File with the read_excel Function

We've now added our data into a DataFrame object that allows us to leverage the data cleaning tools that pandas has to offer!

Reading Error Messages

The error messages shown in Figure 4.1 can be quite confusing and intimidating. Unfortunately, learning to navigate these messages is part of the learning process for using Python to build machine learning models.

While pure Python programmers may love getting so much detail, I find the majority of the error message to be unnecessary and unhelpful. My recommendation is to navigate to the bottom of the error message first. This is where you'll find the human-readable error messages.

Inputting the specific error message into either a generative AI tool or an internet search can also be a helpful approach if you get stuck and aren't able to piece together what may be causing it.

Another consideration that can be important for picking your file format is file size. CSV is a simple file format with no additional compression occurring, making the size of the file larger. For example, our *Retail Sales File 1* file is roughly two times larger as a *.csv* file than a *.xlsx* file. This won't be an important consideration if you're operating in an enterprise setting with data that can be managed and stored outside of a database. However, a smaller file might be preferable when you're sending the data to someone via email, for example.

The last notable difference between *.csv* and *.xlsx* files is the ability for a *.xlsx* file to have multiple tabs. In Figure 4.2, there are five tabs at the bottom of the screen.

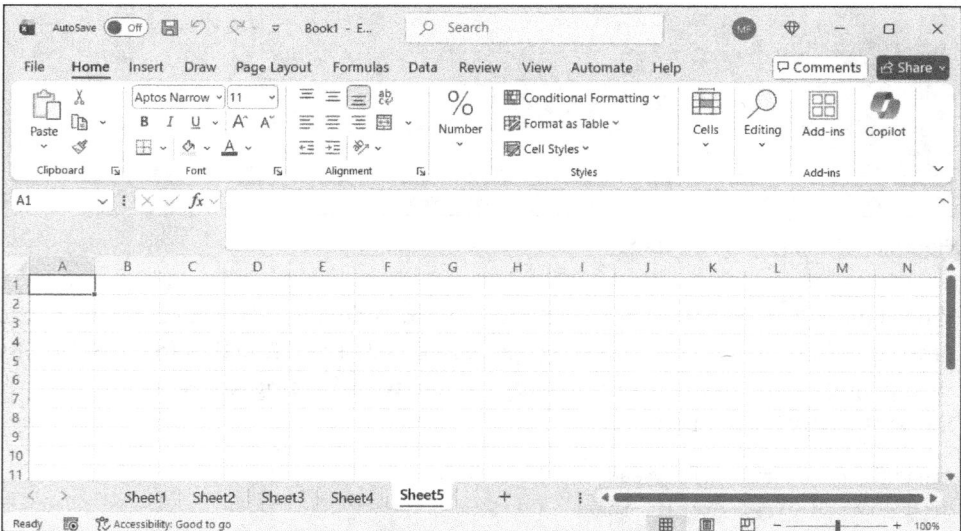

Figure 4.2 Excel File with Five Tabs

There could be data in multiple tabs, so we must specify which tab the function needs to read in the data for. Using the read_excel function from pandas, this is relatively simple to handle with the sheet_name parameter. You're able to pass the position (0 for the first tab, 1 for the second tab, etc.) or name of the sheet (Sheet1, Sheet2, etc.) to this parameter, and the data will then be read into the variable name you assign. In contrast, a *.csv* file can't have multiple tabs, so this parameter is not a consideration.

PDF

While we won't go into extensive detail on PDFs, it's important to state that data isn't always stored in a spreadsheet format. While it is certainly possible to extract data from PDF files, be cautious about agreeing to do so. The technical nuances of pulling data from PDF files are challenging, since PDFs can have structured data (rows and columns) as well as unstructured data (paragraphs of words). Explaining how to properly extract this data could be a book by itself, so it's better that you avoid it if possible.

If there are specific data elements in PDF formats that you want, consider working with the team who creates or manages the PDFs to see if they're able to provide the same data in another format.

4.1.2   Automated

Automated data connections are the preferred choice for machine learning projects. You may not encounter automated data feeds in learning resources like this book, but it's very likely you'll work with automated sources on the job. That said, even some of the world's biggest companies rely on sending Excel files from team to team for their analytics.

There are two main automated data sources—databases and APIs—which we'll discuss in the following sections.

Databases

Databases are the most common automated data feed at companies. Their purpose is to store data. You'll likely hear people refer to them using terms like on-premise data, cloud data, or data lake. While these terms each have different associated products with various pros and cons, it's best to avoid getting overly concerned with the nuances among them, especially when you're just getting started in the machine learning space. If you're looking to be a data engineer, then you certainly need to be up to date on those nuances. For everyone else, this is a lower-priority area to learn.

Every database has its own process for establishing a connection, but this typically involves using a function with parameters. This often includes passing a SQL query through to retrieve the necessary data. For machine learning projects, this is a wonderful thing. You're not responsible for downloading a new file or worrying if the data format has changed and broken your data pipeline. The risk of manual errors is reduced, as well as the cognitive load required to know which steps have to happen in which order. Few things are simpler than just having to press ⌈Ctrl⌉ + ⌈Enter⌉!

APIs

APIs allow you to systematically retrieve data from a source. APIs are almost always used to retrieve data to put it into a database, not to directly pull your data from the API. There are a few reasons for this:

- There's often a cost associated with each pull from an API. Pulling all of the data you might use via an API into a database minimizes the number of times you need to hit the API with a new request.

- APIs aren't great at pulling large volumes of data. If you're pulling a large dataset with a lot of history, it's more effective to pull the data on a recurring basis.

- Access to APIs is generally limited at companies for security purposes. APIs can expose a lot of data, so their access is usually more restricted than databases.

4.1.3   Data Sources for Our Use Cases

Let's take a pause to go over our three use cases and apply what you just learned. All of the source data that we'll discuss can be found on GitHub at *https://github.com/jason-hodson/applied_machine_learning/tree/main*.

Use Case 1: Data Sources

For our first use case, focused on retail sales, we have four different *.csv* files. The data is too large to upload and be stored in GitHub as one file, so it's been split into multiple files. This is also a convenient opportunity for you to learn how to import multiple files at one time. One approach would be to read each file in pandas with its own line of code. Listing 4.3 shows an example of what this looks like.

```
#load pandas library
import pandas as pd

#load os library, used specifically for navigating file paths
import os

#read in the csv file
df_1 = pd.read_csv("Retail Sales File 1.csv", encoding='unicode_escape')
```

Listing 4.3 Reading a Single CSV File

But what happens if you have more than four files, and the names aren't always the same? This is where you need to start thinking like a programmer and looking for places where you're copying and pasting the same code over and over again. Consider the code in Listing 4.4.

```
#create blank data frame
df = pd.DataFrame()

#execute for loop on the set of code
for file_name in os.listdir():

  #specifically only reference the .csv files in the directory
  if file_name.endswith(".csv "):

    #read in the file as a generic object name
    df_temp = pd.read_csv(file_name, encoding = 'unicode_escape')

    #stack each file on top of each other into what was a blank df
    df = pd.concat([df, df_temp], axis = 0)
```

Listing 4.4 Reading Multiple Files from the Same Location

Listing 4.4 introduces some concepts you may or may not have seen before. We won't go into too much detail here, but it's worth providing a brief explanation for those who are newer to Python. In a nutshell, the code is doing the following:

- Creating an empty DataFrame
- Pulling all of the file names from the directory (file path)
- Importing the data of any file that ends in *.csv* and assigning it to df_temp
- Stacking each iteration of the data onto top of each other

Further Resources

If you want more detail on this approach, look into iterating and its associated use cases in Python. Iterating is covered in most Python courses given how critical it is to Python programming. One specific resource to reference is GeekforGeeks' "Iterators in Python" post, available at *https://www.geeksforgeeks.org/python/iterators-in-python/*. This can be a confusing topic, so give yourself some grace if it's a new concept and you're not fully grasping it.

After running the code, you now have a DataFrame called df. Be sure to validate the code to import your data has worked properly, especially when you're doing something new. A helpful function to get your row and column counts is shape. This returns a tuple with the first element being the number of rows in your DataFrame and the second element being the number of columns. Let's run this function for our DataFrame:

```
df.shape
```

The result of running this code on our data is (1347137, 8). The first number represents the number of rows and the second number represents the number of columns.

The last helpful function to mention is head. This shows you a preview of the data. You can use it to display column names and see how certain columns, like dates, come through in the data. By default, it will return five rows of data, but you can input a number to have it return as many or few rows as you wish. Let's run this function for our DataFrame:

```
df.head()
```

The results are shown in Figure 4.3.

	Invoice	StockCode	Description	Quantity	InvoiceDate	Price	Customer ID	Country
0	489434	85048	15CM CHRISTMAS GLASS BALL 20 LIGHTS	12.0	1/12/2009 7:45	6.95	13085.0	United Kingdom
1	489434	79323P	PINK CHERRY LIGHTS	12.0	1/12/2009 7:45	6.75	13085.0	United Kingdom
2	489434	79323W	WHITE CHERRY LIGHTS	12.0	1/12/2009 7:45	6.75	13085.0	United Kingdom
3	489434	22041	RECORD FRAME 7" SINGLE SIZE	48.0	1/12/2009 7:45	2.10	13085.0	United Kingdom
4	489434	21232	STRAWBERRY CERAMIC TRINKET BOX	24.0	1/12/2009 7:45	1.25	13085.0	United Kingdom

Figure 4.3 Result of Using the Head Function

Use Case 2: Data Sources

The data for this use case to improve customer retention for meal deliveries only contains one *.csv* file, which holds the company's order history. We're able to use the simpler approach of reading the *.csv* file in one line, as shown in Listing 4.5.

```
#read in pandas library
import pandas as pd

#read in the .csv file
df = pd.read_csv("order_history_data.csv")
```

Listing 4.5 Read Data

If you only try to use the df.head() function to preview the data, like you did for the first use case's data, you'll notice a few things, as shown in Figure 4.4:

- There are ellipses in the middle of the dataset. This signifies that not all columns are being shown. When you're exploring your data, it's important to look for this to ensure you're aware of what may or may not be shown when using the .head() function.

- A scroll bar appears at the bottom of the page. Be aware that you'll need to scroll to the right to view additional columns.

These issues are common when you're working with datasets that have many columns, and they're further exacerbated by the number of text columns and length of the column names. We'll look to address these issues when we discuss the steps required to prepare this data for the machine learning process.

	Restaurant ID	Restaurant name	Subzone	City	Order ID	Order Placed At	Order Status	Delivery	Distance	Items in order	...	Rating	Review	Cancellation / Rejection reason	Restaurant compensation (Cancellation)	Restaurant penalty (Rejection)	KPT duration (minutes)	Rider wait time (minutes)	Order Ready Marked
0	20320607	Swaad	Sector 4	Delhi NCR	6168884918	11:38 PM, September 10 2024	Delivered	Zomato Delivery	3km	1 x Grilled Chicken Jamaican Tender, 1 x Grill...	...	NaN	NaN	NaN	NaN	NaN	18.35	11.6	Correctly
1	20320607	Swaad	Sector 4	Delhi NCR	6170707559	11:34 PM, September 10 2024	Delivered	Zomato Delivery	2km	1 x Peri Peri Fries, 1 x Fried Chicken Angara ...	...	NaN	NaN	NaN	NaN	NaN	16.95	3.6	Correctly
2	20320607	Swaad	Sector 4	Delhi NCR	6169375019	03:52 PM, September 10 2024	Delivered	Zomato Delivery	<1km	1 x Bone in Peri Peri Grilled Chicken	...	NaN	NaN	NaN	NaN	NaN	14.05	12.2	Correctly
3	20320607	Swaad	Sector 4	Delhi NCR	6151677434	03:45 PM, September 10 2024	Delivered	Zomato Delivery	2km	1 x Fried Chicken Ghostbuster Tender, 1 x Anga...	...	4.0	NaN	NaN	NaN	NaN	19.00	3.3	Correctly
4	20320607	Swaad	Sector 4	Delhi NCR	6167540897	03:04 PM, September 10 2024	Delivered	Zomato Delivery	2km	1 x Peri Peri Krispers, 1 x Fried Chicken Anga...	...	NaN	NaN	NaN	NaN	NaN	15.97	1.0	Correctly

5 rows × 29 columns

Figure 4.4 Head of Data with Many Columns

When we're still exploring the dataset, how can we see more of the data? One option is to read the data in Excel, where you have more flexibility. Since this data is already a *.csv* file and its size is manageable, simply downloading the data from the UI of where you're writing the code is a viable option. We'll explore how to do this in detail later in the book.

Another approach within Python is printing out the column names to see the full list. You can cross reference the results against the output of the `df.head()` results to get a better idea of which columns you're missing. When we print our columns as follows, we get the output shown in Figure 4.5:

```
print(df.columns)
```

```
Index(['Restaurant ID', 'Restaurant name', 'Subzone', 'City', 'Order ID',
       'Order Placed At', 'Order Status', 'Delivery', 'Distance',
       'Items in order', 'Instructions', 'Discount construct', 'Bill subtotal',
       'Packaging charges', 'Restaurant discount (Promo)',
       'Restaurant discount (Flat offs, Freebies & others)', 'Gold discount',
       'Brand pack discount', 'Total', 'Rating', 'Review',
       'Cancellation / Rejection reason',
       'Restaurant compensation (Cancellation)',
       'Restaurant penalty (Rejection)', 'KPT duration (minutes)',
       'Rider wait time (minutes)', 'Order Ready Marked',
       'Customer complaint tag', 'Customer ID'],
      dtype='object')
```

Figure 4.5 Output of Printing Column Names

Use Case 3: Data Sources

While our first two use cases had .csv files, our third use case for crime predictions has an Excel file of past crime data in Los Angeles. As we discussed in Section 4.1.1, this nuance matters. Luckily, we have pandas to help us out.

> **Note**
>
> This use case is based on Los Angeles Police Department crime data, made available via the City of Los Angeles Open Data Portal and redistributed on Kaggle under a CC0 (Public Domain) license. This analysis is for educational purposes only and does not reflect the views of the Los Angeles Police Department or Kaggle.

We'll use the read_excel function to read these files, as shown in Listing 4.6.

```
#import libararies
import pandas as pd
import os

#execute for loop on the set of code
for file_name in os.listdir():

    #specifically only reference the .xlsx files in the directory
    if file_name.endswith(".xlsx"):

        #read in the file as a generic object name
        df_temp = pd.read_excel(file_name)

        #stack each file on top of each other into what was a blank df
        df = pd.concat([df, df_temp], axis = 0)
```

Listing 4.6 Read the Excel Files

When running this function, you may notice it takes quite a while to load the datasets. Let's perform a little test. We'll try saving the combined DataFrame as a .csv file, as pandas is more efficient when reading text files compared to Excel files (see Listing 4.7).

```
#save the dataset to .csv
df.to_csv('crime_data.csv', index=False)
```

Listing 4.7 Write the Combined DataFrame to CSV

You'll notice the code performed this operation pretty quickly. Now the final test: Let's load that .csv file back as a new testing DataFrame (see Listing 4.8). We'll call this df_test instead of overwriting df as a precaution.

```
#test the time to read in the .csv file
df_test = pd.read_csv("crime_data.csv")
```

Listing 4.8 Read In CSV File as a Test

It reads this data *very* quickly! Let's examine why.

When you're saving a file with the *.xlsx* extension, the storage mechanism of this file is efficient. This means the same exact data saved as a *.xlsx* will be smaller than if it is saved as a *.csv* file. When loading this data into Anaconda, there is a 100 mb file limit. The combined *.csv* file is about 250 mb, but the combined *.xlsx* file is only about 140 mb. Regardless, the file had to be split when uploading the data to the project so it can be read in by Python, but we only wanted to have two files instead of the three that a *.csv* file would have required (plus this is a good lesson for educational purposes). The takeaway here is a *.xlsx* file is more efficient at storing data than a *.csv* file is.

However, as you saw, it's much slower to read a *.xlsx* file into Python. This is because pandas is more efficient at reading text data than Excel. Think about this as writing in your native language compared to having to translate from another language. While you can technically communicate with both, you're much more efficient when using your native language.

One final point: Why are you saving to a *.csv* file? This is simply to make your life easier in the future. Rarely are you executing the entirety of a machine learning project at one time, so by saving the combined output as a *.csv* file, you're saving yourself a lot of time in the future by not needing to wait so long for our data to load. Moving forward, we'll just be reading in our *.csv* file each time we spin up a session.

After reviewing the data, you'll see we have a much larger dataset than we did in the first two use cases. When running df.shape, there are over one million rows of data and twenty-eight columns. Similar to our second use case, df.head isn't as useful given the number of columns. However, if you look at the underlying data, you can see that each row is a reported crime with more information about that crime. You can see the full list of columns available in the dataset when running df.columns, as shown in Figure 4.6.

```
Index(['DR_NO', 'Date Rptd', 'DATE OCC', 'TIME OCC', 'AREA', 'AREA NAME',
       'Rpt Dist No', 'Part 1-2', 'Crm Cd', 'Crm Cd Desc', 'Mocodes',
       'Vict Age', 'Vict Sex', 'Vict Descent', 'Premis Cd', 'Premis Desc',
       'Weapon Used Cd', 'Weapon Desc', 'Status', 'Status Desc', 'Crm Cd 1',
       'Crm Cd 2', 'Crm Cd 3', 'Crm Cd 4', 'LOCATION', 'Cross Street', 'LAT',
       'LON'],
      dtype='object')
```

Figure 4.6 Column Names

4.2   Data Exploration

Although this section is not the core topic of the book, it's arguably the most important for successfully applying machine learning. Exploring the data allows you to discover and correct any issues. Every model is only as good as it's input, so the common phrase *garbage in, garbage out* applies here.

As you'll see in our examples, messy data creates a lot of noise. This noise can really hamper the effectiveness of a machine learning model. Taking the time to understand your data will generate a significant return on your time and improve your reputation with your stakeholders. Nothing you do looks worse to a business leader than not accounting for a nuance in the data that they're well aware of as a process subject matter expert.

Exploring the data naturally leads to the data cleaning process described in Section 4.3. The goal of understanding your data is not only to know what it contains but to recognize the type of cleaning it requires.

The Importance of Understanding Your Data

In the middle of writing this book, I started a new job in a different business domain after working in human resources (HR) for more than seven years. Having worked in HR for that length of time, I found that the underlying data was very similar, even when changing companies. HR data generally follows the same structure, so moving from one company to another was more about learning the processes for how the data was generated and understanding how the columns were named. This familiarity made it easier move quickly through the data exploration phase. I already had a good feel for the data, and I knew what to look for and where to look for it.

However, when I moved into a new domain, I was completely lost. I had no mental model for how the data should look. The naming conventions of the columns were completely foreign to me, and I struggled to feel confident in knowing the data I was pulling was the data I thought I was pulling. Seeking out the help of knowledgeable leaders and colleagues helped me overcome this hurdle, but I also often relied on the data exploration techniques in this chapter to better understand the data I was looking at.

Now, let's set up our data types and dive into key data exploration techniques: data visualization, descriptive statistics, and correlation analysis.

4.2.1   Data Types

A data type is a category that indicates the kind of data included in a specific column. For example, a *numeric* data type is a column of data that only includes numerals (e.g., 1, 2, 3). A *string* data type is a column that contains words (e.g., orange, apple, banana).

Speaking from experience, it's best to address the data types of your columns as early as possible, even if it seems like a trivial task. The data type isn't always clear when you load in your data, so you'll often need to perform the alterations to the column explicitly (for example, specify that column A should be an integer). Certain functions or filters only work on certain data types, so this is rarely an optional step.

If you don't address these data type adjustments soon after importing your data, you may run into issues in your code later on, especially if you've created different versions of the same DataFrame—some with the correct data types and others without.

Let's consider a generalized example. Say you have a numeric variable for sales, but when you import your data with pandas, it comes in as a string. If halfway through your code you realize you need to change the sales column to an integer data type, you risk breaking your previous code. You'll also find that uniform data types make it easier to take your code from the exploration phase into your modeling phase. The exploration phase is messy, and you'll only take a subset of your code from the exploration phase into your data pipeline when you build a model. Performing all your critical data exploration tasks at one time, rather than ad hoc as they return errors in your code, will make this transition easier and lower the risk of forgetting important steps for your model.

The starting point for this is using the dtypes function. This returns your columns and their associated data types. From there, there are multiple approaches you can take to adjust data types.

Further Resources

We won't go into detail on the various Python data types, which are beyond the scope of this book. For more information, you can find a great quick reference guide from W3 Schools: *https://www.w3schools.com/python/python_datatypes.asp*.

We'll go through each use case and cover a range of situations you're likely to see when starting out with data for a machine learning project. This will cover basic transformations as well as more complex ones. If you don't already think dates have a special place in data hell, you'll soon understand why they can be so frustrating to work with.

Use Case 1: Updating Data Types

For our retail sales use case, the data provided by Chris imported well overall. If you look at the various columns, you'll see our Price and Quantity columns are both a numeric data type. The main gap is that our InvoiceDate column is an object. Should be simple to fix, right? Let's check the data types of our data:

```
#see the data types of the data frame
df.dtypes
```

Running this function will return the data shown in Table 4.1.

Column	Type
Invoice	Object
StockCode	Object
Description	Object
Quantity	Float64
InvoiceDate	Object
Price	Float64
Customer ID	Float64
Country	Object

Table 4.1 dtypes Function Result

You'll notice that our InvoiceDate column shows as an object. After a quick review of the column's values, it looks like the date structure is day/month/year hour:minute. If you're in the United States, this may be a bit confusing, but nothing that Python can't handle. There is a specific syntax for date parts. You can see this in Listing 4.9 in the format parameter, where the value passed represents the format as %d/%m/%Y %H:%M. This is how you can pass the expected format of the date for pandas to convert your column into a date that the code can understand. However, when running the code in Listing 4.9, we get an error message, which you can see in Figure 4.7 and Figure 4.8.

```
ValueError                               Traceback (most recent call last)
Cell In[12], line 1
----> 1 df['new_invoicedate'] = pd.to_datetime(df['InvoiceDate'], format = '%d/%m/%Y %H:%M')

File /opt/conda/envs/anaconda-panel-2023.05-py310/lib/python3.11/site-packages/pandas/core/tools/datetimes.py:1046, in to_datet
ime(arg, errors, dayfirst, yearfirst, utc, format, exact, unit, infer_datetime_format, origin, cache)
   1044         result = arg.tz_localize("utc")
   1045 elif isinstance(arg, ABCSeries):
-> 1046     cache_array = _maybe_cache(arg, format, cache, convert_listlike)
   1047     if not cache_array.empty:
   1048         result = arg.map(cache_array)

File /opt/conda/envs/anaconda-panel-2023.05-py310/lib/python3.11/site-packages/pandas/core/tools/datetimes.py:250, in _maybe_ca
che(arg, format, cache, convert_listlike)
    248 unique_dates = unique(arg)
    249 if len(unique_dates) < len(arg):
--> 250     cache_dates = convert_listlike(unique_dates, format)
    251     # GH#45319
    252     try:

File /opt/conda/envs/anaconda-panel-2023.05-py310/lib/python3.11/site-packages/pandas/core/tools/datetimes.py:453, in _convert_
listlike_datetimes(arg, format, name, utc, unit, errors, dayfirst, yearfirst, exact)
    451 # `format` could be inferred, or user didn't ask for mixed-format parsing.
    452 if format is not None and format != "mixed":
--> 453     return _array_strptime_with_fallback(arg, name, utc, format, exact, errors)
    455 result, tz_parsed = objects_to_datetime64ns(
    456     arg,
    457     dayfirst=dayfirst,
    (...)
    461     allow_object=True,
    462 )
    464 if tz_parsed is not None:
    465     # We can take a shortcut since the datetime64 numpy array
    466     # is in UTC
```

Figure 4.7 Date Format Error (Part 1)

```
File /opt/conda/envs/anaconda-panel-2023.05-py310/lib/python3.11/site-packages/pandas/core/tools/datetimes.py:484, in _array_st
rptime_with_fallback(arg, name, utc, fmt, exact, errors)
    473 def _array_strptime_with_fallback(
    474     arg,
    475     name,
    (...)
    479     errors: str,
    480 ) -> Index:
    481     """
    482     Call array_strptime, with fallback behavior depending on 'errors'.
    483     """
--> 484     result, timezones = array_strptime(arg, fmt, exact=exact, errors=errors, utc=utc)
    485     if any(tz is not None for tz in timezones):
    486         return _return_parsed_timezone_results(result, timezones, utc, name)

File /opt/conda/envs/anaconda-panel-2023.05-py310/lib/python3.11/site-packages/pandas/_libs/tslibs/strptime.pyx:530, in pandas.
_libs.tslibs.strptime.array_strptime()

File /opt/conda/envs/anaconda-panel-2023.05-py310/lib/python3.11/site-packages/pandas/_libs/tslibs/strptime.pyx:351, in pandas.
_libs.tslibs.strptime.array_strptime()

ValueError: time data "13-12-2009 09:58" doesn't match format "%d/%m/%Y %H:%M", at position 1179. You might want to try:
    - passing `format` if your strings have a consistent format;
    - passing `format='ISO8601'` if your strings are all ISO8601 but not necessarily in exactly the same format;
    - passing `format='mixed'`, and the format will be inferred for each element individually. You might want to use `dayfirst`
alongside this.
```

Figure 4.8 Date Format Error (Part 2)

```
#create new column called new_invoicedate with properly formatted date
df['new_invoicedate'] = pd.to_datetime(df['InvoiceDate'], format = '%d/%m/%Y %H:%M')
```

Listing 4.9 Formatting Data Column

The error message tells us the format doesn't match (`time data "13-12-2009 09:58"` `doesn't match format "%d/%m/%Y %H:%M`). If you go back and look at the columns more carefully, you'll see that some of the data has slashes (/) and some has dashes (-) to separate the date parts. This is certainly not ideal, and it's something that you should bring up to Chris and the team that manages the data storage.

However, you can account for this by replacing the - with a /, as shown in Listing 4.10. Now the code works as you want with no error messages! You can see the new column is formatted as a timestamp in Figure 4.9. This will help you with the visualizations required in Section 4.2.2.

```
#replace '-' with '/' in the column
df['new_invoicedate'] = df['InvoiceDate'].str.replace('-', '/')

#create the new column with the properly formatted date
df['new_invoicedate'] = pd.to_datetime(
  df['new_invoicedate'],
  format = '%d/%m/%Y %H:%M'
  )

#preview the output
df.head()
```

Listing 4.10 Adjusting Date Field Before Converting into a Date

	Invoice	StockCode	Description	Quantity	InvoiceDate	Price	Customer ID	Country	new_invoicedate
0	489434	85048	15CM CHRISTMAS GLASS BALL 20 LIGHTS	12.0	1/12/2009 7:45	6.95	13085.0	United Kingdom	2009-12-01 07:45:00
1	489434	79323P	PINK CHERRY LIGHTS	12.0	1/12/2009 7:45	6.75	13085.0	United Kingdom	2009-12-01 07:45:00
2	489434	79323W	WHITE CHERRY LIGHTS	12.0	1/12/2009 7:45	6.75	13085.0	United Kingdom	2009-12-01 07:45:00
3	489434	22041	RECORD FRAME 7" SINGLE SIZE	48.0	1/12/2009 7:45	2.10	13085.0	United Kingdom	2009-12-01 07:45:00
4	489434	21232	STRAWBERRY CERAMIC TRINKET BOX	24.0	1/12/2009 7:45	1.25	13085.0	United Kingdom	2009-12-01 07:45:00

Figure 4.9 New Timestamp Column Created from Data

Another dynamic to consider with dates is that there is a difference between timestamps and dates. When you perform data visualization to further explore the data, it's likely you don't want to see the data by the timestamp, as seeing it day by day is often more valuable. Datetime, a powerful Python library for dates, allows you to extract the date from the timestamp, as shown in Listing 4.11. You can see the result in Figure 4.10, where we have the same date with the timestamp removed.

```
#import the datetime library, a library used for date and timestamp columns
import datetime

#extract just the date from the timestamp
df['invoicedate'] = df['new_invoicedate'].dt.date

#preview the output
df.head()
```

Listing 4.11 Code to Create Date from Timestamp Using Datetime

	Invoice	StockCode	Description	Quantity	InvoiceDate	Price	Customer ID	Country	new_invoicedate	invoicedate
0	489434	85048	15CM CHRISTMAS GLASS BALL 20 LIGHTS	12.0	1/12/2009 7:45	6.95	13085.0	United Kingdom	2009-12-01 07:45:00	2009-12-01
1	489434	79323P	PINK CHERRY LIGHTS	12.0	1/12/2009 7:45	6.75	13085.0	United Kingdom	2009-12-01 07:45:00	2009-12-01
2	489434	79323W	WHITE CHERRY LIGHTS	12.0	1/12/2009 7:45	6.75	13085.0	United Kingdom	2009-12-01 07:45:00	2009-12-01
3	489434	22041	RECORD FRAME 7" SINGLE SIZE	48.0	1/12/2009 7:45	2.10	13085.0	United Kingdom	2009-12-01 07:45:00	2009-12-01
4	489434	21232	STRAWBERRY CERAMIC TRINKET BOX	24.0	1/12/2009 7:45	1.25	13085.0	United Kingdom	2009-12-01 07:45:00	2009-12-01

Figure 4.10 Create Date from Timestamp Result

Expect Some Troubleshooting

While the example as written makes it appear as if the issue was caught right away, that is far from the truth. I spent the better part of 30 minutes troubleshooting this example before I realized the issue with the different date part delimiters. This is a common occurrence, especially when working with new datasets, and it's part of the reality of preparing your data for a machine learning model. Dates can be very frustrating, but the more practice you have with them, the better you'll get!

Use Case 2: Updating Data Types

Let's use the same df.dtypes code on our second use case order history data. You can see the columns and their associated data types in Figure 4.11.

```
Restaurant ID                                           int64
Restaurant name                                        object
Subzone                                                object
City                                                   object
Order ID                                                int64
Order Placed At                                        object
Order Status                                           object
Delivery                                               object
Distance                                               object
Items in order                                         object
Instructions                                           object
Discount construct                                     object
Bill subtotal                                         float64
Packaging charges                                     float64
Restaurant discount (Promo)                           float64
Restaurant discount (Flat offs, Freebies & others)    float64
Gold discount                                         float64
Brand pack discount                                   float64
Total                                                 float64
Rating                                                float64
Review                                                 object
Cancellation / Rejection reason                        object
Restaurant compensation (Cancellation)                float64
Restaurant penalty (Rejection)                        float64
KPT duration (minutes)                                float64
Rider wait time (minutes)                             float64
Order Ready Marked                                     object
Customer complaint tag                                 object
Customer ID                                            object
dtype: object
```

Figure 4.11 Data Types

We have a mix of text-based columns (object) and number-based columns (int64 and float64), but the column that sticks out the most is Order Placed At. This is the date column of the dataset, which will be important for us to convert into a date given Mary's goal of seeing the probability of a customer placing another order within the next week.

To look further into what the values of this column look like, we'll run the code in Listing 4.12. Figure 4.12 shows its output. This code acts as if we're picking the Order Placed At column to be the sole column in a DataFrame (accomplished with the double brackets) and then uses the .head() function to see the first 10 values of the column.

```
#preview the first 10 values of only the order placed at column
df[['Order Placed At']].head(10)
```

Listing 4.12 First 10 Values of Only the Order Placed at Column

	Order Placed At
0	11:38 PM, September 10 2024
1	11:34 PM, September 10 2024
2	03:52 PM, September 10 2024
3	03:45 PM, September 10 2024
4	03:04 PM, September 10 2024
5	12:28 PM, September 10 2024
6	12:03 AM, September 10 2024
7	10:54 PM, September 09 2024
8	10:51 PM, September 09 2024
9	03:22 PM, September 09 2024

Figure 4.12 Date Values Sample

If you audibly groaned, that's okay, because I certainly did when I was going through the dataset. Unfortunately, other functions won't be able to read the timestamps in their current state. However, this is a great learning opportunity for handling this unique type of date format, and it's a chance to practice leveraging generative AI tools to expedite your coding process. Provide the following prompt to Copilot:

I'm cleaning up a dataset with Python. The timestamp column is formatted in a weird way. Can you help me create some code to get this timestamp formatted in the proper way for other functions to read it? An example of the current format is: 11:38 PM, September 10 2024.

The response from Copilot was helpful in multiple ways. The first is the code response it provided, which is shown in Listing 4.13. The code creates a sample DataFrame with a datetime column formatted like our date column, and then it reformates the date.

```
import pandas as pd

# Example data
data = {'timestamp': ['11:38 PM, September 10 2024', '07:15 AM, July 19 2025']}
df = pd.DataFrame(data)

# Convert the timestamp column
df['timestamp'] = pd.to_datetime(df['timestamp'], format='%I:%M %p, %B %d %Y')

# Now your timestamp column is in standard datetime format
print(df)
```

Listing 4.13 Copilot Answer for Timestamp Formatting Question

This code still needs to be altered to fit our exact data and test how it did, which is shown in Listing 4.14 with the results of the code in Figure 4.13. While the data format conversion is operating as we want, we need to adjust the code to reference the name of our `datetime` column instead of the generated example column `timestamp`.

```
#convert the timestamp column
df['timestamp'] = pd.to_datetime(df['Order Placed At'],
  format='%I:%M %p, %B %d %Y')


#now your timestamp column is in standard datetime format
df[['timestamp']].head(10)
```

Listing 4.14 Adjusting Copilot Answer for Our Data

	timestamp
0	2024-09-10 23:38:00
1	2024-09-10 23:34:00
2	2024-09-10 15:52:00
3	2024-09-10 15:45:00
4	2024-09-10 15:04:00
5	2024-09-10 12:28:00
6	2024-09-10 00:03:00
7	2024-09-09 22:54:00
8	2024-09-09 22:51:00
9	2024-09-09 15:22:00

Figure 4.13 Timestamp Column Updated

The second helpful part of Copilot's response (besides giving an accurate solution) is that it also provided an explanation of how this works (see Figure 4.14). Especially as you're learning, it's important to understand *how* code works. In this case, the % followed by a letter is Python's way of identifying date elements. Compared to the generic results you would get from an internet search on the code, this gives you a response tailored to your specific scenario. Using AI here ends up being a win-win situation. You're able to get to a result faster and learn at the same time.

☑ **What this does:**

- `%l:%M %p` reads the time in 12-hour format with AM/PM.

- `%B %d %Y` reads the full month name, day, and year.

- The result will be a `datetime64` type column, ready for any time-based filtering, grouping, or resampling.

If you're reading this from a CSV or another source, you can plug it directly into the conversion step. Want help chaining this into a larger data-cleaning pipeline?

Figure 4.14 Copilot Explanation of Timestamp Code

Why Not Use Generative AI for the Date Column in Use Case 1?

This is a fair question! My goal was to highlight the differences between working through a problem with an internet search and working through a problem with generative AI. The second approach is where the programming world is going, but it's also important to understand how the standard debugging process works.

Use Case 3: Updating Data Types

As we have for the other use cases, we'll run `df.dtypes` to get the data types of each of our columns for our crime data (see Figure 4.15).

```
DR_NO               int64
Date Rptd          object
DATE OCC           object
TIME OCC            int64
AREA                int64
AREA NAME          object
Rpt Dist No         int64
Part 1-2            int64
Crm Cd              int64
Crm Cd Desc        object
Mocodes            object
Vict Age            int64
Vict Sex           object
Vict Descent       object
Premis Cd         float64
Premis Desc        object
Weapon Used Cd    float64
Weapon Desc        object
Status             object
Status Desc        object
Crm Cd 1          float64
Crm Cd 2          float64
Crm Cd 3          float64
Crm Cd 4          float64
LOCATION           object
Cross Street       object
LAT               float64
LON               float64
dtype: object
```

Figure 4.15 Data Types

As has been the trend, we need to convert our date values into actual dates so we can use those columns for time series or other ways of segmenting data by date. We'll need to do this for Date Rptd, DATE OCC, and TIME OCC. The Date Rptd is the date on which the crime was reported. The DATE/TIME OCC columns are when the crime occurred.

The Date Rptd column should be the easiest to correct as it's already in the proper format; we just need to convert the data type. We'll want to convert DATE OCC and TIME OCC into a full timestamp for ease of use in the future.

First, we'll transform the Date Rptd column in Listing 4.15. This is easy because the date itself is already in the year-month-date format, which is the standard date format for pandas.

```
#convert date rptd column into the pandas datetime format
df['Date Rptd'] = pd.to_datetime(df['Date Rptd'])
```

Listing 4.15 Convert Date Reported Column into Date Time Data Type

Next, we'll create a new column that combines the DATE OCC column and the TIME OCC column into one timestamp column. This is where generative AI can once again be our friend. The nuance with timestamps can be challenging, so let's give Copilot the prompt shown in Figure 4.16.

```
I have two columns of data, one is a date and another is the time. Can you
help me write the Python code to combine these columns to make one
timestamp column? Here is a sample of the data

DATE OCC       TIME OCC 0    2020-03-01    2130 1   2020-02-08
        1800 2   2020-11-04    1700 3   2020-09-09    630 4
        2020-05-02      1800
```

Figure 4.16 Copilot Prompt for Timestamps

The result is a robust code solution (see Figure 4.17) and a thorough explanation of that code (see Figure 4.18).

Now, it's important that we test and validate the solution, rather than simply trusting it. Start by taking the middle section of the code that isn't related to the testing component of Copilot's code, as shown in Listing 4.16.

```
#backfill the time column to ensure it has at least enough values
df['TIME OCC'] = df['TIME OCC'].astype(str).str.zfill(4)

#extract the necessary components for a timestamp
df['TIME OCC'] = df['TIME OCC'].str[:2] + ':' + df['TIME OCC'].str[2:]

#concatenate the date and time together to creat a timestamp
df['TIMESTAMP'] = pd.to_datetime(df['DATE OCC'] + ' ' + df['TIME OCC'])
```

Listing 4.16 Convert Two Columns into One Timestamp Column

Absolutely, Jason! Since your time column uses a compact format like `2130` for 9:30 PM and `630` for 6:30 AM, we'll need to pad it to four digits and then combine it with the date column before converting to a proper timestamp. Here's a clean and efficient way to do it using pandas:

```python
import pandas as pd

# Sample data
data = {
    'DATE OCC': ['2020-03-01', '2020-02-08', '2020-11-04', '2020-09-09', '202
    'TIME OCC': [2130, 1800, 1700, 630, 1800]
}

df = pd.DataFrame(data)

# Step 1: Pad TIME OCC to 4 digits and convert to HH:MM format

df['TIME OCC'] = df['TIME OCC'].astype(str).str.zfill(4)
df['TIME OCC'] = df['TIME OCC'].str[:2] + ':' + df['TIME OCC'].str[2:]

# Step 2: Combine DATE OCC and TIME OCC into a single datetime column

df['TIMESTAMP'] = pd.to_datetime(df['DATE OCC'] + ' ' + df['TIME OCC'])

# Display result

print(df[['DATE OCC', 'TIME OCC', 'TIMESTAMP']])
```

Figure 4.17 Copilot Code Solution for Prompt

What's happening here:

- `str.zfill@` ensures times like `630` become `0630`

- We slice and format it into `HH:MM`

- Then we merge the date and time strings and convert them into proper `datetime` objects

Let me know if you want to localize the timestamp to a specific timezone or extract features like hour, weekday, etc. I can help you build that out too.

Figure 4.18 Copilot Code Explanation

Everything executes properly, but did it do what we wanted it to do? To test this, we'll show a preview of the two columns used to create the new column (see Listing 4.17 and Figure 4.19).

```
#check only the date and time columns alongside the new timestamp column
df[['DATE OCC', 'TIME OCC', 'TIMESTAMP']]
```

Listing 4.17 Select Three Columns Related to the Timestamp Column Creation and Result

DATE OCC	TIME OCC	TIMESTAMP
2020-03-01	21:30	2020-03-01 21:30:00
2020-02-08	18:00	2020-02-08 18:00:00
2020-11-04	17:00	2020-11-04 17:00:00
2020-09-09	06:30	2020-09-09 06:30:00
2020-05-02	18:00	2020-05-02 18:00:00
...	...	...
2024-07-23	14:00	2024-07-23 14:00:00
2024-01-15	01:00	2024-01-15 01:00:00
2024-10-11	23:30	2024-10-11 23:30:00
2024-04-24	15:00	2024-04-24 15:00:00
2024-08-12	23:00	2024-08-12 23:00:00

Figure 4.19 Preview of the New Timestamp Column

This column looks great, so now we can remove the two original columns, as shown in Listing 4.18, since they're no longer needed.

```
#drop the original date and time columns
df = df.drop(columns=['DATE OCC','TIME OCC'])
```

Listing 4.18 Remove Original Two Columns Used to Create New Timestamp Column

You'll notice a common thread across each use case. Dates almost always need some work before they're ready for exploration and modeling.

4.2.2   Data Visualization

Spend some time going through the retail dataset row by row without any visuals. You can do this in Excel or by any other means you prefer. Did you find any patterns? Probably not. That's because humans are much better at identifying patterns and trends in a visual format as opposed to tabular data.

We'll now go through each of the use cases to visualize their respective datasets with the goal of better understanding the data and looking for outliers. Remember, creating a beautiful chart isn't part of that goal. Spending time on formatting for this purpose will only slow you down! You're not creating visuals for others—you're the only one who needs to understand the output. Elements such as titles, axis labels, data labels, and value formatting are not required.

Visualizations in Python

If you're an Excel user who often creates plots, you'll find the underlying concepts for visualizing data in Python are actually quite similar. For basic Python functions and Matplotlib, you need to summarize your data prior to visualizing it. This is similar to Excel, where you often need to put your data into a PivotTable first before you can visualize it.

If you're coming from a Power BI or Tableau background, this data summarization step may feel unfamiliar or like extra work. Software like Power BI or Tableau performs this step automatically on the backend, which makes them easy to use for data visualization. Luckily, pandas has built-in functions that simplify the data aggregation process.

Use Case 1: Visualization to Explore the Data

This dataset has more than one million rows in it, so it's a prime use case for using data visualization to inform your understanding and look for outliers. Since this is sales data, a good place to start is looking for trends with a line graph.

In order to do this, we need to summarize the data by date. There are so many different ways to achieve this in Python. Our approach in Listing 4.19 uses pandas and the groupby() function, with a few other items that we'll discuss next. Figure 4.20 shows the result of this code.

```
#group/summarize the data by date, while summing price
#resetting the index ensures the column names of the output can be referenced
df_summarized = pd.DataFrame(
  df.groupby('invoicedate')['Price'].sum()
  ).reset_index()


#preview the data
df_summarized.head()
```

Listing 4.19 Grouping the Data by Date, Summing the Price

	invoicedate	Price
0	2009-12-01	14450.54
1	2009-12-02	13411.96
2	2009-12-03	13290.25
3	2009-12-04	9687.28
4	2009-12-05	1443.26

Figure 4.20 Result of groupby() While Summing Price

As you string together various functions and methods, it's best to read from the inside to the outside. Here is how to read this function: Using df, we'll groupby() the invoice-date field and add the Price column as part of the groupby() aggregation. groupby() doesn't return a DataFrame, so we need to convert it into a DataFrame and then reset the index to ensure the column Price is actually called Price. Without resetting the index, it may appear that our column is called Price (just elevated above the other column names). However, it won't behave as such if we try to reference Price in our code. Apart from the reset_index(), this is more or less like creating a PivotTable by invoice date that sums price. Figure 4.21 shows what the DataFrame looks like without the reset_index() added on.

invoicedate	Price
2009-12-01	14450.54
2009-12-02	13411.96
2009-12-03	13290.25
2009-12-04	9687.28
2009-12-05	1443.26

Figure 4.21 groupby() Without Resetting the Index

Now that we have our data summarized with the price for each day, we can visualize it. Since it's daily data, we'll use a line plot. Listing 4.20 shows how to create a simple line graph. As mentioned earlier, the objective is to keep this simple, so it's okay to have an ugly graph (and this one is ugly). Some trends should pop out right away; take some time to review Figure 4.22 and jot down what you think stands out.

```
#import matplotlib the most common plotting library in Python
import matplotlib.pyplot as plt

#plot a line graph based on the date and price
df.plot.line(x='invoicedate', y='Price')
```

Listing 4.20 Create a Line Graph of Our Data

Some key observations from this data include:

- There are some large negative numbers. In the context of sales, this could represent returns; however, you should be skeptical about the magnitude of these negative numbers.
- Additionally, there are some significantly large positive numbers. Optimistically, it'd be nice to assume these are great sales days, but skepticism is important when you're working on a machine learning project.

- Ideally, we should be able to get an early sense of how sales are trending, but there is quite a bit of noise that makes it hard to determine the direction of sales.

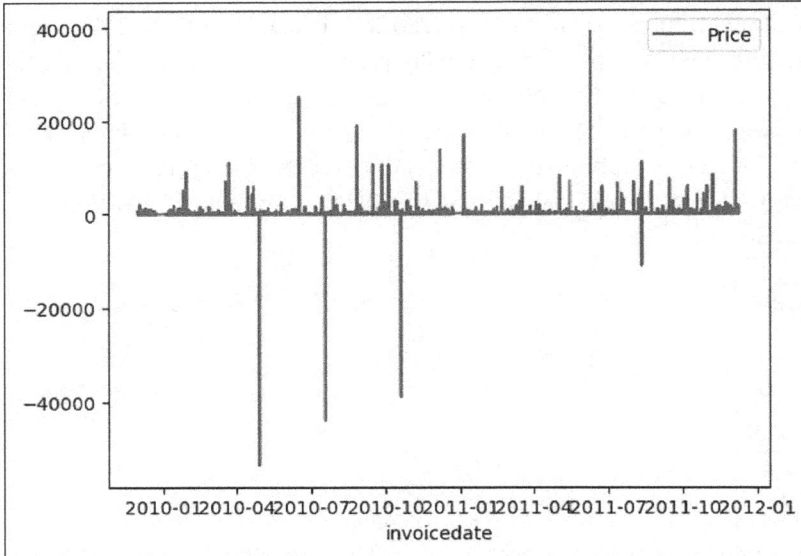

Figure 4.22 Plotting Price Over Time Without Any Data Cleaning

We can safely assume that the negative price values don't make sense, so let's review their rows. Execute the following code to filter to the rows where the price is negative:

```
#check the data where price is negative
df[df['Price'] < 0]
```

We're able to see that there are five rows in Figure 4.23. Notable is the Description column values saying Adjust bad debt. When Chris provided this data, he suggested it was sales data, but now it looks like it could be more complicated than that.

	Invoice	StockCode	Description	Quantity	InvoiceDate	Price	Customer ID	Country	new_invoicedate	invoicedate
179403	A506401	B	Adjust bad debt	1.0	29-04-2010 13:36	-53594.36	NaN	United Kingdom	2010-04-29 13:36:00	2010-04-29
37839	A563186	B	Adjust bad debt	1.0	12/8/2011 14:51	-11062.06	NaN	United Kingdom	2011-08-12 14:51:00	2011-08-12
37840	A563187	B	Adjust bad debt	1.0	12/8/2011 14:52	-11062.06	NaN	United Kingdom	2011-08-12 14:52:00	2011-08-12
26275	A516228	B	Adjust bad debt	1.0	19-07-2010 11:24	-44031.79	NaN	United Kingdom	2010-07-19 11:24:00	2010-07-19
153473	A528059	B	Adjust bad debt	1.0	20-10-2010 12:04	-38925.87	NaN	United Kingdom	2010-10-20 12:04:00	2010-10-20

Figure 4.23 All Rows with Price Less Than 0

After this finding, you go back to Chris for additional clarification, and he tells you this dataset includes ledger items as well as sales data. Debt is included, as we've already found, but credits are also included, and they aren't quite as straightforward to find.

Stop and Explore

With the information you've gotten from Chris, explore the data and see what methods you can use to filter out non-sales-specific data.

To demonstrate, we've downloaded one of the raw files for this use case from GitHub and looked through each of the columns, filtering and sorting them to look for patterns. An example finding: If Quantity is 0, Price is negative, or Customer ID is not a number (NaN), then the data needs to be removed. This is done in Listing 4.21. After performing this task, look at how clean the visual of the data is in Figure 4.24!

```
#apply filters based on the information from Chris and your exploration
df_cleaned = df[
  (df['Quantity'] > 0) &
  (df['Price'] > 0) &
  (df['Customer ID'] > 0)
  ]

#group the dataframe again by date and summing price
df_summarized = pd.DataFrame(df_cleaned.groupby('invoicedate')[ 'Price'].sum())

#reset the index to ensure the columns of the output can be referenced
df_summarized = df_summarized.reset_index()

#plot the data by date and price
df_summarized.plot.line(x='invoicedate ', y='Price')
```

Listing 4.21 Cleaning Data and Then Plotting Again

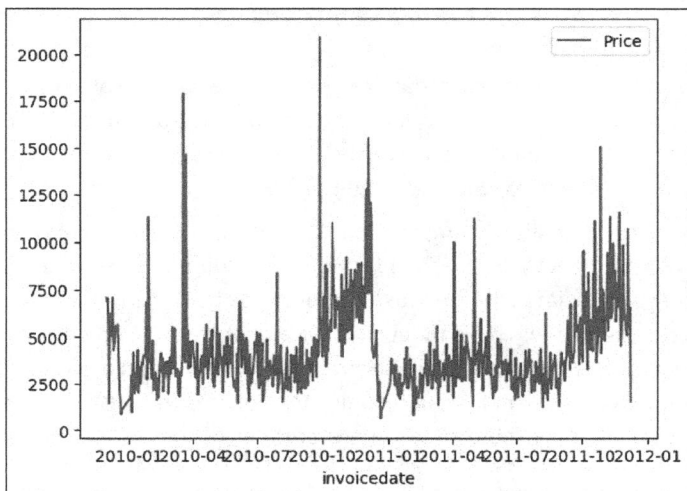

Figure 4.24 Plotting Price Over Time After Cleaning the Data

While we won't get to the details on data cleaning until Section 4.3, the lines are almost always blurred between these processes, so we've taken a first step here for the purposes of our example. Remember that the intended purpose of the data visualization is to understand the data. When you're learning new things about the data that need to be accounted for, then you should make those adjustments as you go.

Do you notice any trends in the data? Take some time to review the graph yourself and see what trends you can find.

The overall trend seems to have some seasonality in it; that is, a recurring pattern of fluctuations. Just from looking at the graph, you can see the sales go up toward the end of each year. There are also a number of large spikes, so we'll want to evaluate those to make sure we're confident they aren't bad data we didn't see before. We'll look at the row-level detail using the following approach, which is similar to the one we used previously:

```
#preview data where the price is greater than 10000
df_cleaned[df_cleaned['Price'] > 10000]
```

This results in the view shown in Figure 4.25.

	Invoice	StockCode	Description	Quantity	InvoiceDate	Price	Customer ID	Country	new_invoicedate	invoicedate
135013	502263	M	Manual	1.0	23-03-2010 15:22	10953.5	12918.0	United Kingdom	2010-03-23 15:22:00	2010-03-23
108640	524159	M	Manual	1.0	27-09-2010 16:12	10468.8	14063.0	United Kingdom	2010-09-27 16:12:00	2010-09-27

Figure 4.25 Price Is Greater Than $10,000

There are only two prices higher than $10,000 and their attributes don't suggest anything is incorrect. These records appear to be outliers of high order counts rather than incorrect data that should be excluded.

The Value of Being Familiar with the Data

As you just experienced with the first use case, data can be messy and nuanced, even when it's less than ten columns. Think about how large most companies' databases are and how much more complex these nuances can become. In my career, I've found some of the most valuable people are the ones who understand the data and its quirks.

I once had a colleague who had been at the company for many years, and he led a team dedicated to providing reports directly to the C-suite. He had accumulated so much knowledge about the company's data that he was promoted to the last level in his career track before vice president (VP). Think about that for a minute: Someone dedicated to ad hoc reporting was one step away from becoming a VP. He was an amazing and beloved employee, which definitely helps, but his value to the company was understanding the data, and he was compensated handsomely for it.

Use Case 2: Visualization to Explore the Data

Given this second use case also has a time component, let's visualize the data by the properly formatted timestamp. Doing so reveals when orders have come in historically. To create this view, we'll aggregate by the `timestamp` field and then count the number of rows to get the number of orders (see Listing 4.22). The results from the code, shown in Figure 4.26, illustrate that this approach doesn't quite give you the insight you need.

```
#count the number of occurances for each value in the timestamp column
df_aggregated = pd.DataFrame(
    df.groupby('timestamp')['Order ID'].count()
    ).reset_index()

#preview the results
df_aggregated.head()
```

Listing 4.22 Aggregating Data by Timestamp

	timestamp	Order ID
0	2024-09-01 00:13:00	1
1	2024-09-01 01:52:00	1
2	2024-09-01 01:54:00	1
3	2024-09-01 02:10:00	1
4	2024-09-01 02:17:00	1

Figure 4.26 Count of Order ID by Timestamp

Sometimes, the difference between a date and a timestamp can be blurred. However, when aggregating the data in this case, it's unlikely we'll need to know to the minute how many orders occurred. Instead, we'll filter by the date of the orders. To do this, we'll create a date column from the timestamp field and then aggregate the results using that date field, as shown in Listing 4.23.

```
#create new column of just the date from the timestamp
df['order_date'] = df['timestamp'].dt.date

#preview the new date column against the original timestamp column
df[['timestamp', 'order_date']].head(10)
```

Listing 4.23 Create Date from Timestamp Field

As you can see, once a field is considered a date, you're able to more easily extract parts from it. In this case, we're using `dt.date` to extract the date from the timestamp field without needing to use % followed by numbers to extract components of the date.

Figure 4.27 shows the result of the code with a view cf the timestamp and the new order_
date field.

timestamp	order_date
2024-09-10 23:38:00	2024-09-10
2024-09-10 23:34:00	2024-09-10
2024-09-10 15:52:00	2024-09-10
2024-09-10 15:45:00	2024-09-10
2024-09-10 15:04:00	2024-09-10
2024-09-10 12:28:00	2024-09-10
2024-09-10 00:03:00	2024-09-10
2024-09-09 22:54:00	2024-09-09
2024-09-09 22:51:00	2024-09-09
2024-09-09 15:22:00	2024-09-09

Figure 4.27 Result of Converting Timestamp to Date

Now we can rerun the code to aggregate the results by the new order_date column. We
get the results by date, as we'd hoped for (see Listing 4.24 and Figure 4.28).

```
#group the data by date while counting the number of orders
df_aggregated = pd.DataFrame(
    df.groupby('order_date')['Order ID'].count()
    ).reset_index()

#preview the results
df_aggregated.head()
```

Listing 4.24 Aggregating Number of Orders by Date

order_date	Order ID
2024-09-01	141
2024-09-02	85
2024-09-03	99
2024-09-04	156
2024-09-05	91

Figure 4.28 Orders by Date

Now we can put the results into a line graph to make it easier to see potential trends by date in the data. This differs from our first use case, where the goal was explicitly predicting future sales by date. Here, we're focused on whether a single customer will purchase again, but it's still wise to understand overall trends before proceeding. Significant variations in the volume of orders over time may suggest time-based factors are important to account for. We'll create a line graph using the same line simple line plot syntax we used in our first use case (see Listing 4.25). Take a moment to review the line graph in Figure 4.29 to identify any trends or make observations about the data.

```
#plot the results by date and count of orders
df_aggregated.plot.line(
   x='order_date',
   y='Order ID'
   )
```

Listing 4.25 Summary of Orders by Date in Line Graph

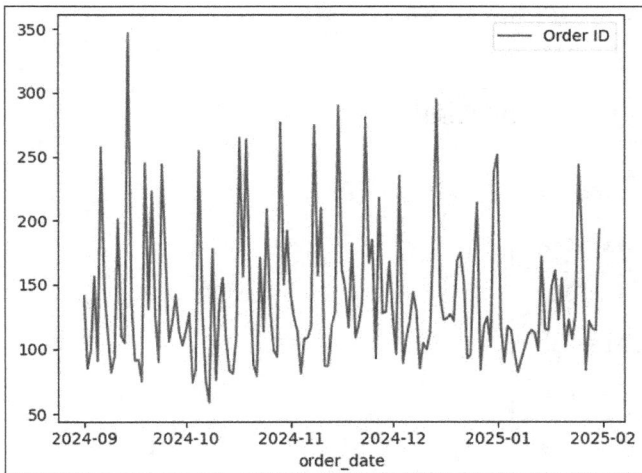

Figure 4.29 Line Graph of Orders by Date

As you may have noticed, this graph has notable spikes occurring with some degree of frequency. This is where understanding your use case and the applicable context is important. These are food delivery orders, so the day of the week seems like it could be relevant. Let's see how the data looks from that view.

To do this, we'll create a new day of the week column and filter data by it. This allows us to see what the order trend looks like by the day of the week. The order of the code in Listing 4.26 is intentional. We need to aggregate the results by the date *first*, before taking the average of each day of the week. This ensures we're calculating the answer to a question like "What is the average number of orders on Monday?" The results are displayed in Figure 4.30.

```
#create a day of the week column, this is a numeric value
df_aggregated['day_of_week'] = pd.to_datetime(
  df_aggregated['order_date']
  ).dt.weekday

#create a day name column
df_aggregated['day_name'] = pd.to_datetime(
  df_aggregated['order_date']
  ).dt.day_name()

#group the data by the day of the week
df_weekdays = pd.DataFrame(
  df_aggregated.groupby(
    ['day_of_week', 'day_name']
    )['Order ID'].mean()
  ).reset_index()

#preview the results
df_weekdays.head(10)
```

Listing 4.26 Aggregated Results by Day of the Week

day_of_week	day_name	Order ID
0	Monday	99.818182
1	Tuesday	131.136364
2	Wednesday	139.863636
3	Thursday	130.863636
4	Friday	154.681818
5	Saturday	186.809524
6	Sunday	134.454545

Figure 4.30 Orders by Day of the Week

In this case, we're calculating both the numeric day of the week (dt.weekday) as well as the name of the day (dt.day_name()). This isn't required, but it does help provide clarity to ensure it's clear whether 0 equates to Sunday or Monday.

The tabular results are helpful; however, to ensure we don't miss any other potential insights, we can also visualize the data with a bar graph, as shown in Listing 4.27, using the .plot function. We'll tell the function which columns to use and then specify which type of plot we'd like to generate. Using the results from Figure 4.30 and Figure 4.31, take a moment to reflect on what this data shows.

```
import pandas as pd
import matplotlib.pyplot as plt

#create bar graph for orders by the day of the week
ax = df_weekdays.plot(
  x='day_name',
  y='Order ID',
  kind='bar'
  )

#display the bar graph
plt.show()
```

Listing 4.27 Bar Graph for Day of the Week Data

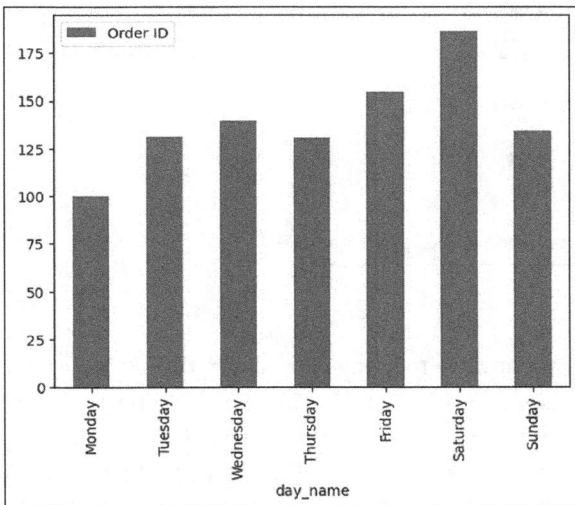

Figure 4.31 Day of the Week Bar Graph

Some key observations from the graph include the following:

- Saturday has the highest order volume, followed by Friday.
- Monday has the lowest order volume.
- Tuesday, Wednesday, Thursday, and Sunday all generally have the same order volume.

This tells us that we should account for the day of the week someone places an order when predicting whether they're going to make another order within the next week.

Now that we have a feel for the time trends of the dataset, let's move on to the customer trends. Given our use case of predicting if someone will make another order within the next week, let's look at the landscape of customer order frequency. We have the customer ID in the dataset, so we can perform a similar activity of aggregating the orders by customer ID. As shown in Listing 4.28, we're summarizing the data at the customer level using the groupby function, resulting in the output shown in Figure 4.32.

```
#create data frame to see how many orders each customer has made
df_customer_frequency = pd.DataFrame(
    df.groupby('Customer ID')['Order ID'].count()
    ).reset_index()

#preview the results
df_customer_frequency.head()
```

Listing 4.28 Aggregate the Number of Orders by Customer ID

	Customer ID	Order ID
	000285ae83ecf06a92b936d4f5b74342edb0e1940e1f00...	3
	00062fa202370fdd3076b794ec9358f36e35469bb8e5ff...	1
	001ab5fc3ee158b4d22e106897cee9b355e6eed50f163e...	1
	001fdf2511dd137361424c9c15ea54774476691ceec5fa...	1
	002afbd83626f0c699892fcb07b2ddf3858911482647b6...	4

Figure 4.32 Orders by Customer ID

At this stage, you might be wondering how we're going to visualize this. Based on the size of the data, there are way too many customers to use a bar graph. A line graph also won't be helpful here. This is a great use case for a *histogram* (a type of bar graph showing the distribution of numerical data, grouped into ranges) to identify the frequency at which customers are ordering.

When building the histogram, as shown in Listing 4.29, we'll define the bins for our use case to ensure the visualization isn't misleading. To do this, we feed it a list of values defined in bin_vals. We're using the range function for this, as it generates a list of numbers starting and ending with the two values provided to it. Take a few moments to review Figure 4.33 and see what the graph may be telling us.

```
#specify how many bins there should be
bin_vals = list(range(0, 26))
```

```
#create histogram for customer order frequency
df_customer_frequency['Order ID'].hist(bins = bin_vals)
```

Listing 4.29 Histogram of Customer Order Frequency

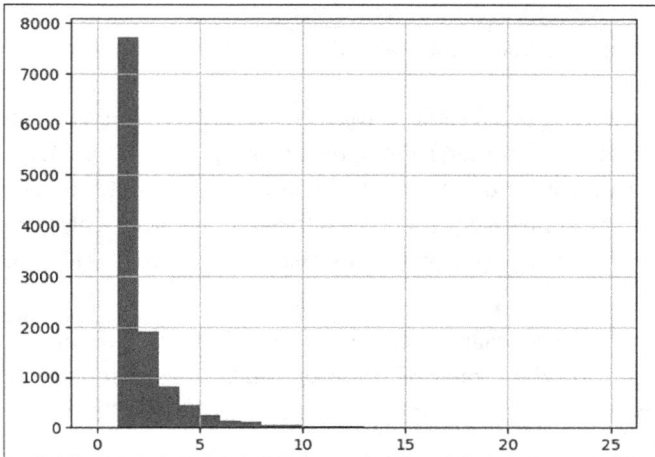

Figure 4.33 Histogram of Customer Order Frequency

You'll notice the first bar (representing one order) jumps out: Almost 8,000 customers only order once. At this time, this shouldn't present any issues from a modeling perspective, but it will be something we want to monitor. Beyond the individuals with only one order, the distribution follows a standard left skew: most customers place orders infrequently, while fewer customers place order frequently.

To quantify these distributions, we can count the number of rows that fit our desired criteria. We want to see how many total customers there are as well as the number customers with more than one order. In Listing 4.30, we've broken apart each component of the calculation into multiple steps by first identifying how many total customers there are and then identifying how many customers there are with more than one order. Finally, we perform the division between the two numbers.

```
#identify how many customers in total there are
all_customers = len(df_customer_frequency)

#identify how many customers at more than one order
one_order_customers = len(
  df_customer_frequency[df_customer_frequency['Order ID'] > 1]
  )

#display total number of customers
print(all_customers)
```

```
#display the customers with more than 1 order
print(one_order_customers)

#calculate and display the percent of customers with more than one order
print(one_order_customers / all_customers)
```

Listing 4.30 Count Number of Customers and Customers with More Than One Order

The result of this code shows there are 11,607 total customers, with 3,894 of them placing more than one order (34% of total customers). One point to keep in mind is Mary's desire to focus on orders placed within one week. If only a third of customers have placed more than one order, it seems likely there aren't many customers ordering within one week of their previous order. This will be the last question we'll explore in this section.

In order to execute the analysis of how many orders occur within seven days of each other, we need to get the orders on the same row to make the data easier to explore. There are a few different ways you could accomplish this; however, follow these steps to get the results in our example:

1. Create a row number in the context of each customer. If you're familiar with SQL, this is the row_number() partition by method.

2. Use this row number to self-join the data together.

3. Analyze and visualize the data.

You can use a combination of Python functions (groupby and cumcount) to create the row number, as shown in Listing 4.31. Figure 4.34 shows a preview of the output.

```
#calculate the row number
df['row_num_join'] = df.sort_values(
  by=['timestamp'], ascending = False)
  .groupby(['Customer ID'])
  .cumcount() + 1

#preview the data where the row_num_join is more than 1
df.loc[
  df['row_num_join'] > 1,
  ['timestamp', 'Customer ID', 'row_num_join']
  ].head(10)
```

Listing 4.31 Python Approach to row_number Partition from SQL

While the results do appear to be accurate, take the time to validate a few records for each customer, since the result only shows us the records for customers with multiple orders. Given how long the customer ID values are, the full value doesn't appear when using .head(). We'll need to print out a few of the values. We can use the index column from Figure 4.34 (bolded numbers to the far left) to pick a few rows (5 and 11, in our example), as shown in Listing 4.32.

```
#see customer values based on index
print(df['Customer ID'][5])
print(df['Customer ID'][11])
```

Listing 4.32 Printing Customer IDs

	timestamp	Customer ID	row_num_join
5	2024-09-10 12:28:00	4103fd4f3ee2166d322e76fabd420ae15f14a2816ca5a3...	2
6	2024-09-10 00:03:00	24d7ca74eb1efe217e88062d2519403ba11d62f5953186...	2
7	2024-09-09 22:54:00	fa1710c1c41dd4f29b810b78f8e7c08a356ca0fdc39307...	2
8	2024-09-09 22:51:00	c521ccf21e7bb2207c3f08d578b6e028d2c3b25986cb35...	2
11	2024-09-08 21:49:00	e1248ba01f5b3ffbab3adde359bfc48adb924a123dd15d...	3
17	2024-09-08 02:55:00	f9e9f7e806c4b5601873c0b9ff2eac2a45b4ab29d61572...	2
27	2024-09-06 23:25:00	693e71ee644f2badf452477f2be8483e5211592ae22cd4...	2
29	2024-09-06 21:23:00	a84ab19e1691226e429b3f4495db99b2937f431d216e18...	2
31	2024-09-06 20:36:00	096a9f7a0c5cf71b57729f2de20cf0080b608dffe8e9c7...	5
32	2024-09-06 20:36:00	ced9e9119fce4573dfcda41998964378f3a2a079d70e50...	2

Figure 4.34 Preview of the Row Number Output

The following customer IDs are printed out:

- 4103fd4f3ee2166d322e76fabd420ae15f14a2816ca5a3be43133260bb1327f4
- e1248ba01f5b3ffbab3adde359bfc48adb924a123dd15daed0063f1c695e5c01

Now, we'll take these values, add them to a list, and filter our data using loc and isin to see what the orders for these two customers look like. This will validate whether our row number approach worked as we expected (see Listing 4.33). The results from the code are shown in Figure 4.35. Give it a review and determine if this is the expected result.

```
#create list of the two customer examples
customer_id_list = [df['Customer ID'][5], df['Customer ID'][11]]

#preview of results for only specific columns and the customers in our list
df
  .loc[
    df['Customer ID'].isin(customer_id_list),
    ['timestamp', 'Customer ID', 'row_num_join']
  ]
    .sort_values(
```

```
         by=['Customer ID' ,'timestamp'],
         ascending = True
    )
```

Listing 4.33 Get Preview of Data for Customers

timestamp	Customer ID	row_num_join
2024-09-02 12:37:00	4103fd4f3ee2166d322e76fabd420ae15f14a2816ca5a3...	1
2024-09-10 12:28:00	4103fd4f3ee2166d322e76fabd420ae15f14a2816ca5a3...	2
2024-11-12 12:37:00	4103fd4f3ee2166d322e76fabd420ae15f14a2816ca5a3...	3
2024-11-22 14:15:00	4103fd4f3ee2166d322e76fabd420ae15f14a2816ca5a3...	4
2024-12-04 13:21:00	4103fd4f3ee2166d322e76fabd420ae15f14a2816ca5a3...	5
2024-09-01 22:01:00	e1248ba01f5b3ffbab3adde359bfc48adb924a123dd15d...	1
2024-09-04 22:34:00	e1248ba01f5b3ffbab3adde359bfc48adb924a123dd15d...	2
2024-09-08 21:49:00	e1248ba01f5b3ffbab3adde359bfc48adb924a123dd15d...	3
2024-09-11 22:30:00	e1248ba01f5b3ffbab3adde359bfc48adb924a123dd15d...	4
2024-09-17 23:20:00	e1248ba01f5b3ffbab3adde359bfc48adb924a123dd15d...	5
2024-09-17 23:22:00	e1248ba01f5b3ffbab3adde359bfc48adb924a123dd15d...	6
2024-09-27 22:33:00	e1248ba01f5b3ffbab3adde359bfc48adb924a123dd15d...	7
2024-09-29 21:50:00	e1248ba01f5b3ffbab3adde359bfc48adb924a123dd15d...	8
2024-10-11 21:51:00	e1248ba01f5b3ffbab3adde359bfc48adb924a123dd15d...	9
2024-10-16 20:53:00	e1248ba01f5b3ffbab3adde359bfc48adb924a123dd15d...	10
2024-10-18 22:21:00	e1248ba01f5b3ffbab3adde359bfc48adb924a123dd15d...	11
2024-11-08 21:17:00	e1248ba01f5b3ffbab3adde359bfc48adb924a123dd15d...	12
2024-11-22 21:06:00	e1248ba01f5b3ffbab3adde359bfc48adb924a123dd15d...	13
2025-01-17 21:45:00	e1248ba01f5b3ffbab3adde359bfc48adb924a123dd15d...	14
2025-01-22 22:28:00	e1248ba01f5b3ffbab3adde359bfc48adb924a123dd15d...	15

Figure 4.35 Preview of Customers with Multiple Orders

You'll find that yes, this is the expected result. Each subsequent order for each customer has an $n + 1$ row number.

Now, we have to create a copy of our data for the join (technically, pandas calls it a *merge*). As part of this process, we need to decide whether to add or subtract from the row number to join either future or past orders with a given order. This can be a tricky decision. For our use case of data evaluation, it makes sense to add the future order to our base data, since we're trying to identify when someone will place a future order. This requires subtracting one from the row number, as shown in Listing 4.34.

```
#select only the necessary columns for the join
df_for_join = df[['timestamp', 'Customer ID', 'row_num_join']]

#adjust the row number to be one less
df_for_join['row_num_join'] = df_for_join['row_num_join'] - 1
```

Listing 4.34 Create New DataFrame and Subtract One from Row Number

Now, we'll execute the join. We'll merge both the customer ID and row number columns using the pd.merge function, as shown in Listing 4.35. The result shown in Figure 4.36 shows that we now have two timestamp columns, with timestamp_x being the order's date and timestamp_y being the next order date.

```
#join the data on customer ID and row number
df_joined = pd.merge(
  df,
  df_for_join,
  how = 'left',
  on = ['Customer ID','row_num_join']
  )
```

Listing 4.35 Merging the Order Data to Get the Future Order Date

Customer ID	timestamp_x	timestamp_y
5d6c2b96db963098bc69768bea504c8bf46106a8a5178e...	2024-09-10 23:38:00	2024-11-14 22:43:00
0781815deb4a10a574e9fee4fa0b86b074d4a0b36175d5...	2024-09-10 23:34:00	2024-10-06 01:52:00
f93362f5ce5382657482d164e368186bcec9c6225fd93d...	2024-09-10 15:52:00	2024-09-25 16:09:00
1ed226d1b8a5f7acee12fc1d6676558330a3b2b742af5d...	2024-09-10 15:45:00	2024-10-20 18:17:00
d21a2ac6ea06b31cc3288ab20c4ef2f292066c096f2c5f...	2024-09-10 15:04:00	2024-09-28 22:57:00
4103fd4f3ee2166d322e76fabd420ae15f14a2816ca5a3...	2024-09-10 12:28:00	2024-11-12 12:37:00
24d7ca74eb1efe217e88062d2519403ba11d62f5953186...	2024-09-10 00:03:00	2024-11-22 22:17:00
fa1710c1c41dd4f29b810b78f8e7c08a356ca0fdc39307...	2024-09-09 22:54:00	2024-09-17 22:07:00
c521ccf21e7bb2207c3f08d578b6e028d2c3b25986cb35...	2024-09-09 22:51:00	2024-10-25 20:09:00
ac2ee09679412d4248884cdfd7637d9b3529eeac0e2f27...	2024-09-09 15:22:00	2024-11-28 12:55:00

Figure 4.36 Future Order Dates Merged onto Dataset

This looks as we'd hoped it would! However, it's a little concerning to only see values in timestamp_y. As we've already found, not all customers have multiple orders. Let's do a quick check to validate how our join was performed. We'll count the rows of the joined dataset, with the dataset filtered with timestamp_y set blank using the .notna() method, as shown in Listing 4.36.

```
#print rows and columns for non-NaN values
print(df_joined[df_joined['timestamp_y'].notna()].shape)
```

```
#print the shape for all values
print(df_joined.shape)
```

Listing 4.36 Counting Rows for Validation of Merge

The results are what we wanted to see! There are about 21,000 rows overall, but only about 8,000 rows when filtering timestamp_y for non-blank values.

Now that we feel good about the dates, we'll create a day difference column and visualize it. This is done in two steps, as shown in Listing 4.37: first calculating the time difference by subtracting timestamp_x from timestamp_y, and then calculating the specific date difference using the dt.days function. The result is shown in Figure 4.37.

```
#create time difference column
df_joined['time_difference'] = df_joined['timestamp_y'] - df_joined['timestamp_x']
```

```
#create day difference column
df_joined['day_difference'] = df_joined['time_difference'].dt.days
```

```
#preview only select columns to validate results
df_joined[[
    'Customer ID',
    'timestamp_x',
    'timestamp_y',
    'time_difference',
    'day_difference'
]].head()
```

Listing 4.37 Create Day Difference Column

Customer ID	timestamp_x	timestamp_y	time_difference	day_difference
5d6c2b96db963098bc69768bea504c8bf46106a8a5178e...	2024-09-10 23:38:00	2024-11-14 22:43:00	64 days 23:05:00	64.0
0781815deb4a10a574e9fee4fa0b86b074d4a0b36175d5...	2024-09-10 23:34:00	2024-10-06 01:52:00	25 days 02:18:00	25.0
f93362f5ce5382657482d164e368186bcec9c6225fd93d...	2024-09-10 15:52:00	2024-09-25 16:09:00	15 days 00:17:00	15.0
1ed226d1b8a5f7acee12fc1d6676558330a3b2b742af5d...	2024-09-10 15:45:00	2024-10-20 18:17:00	40 days 02:32:00	40.0
d21a2ac6ea06b31cc3288ab20c4ef2f292066c096f2c5f...	2024-09-10 15:04:00	2024-09-28 22:57:00	18 days 07:53:00	18.0
4103fd4f3ee2166d322e76fabd420ae15f14a2816ca5a3...	2024-09-10 12:28:00	2024-11-12 12:37:00	63 days 00:09:00	63.0
24d7ca74eb1efe217e88062d2519403ba11d62f5953186...	2024-09-10 00:03:00	2024-11-22 22:17:00	73 days 22:14:00	73.0
fa1710c1c41dd4f29b810b78f8e7c08a356ca0fdc39307...	2024-09-09 22:54:00	2024-09-17 22:07:00	7 days 23:13:00	7.0
c521ccf21e7bb2207c3f08d578b6e028d2c3b25986cb35...	2024-09-09 22:51:00	2024-10-25 20:09:00	45 days 21:18:00	45.0
ac2ee09679412d4248884cdfd7637d9b3529eeac0e2f27...	2024-09-09 15:22:00	2024-11-28 12:55:00	79 days 21:33:00	79.0

Figure 4.37 Date Difference Column

We'll use a histogram again to visualize distribution of day differences. In this case, as shown in Listing 4.38, we'll specify a bin size of 7 days to reflect the 7 days in a week. One challenge here is that we need to account for 0 (same-day orders) within the first bucket. Since we're not trying to make a beautiful visual for our stakeholder, our approach is to set up the first bucket to only include the value 0. This is why our range function starts at -6, which subsequently shifts 63 (9 weeks) down to 57. The 7 in this function specifies we don't want every number, only every seventh number. To ensure we're getting the right number, we'll also print out the number of rows with a date difference of less than or equal to 7. The results are shown in Figure 4.38.

```
#specify the bin size of 7 days
bin_vals = list(range(-6, 57, 7))

#print the bin values to verify
print("bins:", bin_vals)

#print how many orders there were within a week
print("orders within 1 week: ",
    df_joined[df_joined['day_difference'] <= 7].shape[0])

#create histogram of the day differences
df_joined['day_difference'].hist(bins = bin_vals)
```

Listing 4.38 Creating Histogram for Day Difference Between Orders

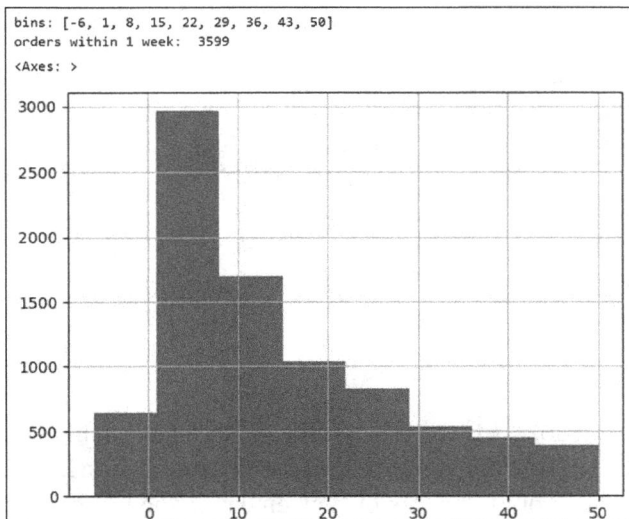

Figure 4.38 Histogram of Day Difference Between Orders

We still see the left skew distribution if we assume all values of 0 should be grouped in with the largest bucket. With 3,599 orders placed within seven days the previous order, that means almost 17% of our 21,000 total orders occur within seven days. Ultimately, this shouldn't present any problems when we get to building our machine learning model.

While this may have felt like a lot of leg work to just analyze the data, it's foundational work that will help you be better informed when building your model. Keep in mind that your stakeholders will almost always ask these types of questions. Business leaders don't think only in the future context. They often consider questions in a historical context, such as "How often does a customer order again within the next week?" Ultimately, a machine learning model uses historical data to build a model to predict the future, but being prepared with a figure like this (17% of orders are from customers who ordered seven or fewer days ago) helps you make a good impression and earn your stakeholders' trust.

Use Case 3: Visualization to Explore the Data

Our crime dataset has a time component, so let's start there by tracking the number of crimes over time. This will give us necessary context on how the overall dataset is trending. To do this, we first need to create a date column using dt.date. Then, we'll create a separate DataFrame showing how many crimes occurred by date using the groupby function (see Listing 4.39). The resulting plot is shown in Figure 4.39. Take a pause and review the graph. Consider how these results might inform a predictive model built on this data.

```
#create the date column
df['DATE'] = df['TIMESTAMP'].dt.date

#create new data frame representing number of crimes on each day
df_by_date = df.groupby('DATE')['DR_NO'].count().reset_index()

#plot the crime by day data frame
df_by_date.plot(kind='line', x = 'DATE', y = 'DR_NO')
plt.show()
```

Listing 4.39 Plot Crime by Day

This plot shows a sharp decline in 2024, which is something we should keep in mind when building a model. For our specific use case, the implications are nuanced. If we were simply trying to predict the amount of crime on a given day in the future, we'd need to consider how our model reflects this recent trend. However, that isn't our use case. We're trying to identify where to allocate resources where crime is most likely to occur—this is important to keep in mind.

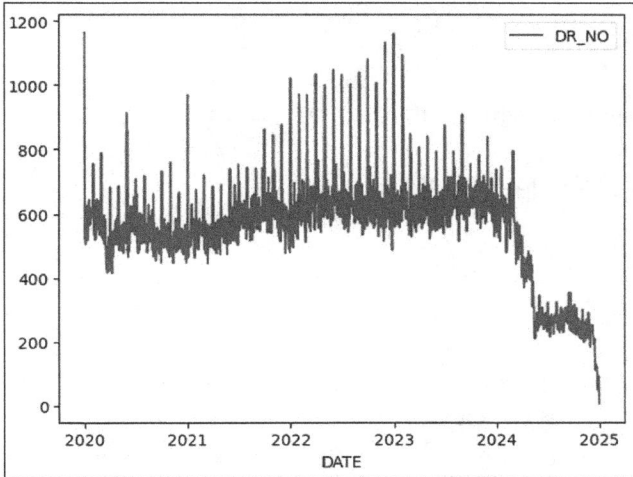

Figure 4.39 Crime by Day Plot

Take this potential scenario: What if the crime distribution by area remained the same (meaning all areas saw an equal decrease in crime) in 2024? This would indicate that the allocation of resources should remain unchanged, as the overall amount of crime is decreasing equally.

Now take this other potential scenario: What if only a few areas saw a near-total reduction of crime? That would certainly change where you allocate resources.

To better understand which of these scenarios aligns with our actual data, let's review the data by area. There are more than twenty unique areas to review, so doing this with a line graph would be challenging. Instead, we'll create a new column that can segment the data into two categories: before or after the start of 2024. We can then compare how much each specific area has changed. Listing 4.40 shows the code to prepare the data. Here, we've created a new column to identify whether a date is before or after the start of 2024 using np.where. We then use the groupby function to summarize the data by area and categorize it as before or after the start of 2024. To perform the percent of total calculation, we pivot the data using pivot.

```
#create new column to identify if date is before or after the start of 2024
df['before_after_2024'] = np.where(
   df['TIMESTAMP'] >= pd.to_datetime('2024-01-01'),
   "2024 or After",
   "Before 2024"
   )

#group data by area and the new 2024 date indicator column
df_area_2024_split = df.groupby(
   ['AREA NAME',
```

```
  'before_after_2024']
  )['DR_NO'].count().reset_index()

#use the pivot function to make the before or after 2024 data as two columns
side by side
df_area_2024_split = df_area_2024_split.pivot(
  index='AREA NAME',
  columns='before_after_2024',
  values='DR_NO'
  ).reset_index()

#create percent change column to use as primary point of comparison
df_area_2024_split['percent_change'] = round(
  (df_area_2024_split['2024 or After'] - df_area_2024_split['Before 2024']) /
  df_area_2024_split['Before 2024']
  , 2)

#sort the values in descending order
df_area_2024_split = df_area_2024_split.sort_values(
  'percent_change',
  ascending = False
  )
```

Listing 4.40 Set Up Data from Before or After 2024 Comparison by Area

Once we have our data prepared, we can plot it. Given the size of the data, we're using a few additional parameters such as ylim, bar_label, and x_ticks, as shown in Listing 4.41. Each of these additions make the chart easier to read. The output of this code is in Figure 4.40.

```
#set width of graph larger to account for number of areas
plt.figure(figsize=(20,8))

#create base bar graph
bar_plt = plt.bar(
  df_area_2024_split['AREA NAME'],
  df_area_2024_split['percent_change']
  )

#specify the limits of the y-axis since we know it's based on a percentage
plt.ylim(-1,0)
```

```
#show the data label on the bar itself
plt.bar_label(bar_plt)

#angle the X-axis label and align to make it easier to read
plt.xticks(rotation=45, ha='right')

#display the bar plot
plt.show()
```

Listing 4.41 Create Bar Graph to Show Change by Group Over Time

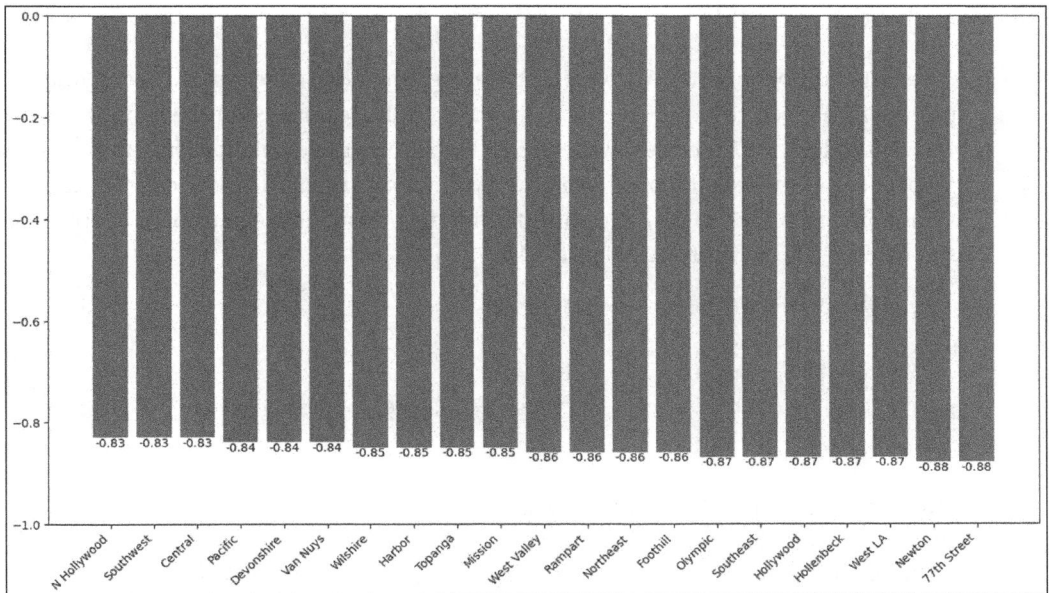

Figure 4.40 Change by Area Before or After 2024

There are changes for all, which is to be expected since we're comparing multiple years to only one year. (This is acceptable because we're comparing all areas across the same time period.) When we look at the data from this view, we can see some variance in the changes in crime rate between each location, but overall the differences are relatively small. This likely means we'll be able to include the entire dataset when training the model, rather than having to exclude data from periods that aren't representative of the future.

We're lucky that this dataset contains a large number of columns that describe the crimes being committed. The real value here is that they're not bunched up into one text-based column. It's much easier to build a model with data that's already separated.

However, this does mean we have a number of columns to visualize. We'll go through each of the following columns to understand how frequently their values show up in the dataset:

- AREA NAME
- Crm Cd Desc
- Premis Desc
- Weapon Desc
- Vict Sex
- Vict Descent
- Vict Age

Given the number of columns, we'll loop through the same code instead of copying and pasting it for each column. To enable the loop, we'll create a list of the column names we want to look at. Then, we'll write the code to group the data and visualize it in a bar graph (see Listing 4.42) by putting a groupby function and plotting function inside of a for loop. Look at each of the code outputs and think about how you'd approach the next steps (see Figure 4.41 through Figure 4.47).

```
#create list of columns to visualize
col_list = [
  'AREA NAME', 'Crm Cd Desc', 'Vict Age',
  'Vict Sex', 'Vict Descent', 'Premis Desc', 'Weapon Desc'
  ]


#for loop through the list of columns
for col in col_list:
  #group the data by the column, using the report number for counting
  group_df = df.groupby(col)['DR_NO'].count().reset_index()

  #create the bar graph
  plt.figure(figsize=(10,6))
  plt.bar(group_df[col], group_df['DR_NO'])
  plt.title(col)
  plt.xticks(rotation=45, ha='right')
  plt.show()
```

Listing 4.42 Create For Loop to Visualize Columns

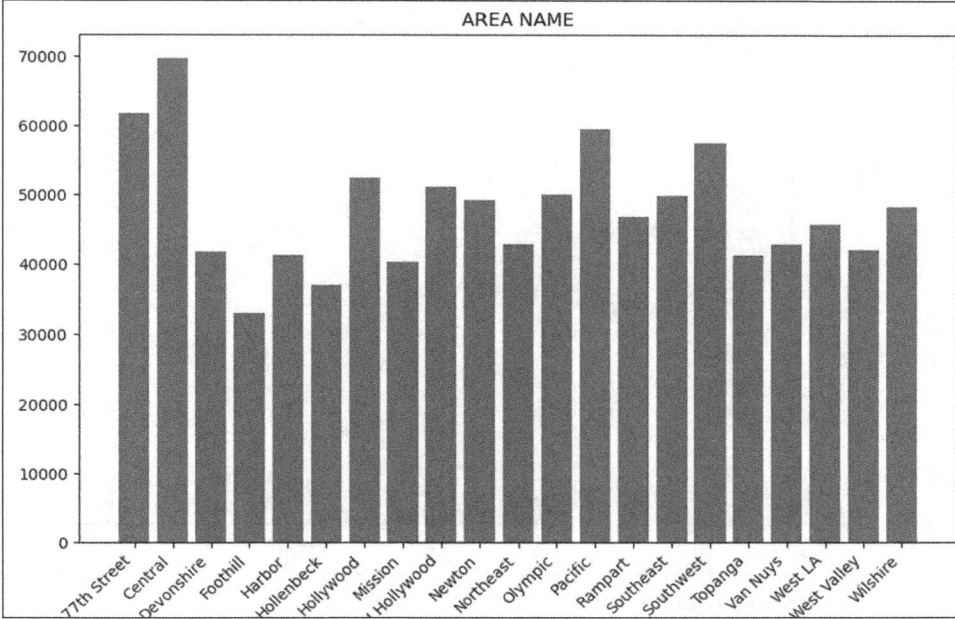

Figure 4.41 Area Name Crime Counts

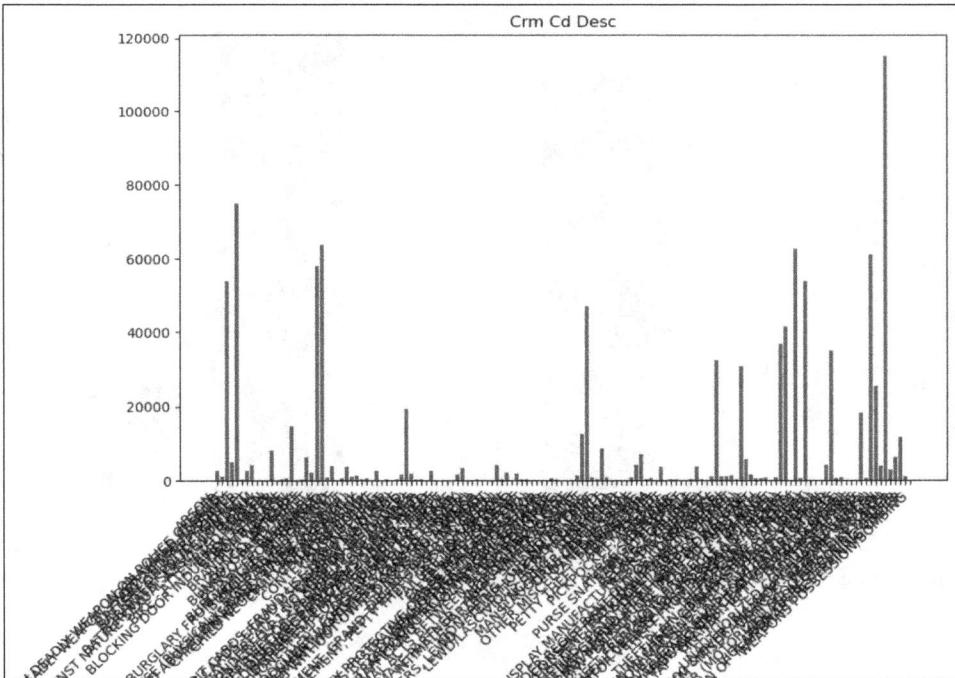

Figure 4.42 Crime Code Counts

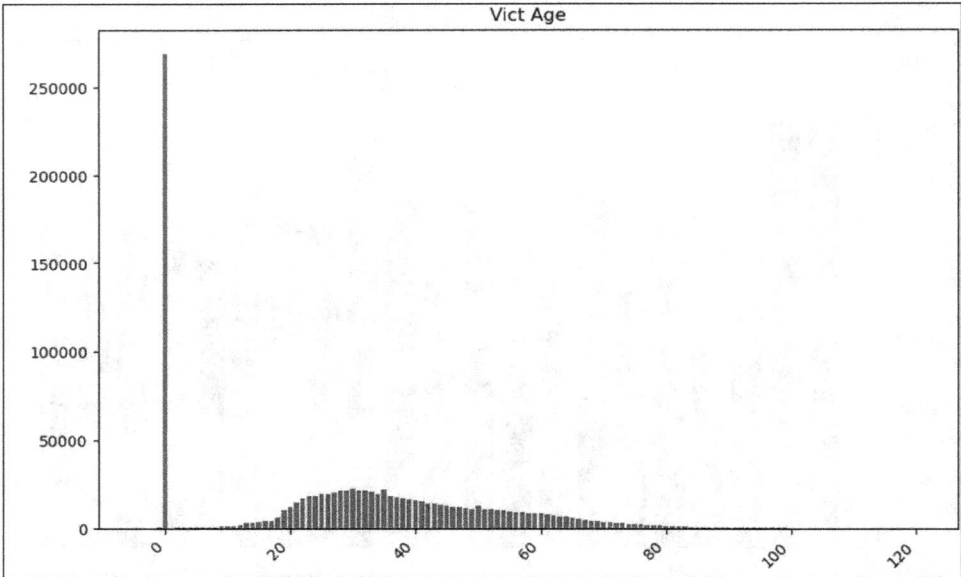

Figure 4.43 Victim Age Counts

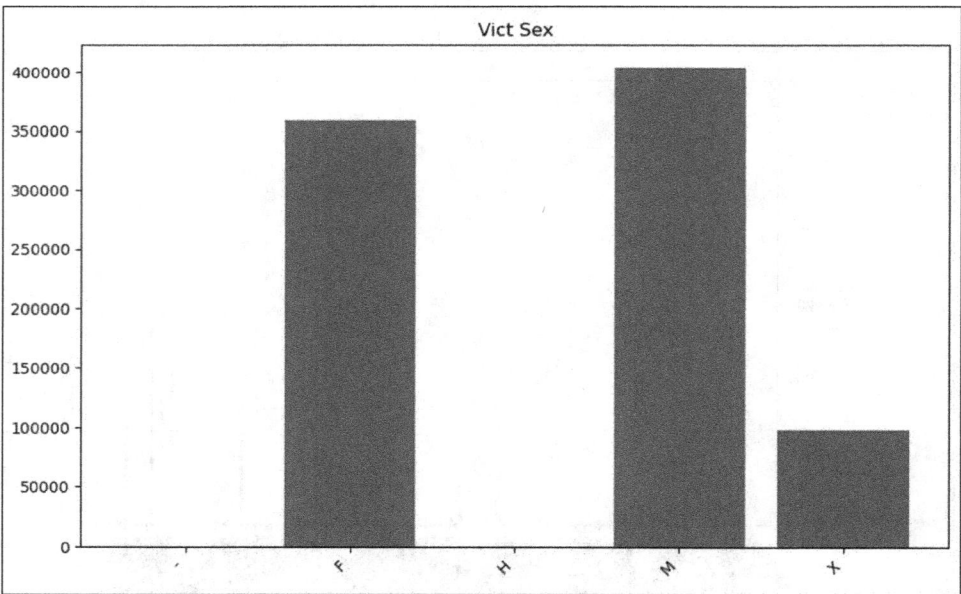

Figure 4.44 Victim Sex Counts

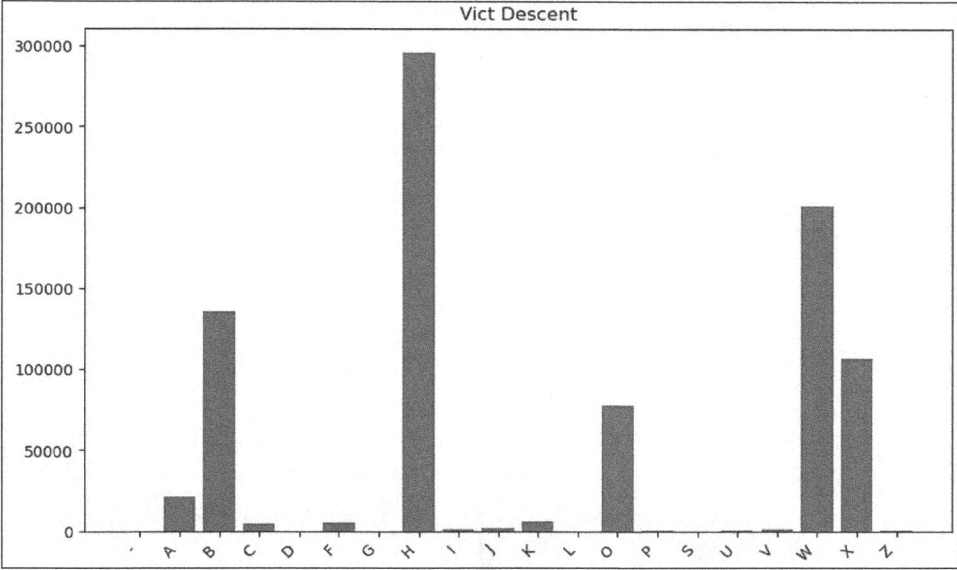

Figure 4.45 Victim Race Counts

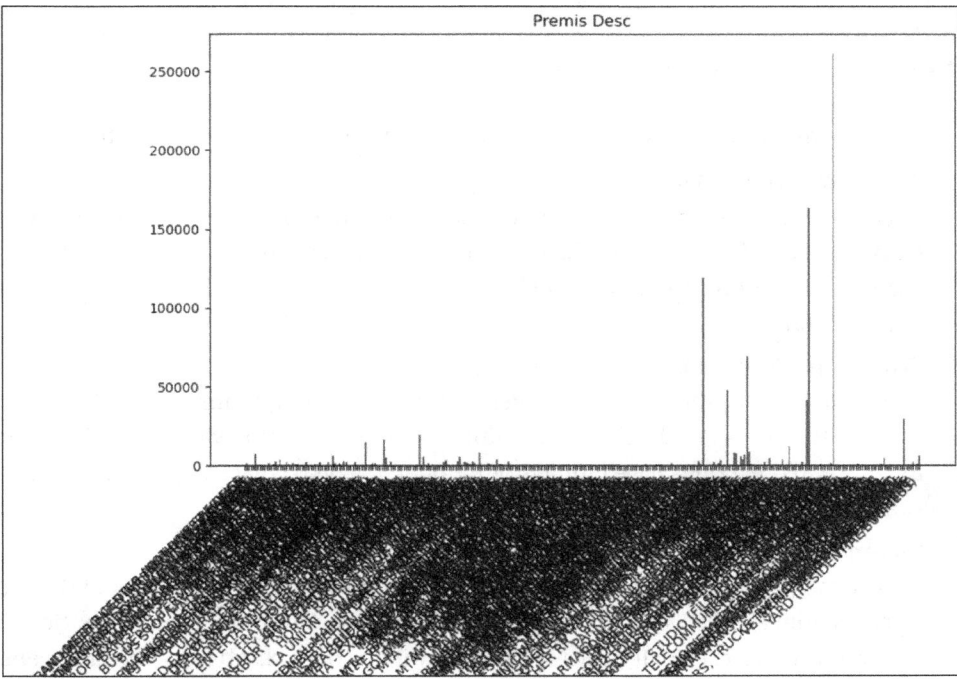

Figure 4.46 Premises Counts

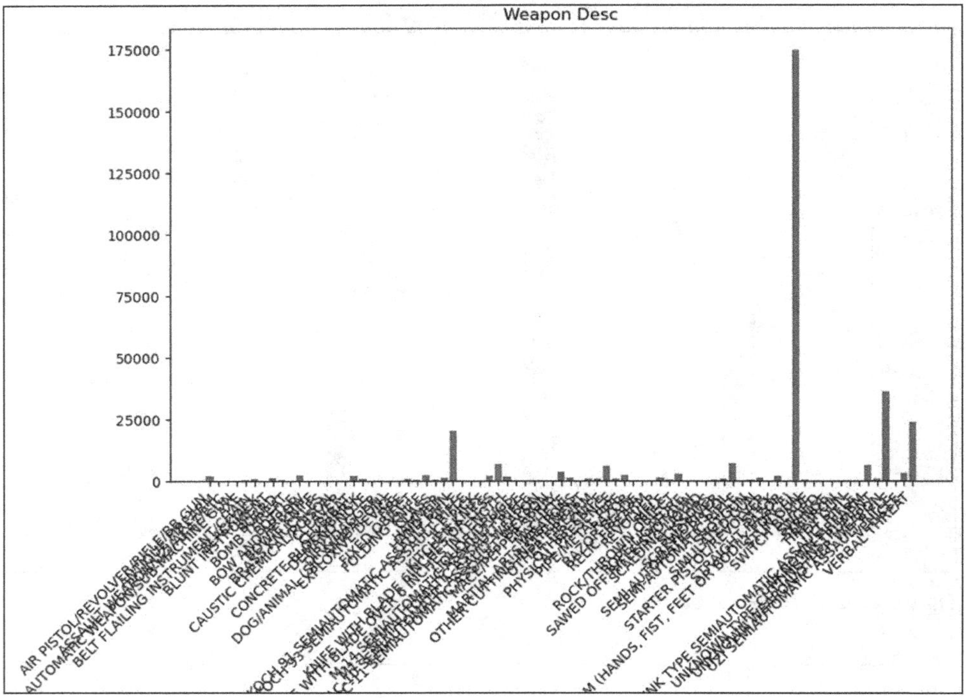

Figure 4.47 Weapon Description Counts

There are some relatively distinct stories that come from each of these graphs:

▶ **Area name (Figure 4.41)**
This will be a key column to our model as we're trying to predict the probability of crime in each of these areas. The distribution of crime is relatively even; however, there are variations, such as areas like the Foothills having much lower crime rates than Central.

▶ **Crime code description (Figure 4.42)**
This column has a lot of unique values that make this graph unreadable. However, we can still get insight from it. This column will require some value grouping if we include it in the model. If we didn't group values, it would create a significant dimensionality problem (a topic we'll discuss in Section 4.3.6).

▶ **Victim age (Figure 4.43)**
This ends up becoming a histogram. It's important to note there are a lot of 0s, and it seems unlikely the LA has a crime epidemic against newborns. The distribution of the graph is interesting: It peaks around 30, with a sharp climb from the late teens, and then gradually decline from the 30s through the 80s. This seems like a viable column that we could use in our model, as long as we account for the likely incorrectly coded 0s.

▶ **Victim sex (Figure 4.44)**

We see that the split of male to female is slightly skewed toward males. This data can likely be used in the model.

▶ **Victim race (Figure 4.45)**

The coding for this column is not clear, so we don't know which races these values belong to. However, we do see there are a few that stick out with higher values, with one being higher than the others. We'll want to consolidate the values for this column and then include it in the model.

▶ **Premises description (Figure 4.46)**

Yikes! This column makes it clear how many unique values there are. However, we can identify a few values that stand out as the highest. This means we can consider consolidating the values before including them in the model.

▶ **Weapon description (Figure 4.47)**

There are quite a few unique values in this column, but only a few stick out. This likely means we can consolidate the values and include them in the model.

4.2.3 Descriptive Statistics

If you don't have a statistics background, no worries! By now, you've hopefully seen you don't need to be an expert in these adjacent topics to build a successful model; we'll share enough information to avoid major gaps in your understanding.

In the context of machine learning projects, descriptive statistics help you understand your data in a numeric sense. Data visualizations allow you to find large trends or outliers that exist in your data, but what if you're looking for something specific, like the average of number sales on a given day? A graph isn't an ideal solution.

From a hierarchy-of-needs perspective, performing this analysis on your target variable is the most important step. Let's return to the sales example from our first use case, for example. If you're predicting sales for a given day, knowing the simple average or median is very helpful when you get to building your model. What happens if your average daily sales total $5,000, but your model predicts $20,000? If you don't know the average daily sales, you don't have the context you need to flag $20,000 as potentially incorrect. This figure would then warrant additional investigation to ensure you've built the model correctly. If you don't catch a mistake like this before going back to your stakeholder, you risk significantly harming your credibility. A stakeholder like Mary may be more forgiving, but you could dig a hole with Chris that's difficult to recover from.

There are a number of perspectives on which statistics are most appropriate to use. This varies by use case, but you should be comfortable using and explaining the following types of descriptive statistics:

- Mean
- Median
- 25th percentile

- 75th percentile
- 10th percentile
- 90th percentile

Mean and median are staples in the descriptive statistics world. Together, they help you understand the skew of the data. When the mean is larger than the median, it likely means there are a small number of rows with large numbers that make the mean higher. The inverse is also true.

Percentiles are extensions of the median (as the median is just the 50th percentile). You can start with the 25th and 75th percentiles, as they tie to the interquartile range that is universally accepted in the statistics community. In business speak, this translates to the upper and lower limit that the data will fall within. By definition, 50% of the data will be within range. You can then leverage the 10th and 90th percentiles to define the data's upper and lower limit (80% of the data falls within these limits).

For our examples, we'll use the NumPy library. Here's the standard for importing the library:

```
import numpy as np
```

NumPy

NumPy stands for Numeric Python. It contains a number of helpful functions, most of which focus on working with numbers. Together with pandas, it's one of the most commonly used libraries in the Python ecosystem.

Use Case 1: Descriptive Statistics

This dataset only has a few numeric columns (quantity and price), so we'll focus on those. In the data visualization exercise, we determined which rows to filter out. In general, you shouldn't run your descriptive statistics until you've taken the initial steps to evaluate and prepare your data. We've already done much of that through our data visualization on this dataset. This use case is a great example, because we'll look at how different the mean is with and without our filter. Listing 4.43 is the code for calculating this; as you can see, we're using the np.mean function to calculate the mean on a number of different data scenarios. Figure 4.48 shows the results of the code.

```
#print mean before summarizing and filtering data
print("Before Summarizing, Without filtering: ",
  np.mean(df['Price'])
  )


#print mean before summarizing the data, but after filtering
print("Before Summarizing, With filtering: ",
```

```
  np.mean(df_cleaned['Price'])
  )

#create a space for the output
print("")

#create grouped dataset by date summing price
df_summarized_no_filters = pd.DataFrame(
  df.groupby('invoicedate')['Price'].sum()
  ).reset_index()

#create grouped dataset after filtering
df_summarized_with_filters = pd.DataFrame(
  df_cleaned.groupby('invoicedate')['Price'].sum()
  ).reset_index()

#print mean after summarizing, but not filtering
print("After Summarizing, Without filtering: ",
  np.mean(df_summarized_no_filters['Price'])
  )

#print mean after summarizing and filtering
print("After Summarizing, With filtering: ",
  np.mean(df_summarized_with_filters['Price'])
  )
```

Listing 4.43 Calculating the Mean of Price with Different Scenarios

```
Before Summarizing, Without filtering:  4.649387727416241
Before Summarizing, With filtering:  3.206561485396916

After Summarizing, Without filtering:  8216.260973509932
After Summarizing, With filtering:  4276.560261589404
```

Figure 4.48 Result of Calculating the Mean of Price with Different Scenarios

Let's walk through each of the scenarios:

- **Before summarizing, without filtering**
  This is the raw data, where there is no aggregation of the individual order records, nor are any filters applied to the dataset as it is originally provided.

- **Before summarizing, with filtering**
  This is the individual order-level data. However, the filters we've discussed and created in the previous code are applied.

- **After summarizing, without filtering**
  This is the aggregation and summarization of the order level data into the day the order was made. This doesn't apply any of our filters to remove the data that's not relevant to our use case.

- **After summarizing, with filtering**
  This is the aggregation and summarization of the order level data by day, with our filters applied.

At the order level, filtering the data resulted in a more than 30% decrease in the average price (4.649 down to 3.207). When summarized at the day level, the decrease is nearly 50% (8,216 down to 4,277). The average shifts significantly!

It's possible there's been some cherry-picking of the data and the outliers are strongly impacting the average, but the median remains unaffected. Listing 4.44 shows how this is calculated by replacing np.mean with np.nanmedian from our previous example in Listing 4.43. Figure 4.49 shows the results of the code.

```
#print median without grouping or filtering
print("Before Summarizing, Without filtering: ",
  np.nanmedian(df['Price'])
  )


#print median with filtering, but not grouping
print("Before Summarizing, With filtering: ",
  np.nanmedian(df_cleaned['Price'])
  )


#create space for the output
print("")


#group data by date while summing price without filters
df_summarized_no_filters = pd.DataFrame(
  df.groupby('invoicedate')['Price'].sum()
  ).reset_index()


#group data by date while summing price with filters
df_summarized_with_filters = pd.DataFrame(
  df_cleaned.groupby('invoicedate')['Price'].sum()
  ).reset_index()


#print median with summarizing, but not filtering
print("After Summarizing, Without filtering: ",
  np.nanmedian(df_summarized_no_filters['Price'])
  )
```

```
#print median with summarizing and filtering
print("After Summarizing, With filtering: ",
  np.nanmedian(df_summarized_with_filters['Price'])
  )
```

Listing 4.44 Calculating the Median of Price with Different Scenarios

```
Before Summarizing, Without filtering:   2.1
Before Summarizing, With filtering:   1.95

After Summarizing, Without filtering:   6377.135
After Summarizing, With filtering:   3701.775
```

Figure 4.49 Results of Calculating the Median of Price with Different Scenarios

We do see some new nuances to consider, particularly the smaller differences at the order level detail, where the aggregation and summarization isn't applied. However, overall, filtering the data still appears to change the underlying distribution of price. Without summarizing the data, we only see about a 7% decrease in the average price (2.1 down to 1.95). This is lower than the same comparison using the mean, which you can interpret to mean there are a small number of price values that are quite large.

What may be potentially confusing, though, is how the row-by-row difference can be so small but the daily aggregation view is still over 40% different (6,377 down to 3,702). This shows you how aggregating your data can change your underlying distributions of the data. By aggregating the data to each day, we're removing all of the noise associated with each product for each order.

In a use case like this, it's often more productive to run your statistics on the aggregated data. The use case means we're looking at overall sales and not at specific products or order volumes. We'll address how using the product-level data can create input for our machine learning model in Chapter 5, but when running descriptive statistics, it's best to avoid these nuances.

Now, let's look at our statistics for the summarized and filtered dataset. You can see the code in Listing 4.45 and the results in Figure 4.50. We're using np.mean, np.nanmedian, and np.percentile to calculate the various statistics on our data to better understand how it's distributed. Take some time to analyze these results and formulate your thoughts on what these numbers mean and how you can interpret them before proceeding.

```
#print the mean
print("Mean: ",
  np.mean(df_summarized_with_filters['Price'])
  )

#print the median
print("Median: ",
```

```
    np.nanmedian(df_summarized_with_filters['Price'])
    )

#print the 25th percentile
print("25th Percentile: ",
    np.percentile(df_summarized_with_filters['Price'], 25)
    )

#print the 75th percentile
print("75th Percentile: ",
    np.percentile(df_summarized_with_filters['Price'], 75)
    )

#print the 10th percentile
print("10th Percentile: ",
    np.percentile(df_summarized_with_filters['Price'], 10)
    )

#print the 90th percentile
print("90th Percentile: ",
    np.percentile(df_summarized_with_filters['Price'], 90)
    )
```

Listing 4.45 Calculating the Descriptive Statistics

```
Mean:    4276.560261589404
Median:   3701.775
25th Percentile:  2876.355
75th Percentile:  5097.8275
10th Percentile:  2152.263
90th Percentile:  7009.086700000004
```

Figure 4.50 Results of Calculating the Descriptive Statistics

The general theme of the skewed data persists, as the higher percentiles have larger gaps between them than the lower percentiles. For example, the difference between the 90th percentile and the 75th percentile is almost 2,000, whereas the difference between the 10th percentile and the 25th percentile is less than 1,000. Additionally, the mean being larger than the median says there are small number of higher sales days that bring up the average but don't bring up the median. We can now feel confident in saying sales on any given day should be between $2,000 and $7,000. When we build the model, our predictions will likely fall within this range.

One nuance to keep in mind with these statistics is the seasonality trend we observed earlier. These summary statistics don't account for seasonality. They only provide a high-level viewpoint of the data and its distribution, not how the data trends over time.

While this isn't an error or concern, it is something to keep in mind as the data journey progresses!

Use Case 2: Descriptive Statistics

In this dataset about food service delivery orders, we have a few different number-based columns we can run our descriptive statistics on. This approach to descriptive statistics differs from the first use case, where we applied them to the variable we are ultimately trying to predict. For this use case, we're using descriptive statistics to better understand our dataset, since these data points will likely become inputs for our model.

We can apply our various statistics on the following numeric columns:

- Bill subtotal
- Packaging charges
- Restaurant discount (Promo)
- Restaurant discount (Flat offs, freebies & others)
- Gold discount
- Brand pack discount
- Total
- Rating
- Restaurant compensation (Cancellation)
- Restaurant penalty (Rejection)
- KPT duration (minutes)
- Rider wait time (minutes)

This is quite the list of columns! Running our descriptive statistics by copying and pasting the same code one by one would be incredibly inefficient. Instead, we'll build a process that runs the descriptive statistics for each of these columns and puts them into a DataFrame.

To accomplish this, we'll be using a for loop to iterate over each of these column names, as shown in Listing 4.46. To capture all of the column's metrics, we'll create a blank Data-Frame where we can add each of the column's metrics. The results are shown in Figure 4.51.

The following steps are occurring in the code shown in Listing 4.46:

1. Create a list of the column names (numeric_cols).
2. Create a blank DataFrame (df_for_metrics). Our metrics will be added here as they're calculated.
3. The for loop's first component calculates each of the descriptive statistics and then saves them to their respective names.
4. Create a dictionary (data).

5. Using data, create a DataFrame that has one row for that specific column's statistics. This is also called data.

6. Lastly, use pd.concat to add the single row of data to our df_for_metrics DataFrame.

7. This will repeat for each column in numeric_cols.

```
#create list of numeric columns
numeric_cols = [
  'Bill subtotal', 'Packaging charges',
  'Restaurant discount (Promo)',
  'Restaurant discount (Flat offs, Freebies & others)',
  'Gold discount', 'Brand pack discount', 'Total', 'Rating',
  'Restaurant compensation (Cancellation)', 'Restaurant penalty (Rejection)',
  'KPT duration (minutes)', 'Rider wait time (minutes)'
  ]

#create blank data frame
df_for_metrics = pd.DataFrame()

#set up for loop to go through each of the numeric columns
for col in numeric_cols:
  mean_val = np.mean(df[col])
  median_val = np.nanmedian(df[col])
  nan_vals = df[col].isna().sum()
  not_nan_vals = df[col].count()

  #put the outputs into a dictionary
  data = {
    "column_name": col, "mean": mean_val,
    "median": median_val, "blanks": nan_vals,
    "not_blank": not_nan_vals
    }

  #create a temporary data frame
  data = pd.DataFrame(data, index=[0])

  #stack the temporary data frame onto the df_for_metrics data
  df_for_metrics = pd.concat(
    [df_for_metrics, data],
    axis = 0,
    ignore_index = True
    )
```

Listing 4.46 Writing Loop for Descriptive Statistics

column_name	mean	median	blanks	not_blank
Bill subtotal	750.076838	629.00	0	21321
Packaging charges	32.564592	28.45	0	21321
Restaurant discount (Promo)	65.091816	80.00	0	21321
Restaurant discount (Flat offs, Freebies & oth...	31.795058	0.00	0	21321
Gold discount	0.099128	0.00	0	21321
Brand pack discount	3.039324	0.00	0	21321
Total	682.616113	597.45	0	21321
Rating	4.356885	5.00	18830	2491
Restaurant compensation (Cancellation)	356.409549	272.58	21188	133
Restaurant penalty (Rejection)	0.000000	0.00	21318	3
KPT duration (minutes)	17.332960	16.33	295	21026
Rider wait time (minutes)	4.825070	3.10	168	21153

Figure 4.51 Results of Descriptive Statistics

Take a few moments to observe the results in Figure 4.51. Here are some key takeaways:

- The billing information is fully populated. We'll likely want to include these data points in our model, at least to start.
- The Rating, Restaurant compensation (Cancellation), and Restaurant penalty (Rejection) columns are mostly blank. We should consider removing these columns from the modeling process, given how often they're not populated.
- The order cost may seem high, but remember these numbers aren't in US dollars. The dataset is based on an Indian company.

Use Case 3: Descriptive Statistics

Compared to the other two use cases, the number of numeric columns for this use case is quite small. The sole numeric column is the age column. We started exploring this column during our data visualization and discovered there are many 0s in the dataset. This will make for a good lesson on statistics and how the mean and median differ.

First, let's compare the mean and median age without any alterations to the underlying data using the median() and mean() functions for Vict Age, as shown in Listing 4.47.

```
#print the median of victim age
print(df['Vict Age'].median())
```

```
#print the mean of the victim age
print(df['Vict Age'].mean())
```

Listing 4.47 Calculate the Median and Mean for Age

The result for median is 30 and the result for the average is 28.9. This is actually quite surprising; given how many 0s showed up in the data, you would expect the average to be lower. Next, let's remove any number that is 0 or less than 0 to see the results (see Listing 4.48).

```
#print the median victim age for ages greater than 0
print(df[df['Vict Age'] > 0]['Vict Age'].median())

#print the mean victim age for ages greater than 0
print(df[df['Vict Age'] > 0]['Vict Age'].mean())
```

Listing 4.48 Calculate the Median and Mean with Only Values Greater Than 0

The results of this are much higher. The median jumped up to 37 and the average jumped up to 39.5. This indicates that we should calculate averages or medians without the 0s, as they're significantly skewing the results for both the median and the average. Sometimes the median adjusts automatically without requiring you to filter the data, but in this situation, there are so many 0s that the median is also skewed quite heavily.

4.2.4   Correlation Analysis

If you haven't done a formal correlation analysis before, it can sound more sophisticated than it really is. The underlying concept involves identifying how closely the trend of two numeric data points follow each other. For example, you could compare the usage of your air conditioning unit to the temperature outside in the summer. As the temperature rises outside, the usage of your air conditioning unit also goes up.

The value of understanding your variable's correlations for a machine learning project goes back to the importance of being informed about the data. As you interpret the results of your model, you can use the correlation analysis as a starting point to see if anything odd may be happening that you didn't expect or weren't aware of.

Correlation Analysis and Regression Models

If you use a regression model, the importance of correlation analysis increases. As you'll learn in Chapter 5, regression models can be less forgiving with the data you provide. For example, if two features are highly correlated—which is called multicollinearity—it can wreak havoc on your model.

So, how do you run a correlation analysis? NumPy has a simple function, `np.corrcoef`, that makes the process quite easy! Let's continue the correlation analysis conversation using our use cases.

Use Case 1: Correlation Analysis

This sales dataset contains only two numeric variables (quantity and price), so the correlation analysis is not extensive. However, it's best practice to complete it anyway, and it shouldn't take much time with only two columns.

If we take the data that hasn't been summarized and calculate a correlation matrix for Price and Quantity using the code in Listing 4.49, we get the result shown in Figure 4.52.

```
#create correlation matrix for price and quantity
np.corrcoef(
    df_cleaned['Price'],
    df_cleaned['Quantity']
    )
```

Listing 4.49 Calculate Correlation Matrix of Price and Quantity

```
array([[ 1.         ,  -0.00490949],
       [-0.00490949,  1.         ]])
```

Figure 4.52 Result of Correlation Matrix

The output is a correlation matrix. The 1s compare the same column to itself and can be ignored, whereas the -0.00490949 shows the correlation between the two variables. A correlation of 0 means there is no relationship. This number is very close to 0, which tells us that on a row-by-row level, this data essentially isn't correlated.

You'll also note this correlation is negative. When a correlation is negative, it signifies an *inverse* relationship between the two columns you're comparing. This means that as the value of one increases, the value of the other decreases.

The next step is to summarize the data, but we haven't yet summarized and aggregated both price and quantity at the same time. Pandas enables this with the agg function, as demonstrated in Listing 4.50. Here, we've summarized the data by date using the groupby function, which is then input into the `np.corrcoef` function to generate a new correlation matrix. After we summarize the data by date, we see a very different result, shown in Figure 4.53.

```
#group the data by date, summing both price and quantity
df_summarized = df_cleaned.groupby(['invoicedate']).agg(
    {'Quantity': 'sum', 'Price': 'sum'}
    )
```

```
#print a preview of the data
print(df_summarized.head())

#print extra line for the output
print("")

#print new correlation matrix for the grouped date
print(np.corrcoef(
    df_summarized['Price'],
    df_summarized['Quantity']
    ))
```

Listing 4.50 Aggregating Data, Then Calculating Correlation Matrix

```
              Quantity      Price
invoicedate
2009-12-01     24437.0    7070.14
2009-12-02     29771.0    6857.51
2009-12-03     48031.0    7069.25
2009-12-04     20069.0    5408.13
2009-12-05      5119.0    1443.26

[[1.          0.51122744]
 [0.51122744  1.         ]]
```

Figure 4.53 Calculating Correlation Matrix After Aggregating Data

Almost like magic, we went from a correlation of essentially 0 to 0.51. A "good" correlation depends on your domain and specific data, but a 0.51 for this use case shows there is a noteworthy relationship between price and quantity. So, how did we go from no relationship to a relationship? Noise.

When you look at granular data, which in this case is itemized sales transactions, you'll encounter a lot of random fluctuations that obscure your true result—that is to say, a lot of noise. Correlation analysis by itself is very bad at sifting through noise. The act of summarizing both quantity and price to the daily level removed a significant amount of noise, which enabled the correlation metric to identify the relationship. If you've worked with someone who is highly experienced in the analytics world, specifically with a math or statistics background, they can tell you that the phase "the data doesn't lie" is rarely true. This is an example of how taking the data through one more step can completely change the outcome.

Use Case 2: Correlation Analysis

The correlation analysis for our second use case, food service deliveries, will be more in depth compared to the first use case given the number of numeric columns. When building a model for this use case, there are two angles to the correlation analysis:

- Are the columns in the dataset correlated? This is important when we get to the modeling stage, as any columns that are too correlated with each other may need to be dealt with.
- Are there indicators that correlate to a customer ordering again within seven days? This will require some additional data prep, but it helps identify the columns that are most predictive of repeat orders within seven days.

Identifying whether the columns in the dataset are correlated is similar to what we did in the previous section, where we created a correlation matrix. As noted earlier, this second use case has many numeric columns, so we'll apply additional visualization on top of the base correlation matrix approach to make the results easier to analyze.

First, we'll create a DataFrame that only contains our numeric variables. We can repurpose our `numeric_cols` from Section 4.2.3 to do this, as shown in Listing 4.51. As you can see, we're creating a list called `numeric_cols`, which restricts how many fields are included in the DataFrame. The resulting DataFrame is previewed in Figure 4.54.

```
#create list of numeric columns
numeric_cols = [
    'Bill subtotal', 'Packaging charges',
    'Restaurant discount (Promo)',
    'Restaurant discount (Flat offs, Freebies & others)',
    'Gold discount', 'Brand pack discount', 'Total', 'Rating',
    'Restaurant compensation (Cancellation)', 'Restaurant penalty (Rejection)',
    'KPT duration (minutes)', 'Rider wait time (minutes)'
    ]

#create new df with only the numeric columns
df_for_correlations = df[numeric_cols]

#preview the data
df_for_correlations.head()
```

Listing 4.51 Create DataFrame for Correlation Matrix

Bill subtotal	Packaging charges	Restaurant discount (Promo)	Restaurant discount (Flat offs, Freebies & others)	Gold discount	Brand pack discount	Total	Rating	Restaurant compensation (Cancellation)	Restaurant penalty (Rejection)	KPT duration (minutes)	Rider wait time (minutes)
715.0	31.75	80.0	0.0	0.0	0.0	666.75	NaN	NaN	NaN	18.35	11.6
1179.0	50.20	175.0	0.0	0.0	0.0	1054.20	NaN	NaN	NaN	16.95	3.6
310.0	11.50	80.0	0.0	0.0	0.0	241.50	NaN	NaN	NaN	14.05	12.2
620.0	27.00	80.0	0.0	0.0	0.0	567.00	4.0	NaN	NaN	19.00	3.3
584.0	25.20	80.0	0.0	0.0	0.0	529.20	NaN	NaN	NaN	15.97	1.0

Figure 4.54 DataFrame of Only Numeric Columns

Next, we create the correlation matrix from this DataFrame. Since we've already selected only the columns that are numeric, this exercise is as simple as providing the DataFrame that the correlation matrix will be created from (see Listing 4.52). To do this, we're using the corr function from pandas. This is an alternative to using the np.corr-coef function. Figure 4.55 shows a preview of the correlation matrix to give you an idea of what it looks like when you don't have any formatting applied.

```
#create correlation matrix
correlation_matrix = df_for_correlations.corr()

#display correlation matrix
print(correlation_matrix)
```

Listing 4.52 Create Correlation Matrix

```
                                                     Bill subtotal  \
Bill subtotal                                             1.000000
Packaging charges                                         0.957385
Restaurant discount (Promo)                               0.482291
Restaurant discount (Flat offs, Freebies & others)        0.230624
Gold discount                                             0.026750
Brand pack discount                                       0.134164
Total                                                    0.958831
Rating                                                   0.059984
Restaurant compensation (Cancellation)                   0.939862
Restaurant penalty (Rejection)                                NaN
KPT duration (minutes)                                   0.376633
Rider wait time (minutes)                                0.189966

                                                     Packaging charges  \
Bill subtotal                                             0.957385
Packaging charges                                         1.000000
Restaurant discount (Promo)                               0.393666
Restaurant discount (Flat offs, Freebies & others)       -0.004306
Gold discount                                             0.020060
Brand pack discount                                       0.101002
Total                                                    0.999105
Rating                                                   0.061174
Restaurant compensation (Cancellation)                   0.952711
Restaurant penalty (Rejection)                                NaN
KPT duration (minutes)                                   0.385651
Rider wait time (minutes)                                0.181590
```

Figure 4.55 Preview of Result from Printing the Correlation Matrix

The challenge with analyzing the results in Figure 4.55 is that we can only see each column and its comparisons independently. This can make it difficult to find trends across the board, especially when you're only looking at numbers without any visual cues to help your eyes differentiate between values.

To address this, we can create a formatted correlation matrix in the form of a heatmap, which represents the strength of the correlations by color, as shown in Listing 4.53. We're using the `heatmap` function, which requires the correlation matrix we calculated in Listing 4.52. Figure 4.56 shows the resulting graph. It's much easier to identify potential trends and themes in the data this way.

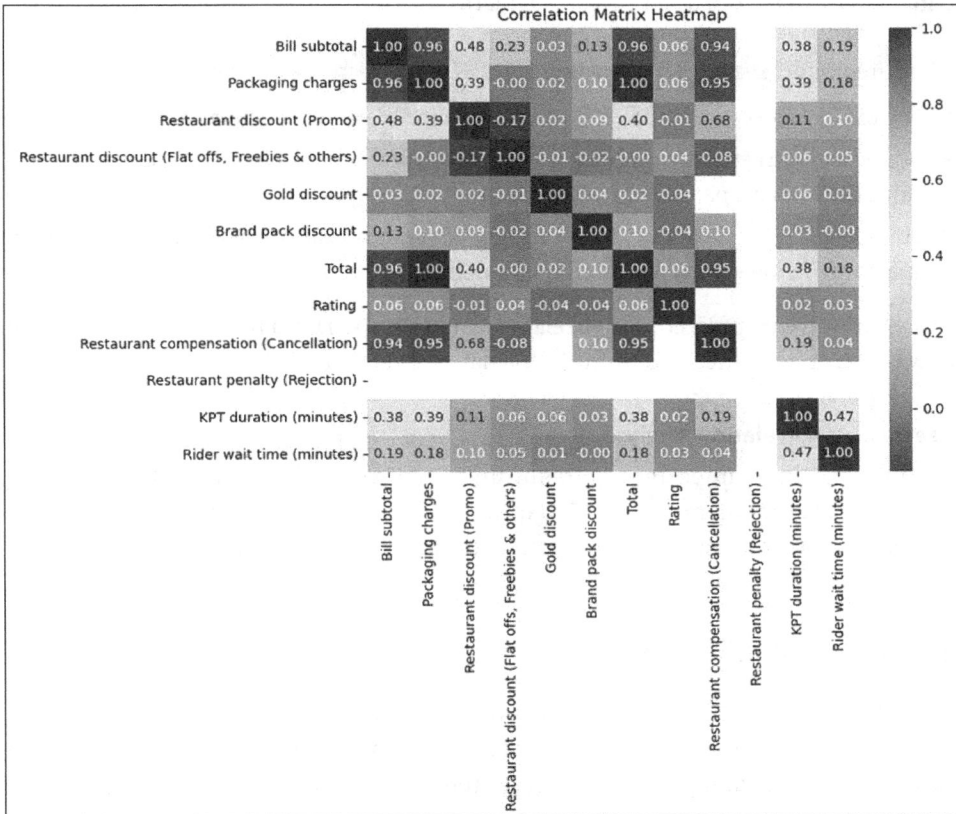

Figure 4.56 Formatted Correlation Matrix

```
#import seaborn which is a robust data visualization package in Python
import seaborn as sns
import matplotlib.pyplot as plt

#set the plot size
plt.figure(figsize=(8, 6))

#create heatmap for the correlation matrix
sns.heatmap(correlation_matrix, annot=True, cmap='coolwarm', fmt=".2f")
```

```
#title the correlation matrix heatmap
plt.title('Correlation Matrix Heatmap')

#display the heatmap
plt.show()
```

Listing 4.53 Create Formatted Correlation Matrix

What are your takeaways from this graph? Here are some key observations:

- The Bill subtotal and Total fields are very similar, with a correlation of 0.96.
- Packaging charges seems to be a flat charge associated with the correlation to Total, since they have a perfect correlation of 1.
- The KPT duration field has some moderate level correlations with the Bill subtotal and Total fields—both have a correlation of 0.38.

Now that we've completed the correlation analysis, we have a better understanding of how the various numeric columns in the data are related.

Use Case 3: Correlation Analysis

The first and most important prerequisite for doing a correlation analysis is having multiple numeric columns. For this dataset, we only have the victim age column. This means we won't be able to conduct a correlation analysis for this use case.

4.3   Data Cleaning (For Now)

It was challenging to decide where to put the data cleaning content in this chapter. Data cleaning is taking the necessary steps to ensure a model can use your data. When compared with data cleaning for data analysis, there is certainly overlap; however, models require specific steps that are not always necessary for data analysis. Two examples of these steps are replacing missing data and dummy coding. While these steps are sometimes used in data analysis projects, they're required for a model to operate as designed.

Data cleaning is a continuous process that spans from the time you receive the data all the way through to when you're building your model and testing it. It's common to build the first iteration of your model and then discover additional data cleaning is required. If it feels like you're constantly cleaning the data, don't worry—that's normal!

Section 4.1 and Section 4.2 both had data cleaning steps within them. This replicates how data cleaning presents itself in the real world. This section will focus on the high-level theory around data cleaning requirements as well as the steps required to get your data ready for the data model process.

4.3.1 Why Isn't Data Already Clean?

Before we dive into specifics, let's discuss why this process is necessary and why it's so time consuming. Businesses are complicated, and the data generated from the various systems and technologies reflect that complexity. There are often two root causes for why data needs cleaning: the business processes themselves and issues with the technology that provides the data.

Business Processes

Data is generated across the world in greatly varying ways. Sometimes it's generated primarily by computers (which tend to be more reliable), while other times it's generated by humans (which tend to be less reliable). To illustrate the impact of business processes on data, let's leverage the HR domain.

HR has many different data sources, most of which revolve around people, given the context of HR. Some of this data is much more reliable than others. Take turnover, for example. Businesses tend to keep accurate records of who has left the company and when. There is a well-defined business process, with many systems capturing and accurately tracking this information. However, say you're trying to build a model that predicts turnover at a company. You have your target variable of turnover, but what other data elements may be helpful when building a model to predict turnover? Consider the following:

- You'll likely want to focus on specific types of turnover. If you focus on voluntary turnover, for example, you'll need to bring in the turnover reason. This is heavily dependent on the individual's manager or HR business partner correctly inputting this information into the system.

- Performance could be a good predictor of someone staying at a company, but how reliable is this information? Many performance management ecosystems at companies are riddled with unreliable ratings due to biases and company politics.

Sometimes it's not always obvious which data is accurate or inaccurate. Sometimes it's impossible to know if data is accurate or inaccurate without an understanding of the business context. Understanding where your data is coming from and how it's generated will help you become a valued partner to your stakeholders and set you apart from others with more technical expertise.

Data Processing

Another important component to understand is how the data is processed and stored. This becomes particularly important when other teams or departments are extracting the data from the source and putting into a database for your consumption. Every step the data takes is an opportunity for an alteration to occur. Paying attention to where the data came from and how it's been handled becomes critical to showing up positively to your stakeholders and building trust with them. For better or worse, if you're the "face of the data," you should be able to answer questions about refresh times and frequency of the data.

Curiosity is incredibly valuable in the data space and helps you be more effective in a machine learning project. Don't be afraid to ask your technology partners "Why?" or dig deeper into a topic that may not seem directly relevant to building a model.

Operational Datasets

Understanding how the data feeds work becomes even more important when you're working with operational datasets and teams that move quickly (and the underlying data changes quickly, too). In one of my roles supporting recruiting functions, the accuracy of the data was heavily dependent on understanding when the data was refreshed.

For example, the data feed was often refreshed with yesterday's data in the early morning hours. Additional processing steps, which were out of our team's control, meant the data often wasn't available to us until 8 a.m. CST. Our data engineer then needed to run the data through the data cleaning process, which took many hours to complete. These nuances are important when considering the application of your model! In this situation, if we were providing a candidate quality score for new applicants, the predictions for that candidate might not be ready until a day and a half after their application submission.

4.3.2   Overview of Cleaning for Regression Models

Regression models have their own nuances to account for. As you'll see in Chapter 5, other types of machine learning models are more forgiving when it comes to the data they're provided. This means there are different considerations you need to make in the data cleaning steps for regression models.

Linear Relationship

The most common type of regression model is linear regression. Using a linear regression model requires your data to have a linear relationship. For example, if you're working an hourly job, your total salary has a linear relationship with how many hours you work.

In the real world, most relationships are not linear. You could easily pick apart the salary and hours worked example and point out the potential for overtime. Luckily, there are many techniques you can use to transform the data into a linear relationship, as well regression techniques—like logistic regression—that don't require a linear relationship. We won't dive into them further, but you can do additional research on these data transformation techniques if you're interested in learning more:

- Logarithmic
- Square root
- Reciprocal
- Exponential

Multicollinearity

Multicollinearity occurs when the variables you're using to build a regression model are correlated. If left unaddressed, multicollinearity can make the results of a regression model unreliable, as the underlying mathematics cannot handle it. The correlation matrix helps identify this. Any time the correlation between the two variables is greater than 0.8, you need to take action to resolve the multicollinearity issue.

There are a number of techniques to resolve the issue. You could simply remove one of the variables if it isn't necessary. Another technique is dimensionality reduction, which we'll discuss in Section 4.3.6. This combines different variables that are similar, providing fewer input variables for your model and helping with multicollinearity.

4.3.3   Inaccurate Data

No matter your modeling technique, you always want to remove inaccurate data. The danger of including inaccurate data isn't a less optimal model—it's a model that appears to be accurate but isn't. This creates a situation where your stakeholder may *think* the model is answering their question, but it's actually answering another question altogether.

This is the most dangerous type of error because it's not detectable by a metric. The best way to identify inaccurate data is to understand the business and the systems from which your data originates. Knowing the business process enables you to identify when the data doesn't align with what could reasonably be expected as an output in the process.

4.3.4   Missing Data

Missing data is an unfortunate reality in the data space, and it can be detrimental to a machine learning project if it's not addressed. How you handle missing values will differ based on whether your data is numeric or categorical. Missing data for categorical columns is easier to handle and can be taken care of with dummy variables (Section 4.3.5). However, numeric columns cannot be blank. The model will return an error telling you the data is missing.

We'll explore options for replacing missing data, demonstrated by our use cases, in the following sections.

Replacing Missing Values

There are a few different ways to approach numeric columns with missing values. The following list, though by no means exhaustive, covers the most commonly used methods:

- **Mean, median, or mode**
  This is one of the most common approaches. You simply take a statistic about the column and replace the missing values with that statistic. It's quick, and it minimizes the impact of your data's underlying distribution. Which of the three metrics you use should depend on the specific use case. Defaulting to the median is often the safest bet because it's not impacted by outliers.

- **Predictive model**
  While it's a bit of inception, you can build a predictive model to replace the missing values of column that are then used to produce a larger predictive model. This option often produces the best results, but you should consider its time implications. You're essentially building two models (or more, if you have multiple columns with missing data) instead of one. To save time, start with a simpler approach in your early iterations of the model and then revisit the possibility of a second predictive model in future iterations.

When to Remove a Column with Missing Data

The best practice for this is highly dependent on your use case. Sometimes, any missing data can be an issue. Other times, 90% of a column's data could be missing, but the column still has value to keep in your model. While relatively subjective, you should use your judgment to decide which data is important to keep in the model.

Another approach can be to run different versions of the model, one with the column in question and one without. The risk to this approach is you're suffering from confirmation bias. When considering which data to provide to your model, it's best to avoid simply throwing in all data and seeing what sticks.

Overall, don't spend too much time worrying about when to remove missing data and when it's okay to leave it in. These are reversible decisions that a model can often help guide. However, you should be intentional about removing data when there are missing values for your target variable. For example, if you're trying to predict the number of sales and there are missing sales values, you should remove those rows, especially if the volume of missing values is small.

Use Case 1: Replacing Missing Values

The dataset for our first use case doesn't have any missing values, so we don't need to take any action.

Use Case 2: Replacing Missing Values

There are a few different decisions we'll need to make with this dataset. In Section 4.1.3, we foreshadowed the missing values issue for the numeric columns in this dataset. We'll focus on correcting the numeric columns first, since the approaches for numeric and text-based columns differ.

If you recall from Section 4.2.3, part of our process to calculate the statistics was identifying how many blank values there were. As a reminder, Figure 4.57 shows that same view of the statistics again.

column_name	mean	median	blanks	not_blank
Bill subtotal	750.076838	629.00	0	21321
Packaging charges	32.564592	28.45	0	21321
Restaurant discount (Promo)	65.091816	80.00	0	21321
Restaurant discount (Flat offs, Freebies & oth...	31.795058	0.00	0	21321
Gold discount	0.099128	0.00	0	21321
Brand pack discount	3.039324	0.00	0	21321
Total	682.616113	597.45	0	21321
Rating	4.356885	5.00	18830	2491
Restaurant compensation (Cancellation)	356.409549	272.58	21188	133
Restaurant penalty (Rejection)	0.000000	0.00	21318	3
KPT duration (minutes)	17.332960	16.33	295	21026
Rider wait time (minutes)	4.825070	3.10	168	21153

Figure 4.57 Descriptive Statistics Results

If you look from the angle of missing values, you can identify three distinct groupings of columns. Each will require a different approach to be taken:

- No missing values
- Mostly missing values
- Minimal missing values

As you'd expect, when columns don't have any missing values, you need to take any additional action for this part of the process.

The next grouping is mostly missing values, which includes the following columns:

- Rating
- Restaurant compensation (Cancellation)
- Restaurant penalty (Rejection)

This is relatively easy to handle, as the best approach is to just remove these columns from our dataset. When there are this many missing values, their value to our use case and model is not meaningful. To remove them, run the code as shown in Listing 4.54. In this code, we're using the drop function, which allows us to provide the names of columns to remove from the dataset.

```
#drop the columns with many missing values
df = df.drop(
  columns=[
    'Rating',
    'Restaurant compensation (Cancellation)',
    'Restaurant penalty (Rejection)'
    ]
  )
```

Listing 4.54 Removing Columns with Many Blank Values

Finally, there are the columns with minimal missing values:

- KPT duration (minutes)
- Rider wait time (minutes)

The cutoff for what is or isn't minimal is straightforward in this context, but it can be fuzzier when you're working with real-world data. Generally, if more than 20% of the data is missing, it's worth considering whether to use that column in the dataset. In this case, less than 1% of the data is blank, so we'll want to preserve these columns.

When considering your approach to fill in the missing values with, think about the decision from two perspectives:

- **How much is missing?**
  If there a lot of data is missing, then an approach like mean, median, or mode may significantly alter the distribution of the column's data. In these cases, leveraging a regression model to predict these values can help maintain the integrity of the underlying distribution. If there isn't a lot of data missing (like in this use case), this risk is much lower. Plus, the speed at which you can leverage a mean, median, or mode makes it a more attractive option.

- **How valuable do you think this will be?**
  This requires an understanding of your use case. Evaluate the data in the context of your use case and determine how important you think this data could be. You can always go back and revisit how you approached the missing data, but it's best to start from a logical place. In this use case, the time it took to get your order as well as how long it took the rider to wait for the order would logically impact whether a customer reorders. The more valuable the column, the more you should consider a regression-like approach.

Combining these considerations is critical. In this use case, the columns are important to our end result. However, there are so few missing values that creating an entire regression process for them would likely not be a valuable use of time.

Before you begin replacing values, you should review the missing values for any potential trends. To do this, count the number of missing values by date to verify the missing values aren't exclusively on the last day of the dataset. This is shown in Listing 4.55, where we've used a combination of groupby to summarize the data at the order level, count to count how many customers ordered on that date, reset_index to ensure our column names can be referenced, and sort_values to order the dataset based on the values in the Customer ID column. You can see a preview of the result in Figure 4.58.

order_date	Customer ID
2024-12-14	11
2024-12-31	10
2024-11-15	8
2024-11-08	7
2024-09-24	7
2024-09-11	5
2024-11-06	5
2025-01-01	5
2024-11-30	5
2024-12-12	4
2024-12-18	4
2024-09-01	4
2024-11-27	4
2024-11-25	4
2024-10-12	4
2024-11-05	4
2024-10-24	4
2024-12-22	4
2024-10-25	4
2025-01-09	4

Figure 4.58 Missing Values by Date

```
#preview the results of the number of KPT duration missing values
df[df['KPT duration (minutes)'].isna()]
   .groupby('order_date')['Customer ID']
   .count()
```

```
  .reset_index()
  .sort_values('Customer ID', ascending = False)
  .head(20)
```

Listing 4.55 Calculate Number of Missing Values by Order Date

As you can see, the dates are scattered throughout the dataset without a notable concentration that would impact how we replace the missing values.

Now, let's test whether we should use a mean or median to replace the missing values from these columns. To generate the mean and median for each one, run the code shown in Listing 4.56. Table 4.2 shows resulting values.

```
#print the mean and median of KPT duration
print(df['KPT duration (minutes)'].mean())
print(df['KPT duration (minutes)'].median())

#print the mean and median of rider wait time
print(df['Rider wait time (minutes)'].mean())
print(df['Rider wait time (minutes)'].median())
```

Listing 4.56 Calculate Mean and Median for Columns with Missing Values

Column with Missing Values	Mean	Median
KPT duration (minutes)	17.33	16.33
Rider wait time (minutes)	4.83	3.1

Table 4.2 Mean and Median Values for Columns with Missing Values

You can see that the mean skews higher, which suggests there are a smaller number of orders with larger values. The median is likely the better option, especially for the rider wait time column given the larger difference between 4.83 and 3.1 minutes. Next, we'll replace these missing values using the code as shown in Listing 4.57. It calculates the median of both columns using the median function and then uses fillna to impute any blank values with their respective column's median.

```
#calculate median of kpt duration and rider wait time
kpt_duration_median = df['KPT duration (minutes)'].median()
rider_wait_median = df['Rider wait time (minutes)'].median()

#replace missing values for kpt duration and rider wait time with the median
df['KPT duration (minutes)'] = df['KPT duration (minutes)']
  .fillna(kpt_duration_median)
```

```
df['Rider wait time (minutes)'] = df['Rider wait time (minutes)']
   .fillna(rider_wait_median)
```

Listing 4.57 Replacing Missing Values with Median

To be safe, we'll verify that this worked properly by checking whether any NaN values remain using the isna().any() function, as shown in Listing 4.58.

```
#check for any remaining missing values in kpt duration or rider wait time
print(df['KPT duration (minutes)'].isna().any())
print(df['Rider wait time (minutes)'].isna().any())
```

Listing 4.58 Checking for NaN Values After Replacing Missing Values

The resulting code returns False for both, meaning we've successfully removed the missing values!

Finally, we'll review the text-based columns for missing values. With this step, we're not necessarily looking to replace the values; however, we do want to decide whether to keep these columns in our data to consider for Section 4.3.5, which covers dummy coding. To do this, we'll loop through all of the columns in our dataset to identify whether they have missing values using a combination of isna and sum to calculate the number of missing values in the column, as shown in Listing 4.59. This reviews the numeric columns again, which is intentional. Rather than going through and identifying which columns to run the loop on, it's more efficient to simply ignore the numeric columns. The result is shown in Figure 4.59.

```
Restaurant ID :  0
Restaurant name :  0
Subzone :  0
City :  0
Order ID :  0
Order Placed At :  0
Order Status :  0
Delivery :  0
Distance :  0
Items in order :  0
Instructions :  20601
Discount construct :  5498
Bill subtotal :  0
Packaging charges :  0
Restaurant discount (Promo) :  0
Restaurant discount (Flat offs, Freebies & others) :  0
Gold discount :  0
Brand pack discount :  0
Total :  0
Review :  21025
Cancellation / Rejection reason :  21135
KPT duration (minutes) :  0
Rider wait time (minutes) :  0
Order Ready Marked :  0
Customer complaint tag :  20852
Customer ID :  0
timestamp :  0
order_date :  0
```

Figure 4.59 Missing Values for All Columns

```
#create list of missing values by column
for col in df.columns:
  print(col, ": ", df[col].isna().sum())
```

Listing 4.59 Identify Number of Missing Values for Each Column

Two columns stick out as worthy columns to remove. The first is Instructions. This column has many different instructions for the values that do exist. It's unlikely to have value to us without natural language processing, which is beyond the scope of this book. The second is the Review column, for the same reasons. We'll drop those columns using the drop function, as shown in Listing 4.60.

```
#drop the instructions and review columns
df = df.drop(columns=['Instructions','Review'])
```

Listing 4.60 Remove Text-Based Columns with Missing Values

Having done this, we're down to 26 columns in our dataset from the original 31 we started with. (The 31 includes the date and timestamp columns we added as part of the data exploration step in Section 4.2.1). This use case is now ready to move to the dummy coding stage, where we'll focus on investigating the text-based columns to see which data will be valuable and reasonable to include in the model.

Use Case 3: Replacing Missing Values

Because this dataset is heavily text-based, the impact of blank values is different. Models can only be given numeric values, so the process of dummy coding (Section 4.3.5) allows us to account for these blanks to ensure the model will run. For this use case, the sole numeric field of the victim's age has no missing values. However, we can look at other fields with missing values to decide whether we'll include them as part of our dummy coding process.

We'll run the code in Listing 4.61 to loop through all of the columns and identify how many blank values there are using a combination of isna and sum in a for loop. The results are displayed in Figure 4.60. Take a few moments to review Figure 4.60 to see which (if any) columns you'd want to remove from the data based on how many blank values there are.

```
#loop through each column of the data frame
for col in df.columns:
  #print out the total number of na/blank values in the dat frame
  print(col, ": ", df[col].isna().sum())
```

Listing 4.61 Check How Many na/blank Values Are in Each Column

The results indicate that it's best not to remove any of these columns. If some of these columns were numeric, we might have a different perspective. However, since these are text-based columns that need to be dummy coded, it's better to keep them all in. Even for a column like Weapon Used Cd, whether or not a field is populated can provide valuable information for our predictions.

```
DR_NO :  0
Date Rptd :  0
AREA :  0
AREA NAME :  0
Rpt Dist No :  0
Part 1-2 :  0
Crm Cd :  0
Crm Cd Desc :  0
Mocodes :  151591
Vict Age :  0
Vict Sex :  144604
Vict Descent :  144616
Premis Cd :  16
Premis Desc :  588
Weapon Used Cd :  677654
Weapon Desc :  677654
Status :  1
Status Desc :  0
Crm Cd 1 :  11
Crm Cd 2 :  935717
Crm Cd 3 :  1002532
Crm Cd 4 :  1004783
LOCATION :  0
Cross Street :  850626
LAT :  0
LON :  0
TIMESTAMP :  0
DATE :  0
```

Figure 4.60 Blank Values in Columns

4.3.5   Dummy Coding

For a data transformation so commonly used in machine learning, it's fair to question the branding impact of using the term *dummy* in its name. But don't let the name fool you: if you aren't familiar with this concept, it's one you need to learn. The return on investment (ROI) of knowledge in this space is significant, given the frequency at which dummy coding is used.

In its simplest form, dummy coding takes categorical (text data) and converts it into numbered data, because models require data to be numeric. Figure 4.61 shows a simple example of what dummy coding looks like in practice. Before you continue on, think about this example and the limitations that dummy coding can have.

ID	Fruit		ID	Apple	Orange	Banana
1	Apple	Dummy Code	1	1	0	0
2	Orange		2	0	1	1
3	Apple		3	1	0	0
4	Banana		4	0	0	1
5	Banana		5	0	0	1

Figure 4.61 Example of Dummy Coded Data

We're now migrating out of the data exploration phase and into the model data pipeline phase. To keep your code clean, don't create a new notebook. You should bring over only the necessary code for data importing and cleaning based on what you learned in the data exploration phase. The data exploration notebook will likely be messy, but the data pipeline code should be cleaner, more stable, and reproducible.

Reproducible Code

If you haven't done much coding before, the concept of *reproducible code* is an important one to know. If you write code and come back to it in two months, you want that code to produce the same results. You might assume running the same code will always produce the same outcome. One common reason it doesn't is that the process of writing code isn't always linear. You often delete and rewrite sections of code, so it's not uncommon to write parts out of order, making it impossible to reproduce the same result.

In the following sections, we'll start by digging into an important aspect of dummy coding: cardinality. We'll then perform dummy coding for our three use cases.

Cardinality

This is one of those fancy terms you can use when you're trying to make a point and sound smart. Cardinality simply refers to the number of unique values in the given column. A high-cardinality column for a dataset about the United States would be city, as there are a lot of unique city values. A medium-cardinality column for the same dataset would be state. There are 50 states, far fewer than the number of cities. A low-cardinality column then would be region (South, Northeast, Midwest, etc.). There are a relatively small number of unique regions in the United States.

Why does this matter? In dummy coding, each unique value in a column becomes its own column. There are thousands of unique US cities, so creating a column for each city would result in way too many columns. This creates technical problems for a model and significantly increases its processing times. From a use case perspective, data with high cardinality is *generally* less valuable. If your objective is to identify a city's geographic location, the data won't represent the similarities and differences between New York City, Boston, and Los Angeles. However, in the region column, New York City and Boston would be grouped together and Los Angeles would appear in a different region. This distinction makes the region column more valuable for your model.

But what happens when your data has a relatively small number of columns and you need to leverage a column with high cardinality? Section 4.3.6 covers dimensionality reduction, which you can use to consolidate the number of columns created from the dummy coding process.

Use Case 1: Dummy Coding

During the data exploration phase for this dataset, we found that we need to clean up the data and filter some of it out. The code in Listing 4.62 is carried over from the data exploration notebook.

```python
#load in necessary libraries
import pandas as pd
import os
import datetime

#create blank dataframe
df = pd.DataFrame()

#create loop to go through all files in the current directory
for file_name in os.listdir():

    #set up if statement to look at only .csv files
    if file_name.endswith(".csv"):

    #read in the csv file into a temp df
    df_temp = pd.read_csv(file_name, encoding='unicode_escape')

    #stack the data frames on top of each other
    df = pd.concat([df, df_temp], axis = 0)

#replace the hyphens with backslashes for the date
df['new_invoicedate'] = df['InvoiceDate'].str.replace('-','/')

#convert the date column to a pandas date time
df['new_invoicedate'] = pd.to_datetime(
    df['new_invoicedate'],
    format = '%d/%m/%Y %H:%M'
    )

#create a date for from the timestamp
df['invoicedate'] = df['new_invoicedate'].dt.date

#keep only the data identified as legitimate transaction records
df_cleaned = df[
    (df['Quantity'] > 0) &
    (df['Price'] > 0) &
    (df['Customer ID'] > 0)
    ]
```

Listing 4.62 Code to Import Data and Apply Necessary Adjustments Previously Identified

Let's do some light exploration to find columns to dummy code. First, we'll look at our data's columns and the values within them by running the following code:

```
#preview data
df_cleaned.head()
```

The result is shown in Figure 4.62.

	Invoice	StockCode	Description	Quantity	InvoiceDate	Price	Customer ID	Country	new_invoicedate	invoicedate
0	489434	85048	15CM CHRISTMAS GLASS BALL 20 LIGHTS	12.0	1/12/2009 7:45	6.95	13085.0	United Kingdom	2009-12-01 07:45:00	2009-12-01
1	489434	79323P	PINK CHERRY LIGHTS	12.0	1/12/2009 7:45	6.75	13085.0	United Kingdom	2009-12-01 07:45:00	2009-12-01
2	489434	79323W	WHITE CHERRY LIGHTS	12.0	1/12/2009 7:45	6.75	13085.0	United Kingdom	2009-12-01 07:45:00	2009-12-01
3	489434	22041	RECORD FRAME 7" SINGLE SIZE	48.0	1/12/2009 7:45	2.10	13085.0	United Kingdom	2009-12-01 07:45:00	2009-12-01
4	489434	21232	STRAWBERRY CERAMIC TRINKET BOX	24.0	1/12/2009 7:45	1.25	13085.0	United Kingdom	2009-12-01 07:45:00	2009-12-01

Figure 4.62 Preview of Data Before Any Additional Adjustments

There are three potential columns to dummy code: Country, Description, and Customer ID. Country is an easy decision for dummy coding because of its lower cardinality. The lower the cardinality of a column, the less computational effort required.

The dataset still has over 800,000 rows, so we need some assistance in identifying the cardinality of the respective columns. We can use the nunique() function to do this, as shown in Listing 4.63. The result of the code is shown in Figure 4.63.

```
#print unique valeus in description column
print("Unique values in description column: ",
  df_cleaned['Description'].nunique()
  )


#print unique values in customer ID column
print("Unique values in customer ID column: ",
  df_cleaned['Customer ID'].nunique()
  )


#print unique values in country column
print("Unique values in country column: ",
  df_cleaned['Country'].nunique()
  )
```

Listing 4.63 Count Unique Values of Columns

```
Unique values in description column:  5283
Unique values in customer ID column:  5878
Unique values in country column:   41
```

Figure 4.63 Results of Counting Unique Values in Each Column

Because of the limited columns available in this dataset, we'll include both the description and customer IDs in the dummy coding process. Machine learning algorithms crave data, so giving the algorithm more data in a situation like this—where our dataset doesn't have hundreds of columns—is the wise move. This decision means we're going to create more than 11,000 new columns, which will necessitate using one of the dimensionality reduction techniques outlined in Section 4.3.6.

With this use case, we're going to dummy code each of these columns independently. This will make our dimensionality reduction steps easier to execute. Don't worry, we'll put all of the columns back together at the end!

Let's start with the lowest cardinality column, Country. As shown in Listing 4.64, we're using get_dummies to execute the dummy coding process. We provide this function to our DataFrame with only the Country column, which then creates a new column for each unique value in the Country column. When you run this code yourself, you'll see a lot of blanks. Figure 4.64 shows the preview window scrolled to the right to display the first five rows, which are blank and belong to the United Kingdom. With this approach, each row will only have one column populated with a 1 for each respective column being dummy coded.

```
#dummy code the country column
df_country = pd.get_dummies(
    df_cleaned[['Country']],
    dtype=float
    )

#print the number of columns and rows
print(df_country.shape)

#preview the data
df_country.head()
```

Listing 4.64 Dummy Coding the Country Field

(805549, 41)

...	Country_Singapore	Country_Spain	Country_Sweden	Country_Switzerland	Country_Thailand	Country_USA	Country_United Arab Emirates	Country_United Kingdom	Country_Unspecified	Country_West Indies
	0.0	0.0	0.0	0.0	0.0	0.0	0.0	1.0	0.0	0.0
	0.0	0.0	0.0	0.0	0.0	0.0	0.0	1.0	0.0	0.0
	0.0	0.0	0.0	0.0	0.0	0.0	0.0	1.0	0.0	0.0
	0.0	0.0	0.0	0.0	0.0	0.0	0.0	1.0	0.0	0.0
	0.0	0.0	0.0	0.0	0.0	0.0	0.0	1.0	0.0	0.0

Figure 4.64 Row and Column Count, as Well as Data Preview

Next, we'll dummy code the description column, as shown in Listing 4.65.

```
#dummy code the description column
df_description = pd.get_dummies(
  df_cleaned[['Description']],
  dtype=float
  )
```

Listing 4.65 Attempting to Dummy Code the Description Column

When you run this code yourself, you'll likely see the time to execute. If you're using Anaconda Cloud, you'll likely get the following error message about your kernel restarting as shown in Figure 4.65. Before you read on, ask yourself what this could mean.

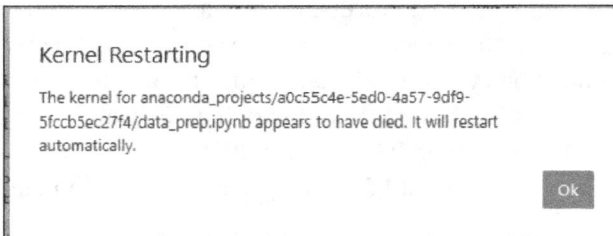

Kernel Restarting

The kernel for anaconda_projects/a0c55c4e-5ed0-4a57-9df9-5fccb5ec27f4/data_prep.ipynb appears to have died. It will restart automatically.

Ok

Figure 4.65 Kernel Restarting Popup Message

This indicates the processing power required to execute this type of action exceeds what Anaconda provides us for free. While the statement "you get what you pay for" is certainly relevant, it also serves as a good lesson. The action of dummy coding a column with cardinality this high is computationally expensive.

So, how do we proceed without missing out on this potentially valuable data? If we go back to what the root cause, the cardinality of the description column is too high. Let's reduce the cardinality with intentionality!

One approach is to take the top *n* number of values and categorize the rest as other. To do this, we can go back to our groupby() data summarization function from Section 4.2.2, as shown in Listing 4.66. Here, we've used the groupby function to summarize our data by the Description column and count the number of rows. Using Invoice here is arbitrary, as we only need to select a column we know will be populated. We then sort the data to make the preview easier to review. The results are shown in Figure 4.66.

```
#aggregate the data by description, counting the number of records
df_summarized = pd.DataFrame(
  df_cleaned.groupby('Description')['Invoice'].count()
  ).reset_index()

#sort the values by the descriptions that have the most invoices
df_summarized = df_summarized.sort_values(
  by='Invoice',
```

```
  ascending=False
  )
```

```
#preview the data
df_summarized.head()
```

Listing 4.66 Identify the Most Common Values of the Description Column

	Description	Invoice
5047	WHITE HANGING HEART T-LIGHT HOLDER	5181
3769	REGENCY CAKESTAND 3 TIER	3428
292	ASSORTED COLOUR BIRD ORNAMENT	2777
2392	JUMBO BAG RED RETROSPOT	2702
3844	REX CASH+CARRY JUMBO SHOPPER	2141

Figure 4.66 Top 5 Most Common Values of the Description Column

Now we need to decide what the cutoff should be. Ideally, the results will be based off the top *n* number of values instead of a frequency cutoff. This ensures that as the data grows and changes, our data pipeline doesn't break.

One approach is to export your data to a *.csv* file to explore it. You're able to more easily filter, sort, and scroll through the columns in Excel as compared to Python. Another trick directly within Python is to use a combination of the head and tail functions, as follows:

```
#use head and tail to select specific range to preview
df_summarized.head(100).tail(10)
```

These functions are essentially the top and bottom *n* of your DataFrame, so combining them together enables you to see the last ten values of the top 100, as shown in Figure 4.67.

	Description	Invoice
2405	JUMBO STORAGE BAG SKULLS	1041
3928	ROUND SNACK BOXES SET OF 4 FRUITS	1036
3688	RED RETROSPOT CAKE STAND	1024
4615	STRAWBERRY CHARLOTTE BAG	1017
4570	SPOTTY BUNTING	1017
2107	GREY HEART HOT WATER BOTTLE	1003
2396	JUMBO BAG SPACEBOY DESIGN	1000
2900	NO SINGING METAL SIGN	998
951	CHARLOTTE BAG SUKI DESIGN	991
2081	GREEN REGENCY TEACUP AND SAUCER	990

Figure 4.67 91st Through 100th Most Common Values in the Description Column

A hundred unique values is still a lot, so what about using the top 50 description values instead? Execute the following code to see the result in Figure 4.68:

```
#use head and tail to select specific range to preview
df_summarized.head(50).tail(10)
```

	Description	Invoice
3128	PAPER CHAIN KIT VINTAGE CHRISTMAS	1330
5270	ZINC METAL HEART DECORATION	1317
3298	PINK CREAM FELT CRAFT TRINKET BOX	1281
163	72 SWEETHEART FAIRY CAKE CASES	1270
1128	COOK WITH WINE METAL SIGN	1270
1635	FELTCRAFT 6 FLOWER FRIENDS	1265
2390	JUMBO BAG PINK VINTAGE PAISLEY	1257
4654	SWEETHEART CERAMIC TRINKET BOX	1244
3491	PLASTERS IN TIN SPACEBOY	1230
204	ALARM CLOCK BAKELIKE RED	1226

Figure 4.68 41st Through 50th Most Common Values in the Description Column

The difference from the 41st value is greater when compared with the 50th value, so let's go with the top 50 description values. You may be thinking there is a high degree of subjectivity here, and you'd be correct. A good rule of thumb is to consider the balance between making progress (avoiding analysis paralysis) and having a defensible decision. This isn't a decision you need to dwell on for three hours. You can always update it in future iterations. From a defensibility standpoint, there are two primary considerations:

- The first is that round numbers are less specific, and therefore a stakeholder like Chris is less likely to call out the decision and dig into it. Imagine if you picked a number like 37. This could appear more random to Chris and warrant additional questions that aren't necessary or valuable. This is part of stakeholder management.

- The second is the distribution of the data. When you look at the top 100, the difference in frequency between the 90th value and the 100th value was minimal, so why add additional values in that will require additional processing power?

To execute the process of only keeping the names of the top 50 most common values and grouping all other values, we'll take the top 50 descriptions and join (merge) them onto our DataFrame (see Listing 4.67). With the merge() function, it will include any new columns not being used to bring the data together. The second line is included to rename the description column from Description to Description Grouped. This allows us to identify the original column from the newly grouped column after the merge is executed.

```
#select the top 50 description values
df_description_top_50 = df_summarized.head(50)

#Rename colummn to description grouped
df_description_top_50['Description Grouped'] = df_description_top_50['Description']

#merge onto the original dataset
df_cleaned = df_cleaned.merge(
  df_description_top_50[['Description', 'Description Grouped']],
  left_on = 'Description',
  right_on = 'Description'
  )
```

Listing 4.67 Identify 50 Most Common Values in the Description Field and Join Back onto the Main Data

Now we can return to the dummy coding process. Run the following lines of code to preview the data and get the result shown in Figure 4.69:

```
#preview data
df_cleaned.head()
```

	Invoice	StockCode	Description	Quantity	InvoiceDate	Price	Customer ID	Country	new_invoicedate	invoicedate	Description Grouped
0	489434	21232	STRAWBERRY CERAMIC TRINKET BOX	24.0	1/12/2009 7:45	1.25	13085.0	United Kingdom	2009-12-01 07:45:00	2009-12-01	STRAWBERRY CERAMIC TRINKET BOX
1	489461	21232	STRAWBERRY CERAMIC TRINKET BOX	48.0	1/12/2009 10:49	1.25	17865.0	United Kingdom	2009-12-01 10:49:00	2009-12-01	STRAWBERRY CERAMIC TRINKET BOX
2	489536	21232	STRAWBERRY CERAMIC TRINKET BOX	3.0	1/12/2009 12:13	1.25	16393.0	United Kingdom	2009-12-01 12:13:00	2009-12-01	STRAWBERRY CERAMIC TRINKET BOX
3	489537	21232	STRAWBERRY CERAMIC TRINKET BOX	20.0	1/12/2009 12:14	1.25	14040.0	United Kingdom	2009-12-01 12:14:00	2009-12-01	STRAWBERRY CERAMIC TRINKET BOX
4	489546	21232	STRAWBERRY CERAMIC TRINKET BOX	12.0	1/12/2009 12:30	1.25	14156.0	EIRE	2009-12-01 12:30:00	2009-12-01	STRAWBERRY CERAMIC TRINKET BOX

Figure 4.69 Data Preview After Join of Description Top 50 Data

Then, as shown in Listing 4.68, we use the get_dummies function with only the Description Grouped column to create a column for each distinct value in our newly created Description Grouped column. The results are shown in Figure 4.70.

```
#dummy code the grouped description column
df_description = pd.get_dummies(
  df_cleaned[['Description Grouped']],
  dtype=float
  )
```

```
#print the number of rows and columns
print(df_description.shape)

#preview the data
df_description.head()
```

Listing 4.68 Dummy Code the New Description Grouped Column

	Description Grouped_6 RIBBONS RUSTIC CHARM	Description Grouped_60 TEATIME FAIRY CAKE CASES	Description Grouped_72 SWEETHEART FAIRY CAKE CASES	Description Grouped_ALARM CLOCK BAKELIKE RED	Description Grouped_ASSORTED COLOUR BIRD ORNAMENT	Description Grouped_BAKING SET 9 PIECE RETROSPOT	Description Grouped_CHOCOLATE HOT WATER BOTTLE	Description Grouped_COOK WITH WINE METAL SIGN
0	0.0	0.0	0.0	0.0	0.0	0.0	0.0	0.0
1	0.0	0.0	0.0	0.0	0.0	0.0	0.0	0.0
2	0.0	0.0	0.0	0.0	0.0	0.0	0.0	0.0
3	0.0	0.0	0.0	0.0	0.0	0.0	0.0	0.0
4	0.0	0.0	0.0	0.0	0.0	0.0	0.0	0.0

(87947, 50)

5 rows × 50 columns

Figure 4.70 Preview of Description Grouped Being Dummy Coded

> **Tip**
> You may have noticed how messy these column names are. At this stage, you should avoid cleaning up the column names. This will introduce manual, potentially unsustainable code to your data pipeline.

The customer ID column will experience the same kernel restarting issue, so we'll follow the same process of identifying the most common values in the column that we did for the description. For simplicity's sake, we'll leverage the top 100 customer IDs (see Listing 4.69 and Figure 4.71).

```
#count the number of instances each customer appears in the data
df_summarized = pd.DataFrame(
    df_cleaned.groupby('Customer ID')['Invoice'].count()
    ).reset_index()

#sort from most to least
df_summarized = df_summarized.sort_values(
    by='Invoice',
    ascending=False
    )

#use head and tail to preview specific range of the data
df_summarized.head(100).tail(10)
```

Listing 4.69 Identify the Frequency of Each Customer ID Showing Up in the Data

	Customer ID	Invoice
1514	14085.0	111
4493	17589.0	111
4482	17576.0	110
5033	18225.0	110
4287	17340.0	110
973	13451.0	109
4930	18109.0	109
3134	15998.0	109
5031	18223.0	108
2166	14849.0	108

Figure 4.71 90th Through 100th Most Common Values in the Customer ID Column

Now we need to join the most common customer ID values onto the main data, as shown in Listing 4.70.

```
#select top 100 values
df_customer_top_100 = df_summarized.head(100)

#rename the column
df_customer_top_100['Customers Grouped'] = df_customer_top_100['Customer ID']

#merge onto original dataset
df_cleaned = df_cleaned.merge(
  df_customer_top_100[['Customer ID', 'Customers Grouped']],
  left_on = 'Customer ID',
  right_on = 'Customer ID'
  )
```

Listing 4.70 Add the Top 100 Customer IDs Back Onto the Main Data

When you attempt to dummy code the data per Listing 4.71, you'll notice something weird happens. The dummy coding doesn't appear to work (see Figure 4.72). Before you continue on, spend some time thinking about why this is occurring.

```
#dummy code the customer data
df_customers = pd.get_dummies(
  df_cleaned[['Customers Grouped']],
  dtype=float,
  dummy_na = True
  )
```

```
#print the number of rows and columns
print(df_customers.shape)

#preview the data
df_customers.head()
```

Listing 4.71 Dummy Code the Customer ID Column

	Customers Grouped
0	17865.0
1	17865.0
2	17865.0
3	17865.0
4	17865.0

Figure 4.72 Preview of the Customer ID Column ... Something Is Wrong

If you guessed that it's related to the customer ID being a numeric column, you're correct! We need to convert the customer ID column into a categorical column to make the dummy coding work. Let's pause on this point and think about this dynamic. Our code views the customer ID as a number, suggesting customer 1 is 5 values less than customer 6. Sometimes, there's hidden value in this. For example, if your customer IDs are generated sequentially, a lower number can indicate a customer who started shopping with the company earlier, while a higher number indicates a new customer. This can be a grey area, but for this use case, we'll convert the column to a categorical variable and dummy code it. As shown in Listing 4.72, we'll use the get_dummies function again, but we need to ensure our column is a string data type for the function to work properly. We do this by using the astype function. You'll get the result shown in Figure 4.73.

```
#updated code for dummy coding the customer column
df_customers = pd.get_dummies(
  df_cleaned[['Customers Grouped']].astype(str),
  dtype=float,
  dummy_na = True
  )

#print the number of rows and columns
print(df_customers.shape)

#preview the data
df_customers.head()
```

Listing 4.72 Dummy Code Customer ID Column While Converting to String Data Type

```
(20321, 101)
```

	Customers Grouped_12471.0	Customers Grouped_12540.0	Customers Grouped_12682.0	Customers Grouped_12748.0	Customers Grouped_12867.0	Customers Grouped_12921.0	Customers Grouped_12949.0	C Grouped
0	0.0	0.0	0.0	0.0	0.0	0.0	0.0	
1	0.0	0.0	0.0	0.0	0.0	0.0	0.0	
2	0.0	0.0	0.0	0.0	0.0	0.0	0.0	
3	0.0	0.0	0.0	0.0	0.0	0.0	0.0	
4	0.0	0.0	0.0	0.0	0.0	0.0	0.0	

5 rows × 101 columns

Figure 4.73 Customer ID Column Dummy Coded After Converting to String

Adjusting to a string seems to have worked, and now we're ready to go see how dimensionality reduction can help us, right? Nope. Did you notice any potential problems with how we executed the dummy coding? Spend some time reviewing the previous steps before reading on.

Each of our row counts are different! There's no way to know how to bring this data back together. An interesting feature (or arguably, bug) of the merge function is that the default for the how parameter is an inner join. This means only the matches were included in the dummy coded dataset, since an inner join will only return rows where there are matches in both datasets.

One option is to update our dummy code functions to include a unique identifier, like invoice, that we can then use to join the datasets together. While this will work, it requires using the pandas merge function, which is less efficient from a computation perspective. A more efficient approach is to ensure the row order and quantity is maintained across each dummy coding process. This allows us to leverage the concat function to put each of these datasets side by side, which is a much simpler task for Python.

First, we rerun the merge with the how parameter included (see Listing 4.73 and Listing 4.74).

```
#select the top 50 values
df_description_top_50 = df_summarized.head(50)

#rename column
df_description_top_50['Description Grouped'] = df_description_top_50['Description']

#updated merge using the how parameter
df_cleaned = df_cleaned.merge(
  df_description_top_50[['Description', 'Description Grouped']],
  how = 'left',
```

```
left_on = 'Description',
right_on = 'Description'
)
```

Listing 4.73 Rerunning the Join of Description Grouped Using the how Parameter

```
#selecting top 100 values
df_customer_top_100 = df_summarized.head(100)

#renaming columns
df_customer_top_100['Customers Grouped'] = df_customer_top_100['Customer ID']

#updated merge using how parameter
df_cleaned = df_cleaned.merge(
  df_customer_top_100[['Customer ID', 'Customers Grouped']],
  how = 'left'
  left_on = 'Customer ID',
  right_on = 'Customer ID'
)
```

Listing 4.74 Rerunning the Join of the 100 Customer ID Data Using the how Parameter

After running through the get_dummies(), you can see both outputs are the same—about 800,000 records. This will make it easier to combine all of the DataFrames once the dimensionality reduction exercise is complete.

In Listing 4.75 we're rerunning the get_dummies function on the Description Grouped data after the changes we've made. Figure 4.74 shows the preview of this code's output .

```
#dummy code the description column
df_description = pd.get_dummies(
  df_cleaned[['Description Grouped']],
  dtype=float
  )

#print number of rows and columns
print(df_description.shape)

#preview the data
df_description.head()
```

Listing 4.75 Rerunning the Dummy Coding for the Description Column

4.3 Data Cleaning (For Now)

(805549, 50)

	Description Grouped_6 RIBBONS RUSTIC CHARM	Description Grouped_60 TEATIME FAIRY CAKE CASES	Description Grouped_72 SWEETHEART FAIRY CAKE CASES	Description Grouped_ALARM CLOCK BAKELIKE RED	Description Grouped_ASSORTED COLOUR BIRD ORNAMENT	Description Grouped_BAKING SET 9 PIECE RETROSPOT	Description Grouped_CHOCOLATE HOT WATER BOTTLE	Description Grouped_COOK WITH WINE METAL SIGN	Description Grouped_FELTCRAFT 6 FLOWER FRIENDS	Description Grouped_GIN + TONIC DIET METAL SIGN	
0	0.0	0.0	0.0	0.0	0.0	0.0	0.0	0.0	0.0	0.0	...
1	0.0	0.0	0.0	0.0	0.0	0.0	0.0	0.0	0.0	0.0	...
2	0.0	0.0	0.0	0.0	0.0	0.0	0.0	0.0	0.0	0.0	...
3	0.0	0.0	0.0	0.0	0.0	0.0	0.0	0.0	0.0	0.0	...
4	0.0	0.0	0.0	0.0	0.0	0.0	0.0	0.0	0.0	0.0	..

5 rows × 50 columns

Figure 4.74 Preview of the Rerun to Dummy Coding the Description Column

In Listing 4.76, we're rerunning the get_dummies function on the Customer Grouped column after the changes we've made. Figure 4.75 shows a preview of the code's output.

```
#dummy code the customer data
df_customers = pd.get_dummies(
    df_cleaned[['Customers Grouped']].astype(str),
    dtype=float,
    dummy_na = True
    )

#print number of rows and columns
print(df_customers.shape)

#preview the data
df_customers.head()
```

Listing 4.76 Rerunning the Dummy Coding for the Customer ID Column

(805549, 102)

	Customers Grouped_12415.0	Customers Grouped_12471.0	Customers Grouped_12681.0	Customers Grouped_12682.0	Customers Grouped_12748.0	Customers Grouped_12867.0	Customers Grouped_12921.0	Customers Grouped_13001.0	Customers Grouped_13050.0	Cus Grouped_1
0	0.0	0.0	0.0	0.0	0.0	0.0	0.0	0.0	0.0	
1	0.0	0.0	0.0	0.0	0.0	0.0	0.0	0.0	0.0	
2	0.0	0.0	0.0	0.0	0.0	0.0	0.0	0.0	0.0	
3	0.0	0.0	0.0	0.0	0.0	0.0	0.0	0.0	0.0	
4	0.0	0.0	0.0	0.0	0.0	0.0	0.0	0.0	0.0	

5 rows × 102 columns

Figure 4.75 Preview of the Rerun of Dummy Coding the Description Column

This use case is a great example of how iterative the process can be. Rarely do you run code properly the first time. There were even errors that required fixing while putting together this example. Remember that it's normal to iterate within this process!

Use Case 2: Dummy Coding

The dummy coding process for this use case will build upon what you learned in the first use case. There are columns in this dataset that have high dimensionality (many unique values), so we won't be able to simply apply dummy coding across the entirety of our text-based columns as they are.

Given the number of columns we have, let's first identify how many unique values there are in each one. We used a similar approach to confirm the number of null values before replacing them. The difference here is we need to count the number of *unique* values, which we do by identifying how many unique values there are and then calculating the length of the list. As shown in Listing 4.77, we're calculating the number of unique values in each column by looping through them, using a combination of the unique and len functions. The results are shown in Figure 4.76.

```
#print unique values for each column
for col in df.columns:
  print(col, ": ", len(df[col].unique()))
```

Listing 4.77 Count Number of Unique Values in Each Column

```
Restaurant ID :   21
Restaurant name :   6
Subzone :   8
City :   1
Order ID :   21321
Order Placed At :   19114
Order Status :   6
Delivery :   1
Distance :   22
Items in order :   6123
Discount construct :   528
Bill subtotal :   1807
Packaging charges :   2727
Restaurant discount (Promo) :   700
Restaurant discount (Flat offs, Freebies & others) :   318
Gold discount :   29
Brand pack discount :   415
Total :   3147
Cancellation / Rejection reason :   6
KPT duration (minutes) :   2415
Rider wait time (minutes) :   311
Order Ready Marked :   3
Customer complaint tag :   6
Customer ID :   11607
timestamp :   19114
order_date :   153
```

Figure 4.76 Unique Values for Dummy Coding

To simplify this process, we'll keep all the columns. We can then use the data to decipher whether columns are numeric or text based. There are a few categories these columns will fit into:

- Dummy code as is.
- Dummy code with possible summarization.
- Don't dummy code and remove.

The dummy-code-as-is list is unfortunately small, with only two columns fitting this description. Restaurant name sticks out as an easy item for dummy coding since it only has six unique values. Subzone also fits into this category with only eight unique values.

The list of columns to investigate further includes Distance, Customer complaint tag, Cancellation / Rejection reason, and Order Ready Marked.

4

Distance is a text-based column, but its name suggests it may have some numeric value. This flags further investigation because there are only 22 unique values, meaning the granularity of the values may be quite high level. Rather than dummy coding all 22 unique values, there may be an opportunity to simplify this column or convert it into a numeric field to avoid dummy coding all together.

To investigate the column, we'll identify the unique number of values using the code shown in Listing 4.78. The results from this code are shown in Figure 4.77. Take a moment to consider how you'd approach the next steps with this column. Would you keep it as is, group some of the values together and dummy code them, or convert it into a numeric column?

```
#look at unique values in distance
df['Distance'].unique()
```

Listing 4.78 Unique Values in the Distance Column

```
array(['3km', '2km', '<1km', '1km', '6km', '4km', '5km', '8km', '7km',
       '10km', '9km', '11km', '16km', '14km', '15km', '17km', '12km',
       '18km', '13km', '19km', '21km', '20km'], dtype=object)
```

Figure 4.77 Unique Values in the Distance Column

Let's convert this into a numeric column. Numeric columns are valuable in predictive modeling because they maintain the relationship between the values. In this instance, there could be a specific cutoff or threshold in the distance of the delivery that translates to repeat orders. Having all the values in one column makes it easier for the model to identify and utilize this trend.

To do this, we need to create a function that addresses two main nuances. First, we need to address the <1km value since it isn't inherently a number itself. The second step is removing the km from all of the numbers (ideally, without having to explicitly reference each interval). To do this, we'll write a function and then apply it to the distance column. First, the function is created, as shown in Listing 4.79. We're using def, which creates a function and Python's approach to an if statement. This creates a function we can apply to our distance column.

```
#define function to convert the distance from a string to numeric column
def convert_distance(col):
  if col == '<1km':
    return 0.5
```

```
  else:
    return col[:-2]
```

Listing 4.79 Convert Text Distance into Numeric Distance

Next, we convert the value using the apply function. We also want to double-check that this worked as we expected by using groupby to view the data for each unique combination of the original Distance column and new DistanceNumeric columns. This ensures the original text-based values are properly converted into numeric columns. Listing 4.80 shows the code to achieve this and Figure 4.78 shows the result.

```
#apply the function created to convert distance column from string to numeric
df['DistanceNumeric'] = df['Distance'].apply(convert_distance)

#check the results
df.groupby(['Distance', 'DistanceNumeric'])['timestamp'].count()
```

Listing 4.80 Apply Function to Convert to Numeric Column and Confirm Results

```
Distance   DistanceNumeric
10km       10                    384
11km       11                    253
12km       12                    115
13km       13                     63
14km       14                     82
15km       15                     77
16km       16                     87
17km       17                     26
18km       18                     25
19km       19                     19
1km        1                    3346
20km       20                      3
21km       21                      2
2km        2                    3558
3km        3                    3212
4km        4                    2410
5km        5                    2140
6km        6                    2112
7km        7                    1282
8km        8                     709
9km        9                     767
<1km       0.5                   649
```

Figure 4.78 Confirming Results of Converting to Numeric Column

All of the results return as expected!

You may be wondering why this is in the dummy coding section if we just actively avoided dummy coding. That's a fair question. Avoiding dummy coding is very common when preparing your data, so this example gives you a realistic representation of how the data preparation process usually looks. A key takeaway from the distance column is to carefully consider the true meaning of what the data represents. For instance, this column represents a number. You'll see the value in keeping distance as one column instead of dummy coding it and turning it into 20 columns once we start building the models.

As a final housekeeping item with this column, we'll remove the original column using the drop function, as shown in Listing 4.81.

```
#drop the original distance column
df = df.drop(columns=['Distance'])
```

Listing 4.81 Remove Original Distance Column

Next, we'll look into the customer complaint tag. The first step is the same as the distance column: we'll evaluate the unique values (see Listing 4.82). The resulting values shown in Figure 4.79 are standardized, meaning we don't need to apply any additional adjustments before dummy coding them.

```
#look at unique values in customer complaint tag column
df['Customer complaint tag'].unique()
```

Listing 4.82 Identify Unique Values in Customer Complaint Tag

```
array([nan, 'Poor taste or quality', 'Non-refunded complaint',
       'Poor packaging or spillage', 'Item(s) missing or not delivered',
       'Wrong item(s) delivered'], dtype=object)
```

Figure 4.79 Unique Values in Customer Complaint Tag Column

We can use the same process for the remaining two columns of Cancellation / Rejection reason and Order Ready Marked. Listing 4.83 shows the code and Figure 4.80 and Figure 4.81 show the results. Ultimately, these columns are standardized, so no additional work is required before we can dummy code them.

```
#check unique values in cancellation/rejection reason
df['Cancellation / Rejection reason'].unique()
```

```
#check unique values in order ready marked column
df['Order Ready Marked'].unique()
```

Listing 4.83 Identify Unique Values for Cancellation/Rejection Reason and Order Ready Marked

```
array([nan, 'Cancelled by Zomato', 'Cancelled by Customer',
       'Merchant device issue', 'Kitchen is full', 'Items out of stock'],
      dtype=object)
```

Figure 4.80 Cancellation Reason/Rejection Reason Unique Values

```
array(['Correctly', 'Incorrectly', 'Missed'], dtype=object)
```

Figure 4.81 Order Ready Marked Unique Values

Before we start dummy coding, we'll drop the additional columns we know we won't need anymore. We'll keep identifying columns that may be needed later, but at this step we'll remove the junk columns (see Listing 4.84). This reduces the memory required for the server to process the dummy coding. We're now down to 18 columns in the dataset.

```
#create list of columns to drop
cols_to_drop = [
  'Restaurant ID',
  'City',
  'Order ID',
  'Delivery',
  'Order Placed At',
  'Items in order',
  'Discount construct',
  'Order Status']


#drop columns
df = df.drop(columns = cols_to_drop)
```

Listing 4.84 Remove Junk Columns Before Dummy Coding

We can take a simpler approach here than we did for the first use case. In that use case, we had a small number of high-dimensionality columns to dummy code, so we did each one individually, which made for easier dimensionality reduction. However, this data has low-dimensionality columns, so we'll be able to pass the columns we want to dummy code, and the function will return use our entire dataset intact with the new dummy coded values. This saves us the additional steps of stitching the data back together.

Listing 4.85 shows the dummy coding. This results in our dataset becoming 40 columns, which is still very manageable (especially when you consider the number of columns we had in the first use case).

```
#create list of columns to dummy code
cols_to_dummy = [
  'Restaurant name',
  'Subzone',
  'Cancellation / Rejection reason',
  'Order Ready Marked',
  'Customer complaint tag'
  ]
```

```
#dummy code columns
df_dummied = pd.get_dummies(df, columns=cols_to_dummy)
```

Listing 4.85 Dummy Code Data

The concludes the dummy coding portion for this use case! Compared to the first use case, this offers a notably different perspective on how dummy coding should be considered when applied to real data.

Use Case 3: Dummy Coding

As a heads up, this section is quite lengthy. Given the number of text-based columns with high dimensionality in this dataset, we'll need to go through each column and decide two things:

- Will we keep this column or not?
- Do we need to reduce the number of unique values before dummy coding?

Unfortunately, this shouldn't be done exclusively with a loop. We'll end up with a better result by going through each column to determine when values should be kept as standalone, grouped with other values, or placed in a broad Other category.

As a starting point, we'll go through the area names, as shown in Listing 4.86. We're using the groupby function to identify how many times each area name is in the data. We use the DR_NO column to count, as we know it should be populated. We then sort the values using sort_values so the most frequent values are displayed at the top. Figure 4.82 shows the result.

```
#count number of unique values in the column
print(df['AREA NAME'].nunique())


#create new data frame that counts the number of crimes for each area name
#sorts the values from largest to smallest
df_review = df.groupby(
  'AREA NAME')['DR_NO']
  .count()
  .reset_index()
  .sort_values('DR_NO', ascending = False)


#displays the entirety of the data frame
df_review
```

Listing 4.86 Area Name Unique Values and Counts

AREA NAME	DR_NO
Central	69671
77th Street	61752
Pacific	59500
Southwest	57430
Hollywood	52438
N Hollywood	51106
Olympic	50062
Southeast	49926
Newton	49174
Wilshire	48238
Rampart	46823
West LA	45723
Northeast	42948
Van Nuys	42877
West Valley	42141
Devonshire	41744
Harbor	41380
Topanga	41366
Mission	40341
Hollenbeck	37078
Foothill	33129

Figure 4.82 Count of Crimes by Area Name

There are twenty-one unique areas, which is a lot. This will present challenges for regression-based models. However, given our use case—to understand where crime is occurring so we can allocate resources there—grouping these values won't make sense. We'll leave the area name column as is.

Next, we'll evaluate the Crm Cd Desc (crime code description) column in the same way, as shown in Listing 4.87. Figure 4.83 shows the results.

```
#count number of unique values in the column
print(df['Crm Cd Desc'].nunique())

#create new data frame that counts the number of crimes for each crime code
#sorts the values from largest to smallest
df_review = df.groupby(
```

```
'Crm Cd Desc')['DR_NO']
.count()
.reset_index()
.sort_values('DR_NO', ascending = False)

#displays the first 20 rows of the data frame
df_review.head(20)
```

Listing 4.87 Unique Values and Counts for the Crime Code Description Field

Crm Cd Desc	DR_NO
VEHICLE - STOLEN	115184
BATTERY - SIMPLE ASSAULT	74817
BURGLARY FROM VEHICLE	63511
THEFT OF IDENTITY	62534
VANDALISM - FELONY ($400 & OVER, ALL CHURCH VA...	61084
BURGLARY	57874
THEFT PLAIN - PETTY ($950 & UNDER)	53716
ASSAULT WITH DEADLY WEAPON, AGGRAVATED ASSAULT	53525
INTIMATE PARTNER - SIMPLE ASSAULT	46710
THEFT FROM MOTOR VEHICLE - PETTY ($950 & UNDER)	41311
THEFT FROM MOTOR VEHICLE - GRAND ($950.01 AND ...	36943
THEFT-GRAND ($950.01 & OVER)EXCPT,GUNS,FOWL,LI...	35144
ROBBERY	32314
SHOPLIFTING - PETTY THEFT ($950 & UNDER)	30898
VANDALISM - MISDEAMEANOR ($399 OR UNDER)	25368
CRIMINAL THREATS - NO WEAPON DISPLAYED	19276
TRESPASSING	18425
BRANDISH WEAPON	14531
INTIMATE PARTNER - AGGRAVATED ASSAULT	12657
VIOLATION OF RESTRAINING ORDER	11748

Figure 4.83 Count of Crimes by Crime Code

There are 140 unique values, which is way too many for us to keep. We need to decide what our cutoff will be, which becomes relatively subjective. The drop off in the data appears to start happening after VANDALISM – MISDEAMEANOR ($399 OR UNDER). This means we want to keep the top 15 values, then group everything else into the "other" column.

Given the length of each crime code description, we'll use our DataFrame to pull the top fifteen values when it creates this new column (see Listing 4.88). This is done using np.where, which acts like a simple if statement. The first parameter for this function is the logical criteria. We use isin for this, passing in a list of the top 15 crime codes. This means only the values for the top 15 crime codes appear in the columns, since we specify to return Crm Cd Desc values when true. Any other values will return as Other.

```
#create list of the top 15 unique values based on count of the Crm Cd Desc
top_15_crime_codes = list(df_review.head(15)['Crm Cd Desc'].unique())

#create new column that either shows the top 15 crime code or other
df['crime_description'] = np.where(
    df['Crm Cd Desc'].isin(top_15_crime_codes),
    df['Crm Cd Desc'],
    'Other'
    )
```

Listing 4.88 Create Column That Only Keeps the Top 15 Most Common Values in the Crime Code

If we run the same grouping code on the new grouped column for the crime code descriptions, we get a list of 16 unique values (the 15 we selected, plus one more for Other), as shown in Figure 4.84.

crime_description	DR_NO
Other	213914
VEHICLE - STOLEN	115184
BATTERY - SIMPLE ASSAULT	74817
BURGLARY FROM VEHICLE	63511
THEFT OF IDENTITY	62534
VANDALISM - FELONY ($400 & OVER, ALL CHURCH VA...	61084
BURGLARY	57874
THEFT PLAIN - PETTY ($950 & UNDER)	53716
ASSAULT WITH DEADLY WEAPON, AGGRAVATED ASSAULT	53525
INTIMATE PARTNER - SIMPLE ASSAULT	46710
THEFT FROM MOTOR VEHICLE - PETTY ($950 & UNDER)	41311
THEFT FROM MOTOR VEHICLE - GRAND ($950.01 AND ...	36943
THEFT-GRAND ($950.01 & OVER)EXCPT,GUNS,FOWL,LI...	35144
ROBBERY	32314
SHOPLIFTING - PETTY THEFT ($950 & UNDER)	30898
VANDALISM - MISDEAMEANOR ($399 OR UNDER)	25368

Figure 4.84 New Crime Code Grouped Column Results

Next, we'll evaluate the Vict Sex column in the same way (see the code in Listing 4.89 and the results in Figure 4.85).

```
#count number of unique values in the column
print(df['Vict Sex'].nunique())

#create new data frame that counts the number of crimes for each value
#sorts the values from largest to smallest
df_review = df.groupby(
  'Vict Sex')['DR_NO']
  .count()
  .reset_index()
  .sort_values('DR_NO', ascending = False)

#displays all rows of the data frame
df_review
```

Listing 4.89 Show Unique Values and Counts by the Victim Sex Column

Vict Sex	DR_NO
M	403834
F	358543
X	97751
H	114
-	1

Figure 4.85 Count of Values for the Victim Sex Column

The results of this column should be viewed from the broader context of our data. If we dummy coded this column as is, we'll only get five new columns. By itself, that isn't a concern. However, given how many columns we'll be creating when dummy coding, we should cut down the number of columns as much as we can without harming our ability to capture predictive power. Keep in mind that the blank values aren't included here. From a modeling perspective, it makes the most sense to keep the male and female values, while assigning everything else to Other. This will result in three unique columns created when we dummy code the data. In Listing 4.90, we've created a list of gender values to keep. We then use the same approach from Listing 4.88 where we're using np.where and isin to create a column that only returns the gender values we want to keep.

```
#create list of male and female values
gender_list_keep = ['M', 'F']
```

```
#create new column that groups the values that aren't M or F into an Other value
df['victim_gender'] = np.where(
    df['Vict Sex'].isin(gender_list_keep),
    df['Vict Sex'],
    'Other'
    )
```

Listing 4.90 Create Column That Consolidates the Values of the Victim Sex

We'll evaluate the Vict Descent (race/ethnicity) field next, as we've done previously (see Listing 4.91 and Figure 4.86).

Vict Descent	DR_NO
H	296365
W	201424
B	135810
X	106659
O	77996
A	21336
K	5990
F	4838
C	4631
J	1586
V	1193
I	1015
Z	577
P	288
U	221
D	91
L	77
G	74
S	58
-	2

Figure 4.86 Count of Unique Values for Victim Race/Ethnicity

```
#count number of unique values in the column
print(df['Vict Descent'].nunique())
```

```
#create new data frame that counts the number of crimes for each value
```

```
#sorts the values from largest to smallest
df_review = df.groupby(
   'Vict Descent')['DR_NO']
   .count()
   .reset_index()
   .sort_values('DR_NO', ascending = False)

#displays all rows of the data frame
df_review
```

Listing 4.91 Unique Values and Count by Each Value

The 20 unique values will need to be consolidated. There are two potential cutoff points: either after X or 0. In this situation, we'll cut off the values and assign them to Other after 0, given there are 75,000 crimes associated with this value. This will result in six unique values in a consolidated column. In Listing 4.92, we've used the same approach from Listing 4.88 and Listing 4.90 for the Vict Descent column.

```
#list of values to keep
race_list_keep = ['H', 'W', 'B', 'X', 'O']

#create new race column that is consolidated
df['victim_race'] = np.where(
   df['Vict Descent'].isin(race_list_keep),
   df['Vict Descent'],
   'Other'
   )
```

Listing 4.92 Create New Column That Consolidates the Race/Ethnicity Values

Now, let's look at the Premis Desc (premises description) column. This was one of the graphs we created in Section 4.2.2 with a bunch of text smashed together, making it unreadable. The code and resulting value counts are shown in Listing 4.93 and Figure 4.87.

```
#count number of unique values in the column
print(df['Premis Desc'].nunique())

#create new data frame that counts the number of crimes for each value
#sorts the values from largest to smallest
df_review = df.groupby(
   'Premis Desc')['DR_NO']
   .count()
   .reset_index()
   .sort_values('DR_NO', ascending = False)
```

```
#displays top 20 values of the data frame
df_review.head(20)
```

Listing 4.93 Unique Values and Value Counts for the Premis Desc Field

Premis Desc	DR_NO
STREET	261259
SINGLE FAMILY DWELLING	163640
MULTI-UNIT DWELLING (APARTMENT, DUPLEX, ETC)	119008
PARKING LOT	69137
OTHER BUSINESS	47643
SIDEWALK	40859
VEHICLE, PASSENGER/TRUCK	29300
GARAGE/CARPORT	19361
DRIVEWAY	16083
DEPARTMENT STORE	14431
RESTAURANT/FAST FOOD	12312
PARKING UNDERGROUND/BUILDING	8228
OTHER PREMISE	8039
MARKET	7883
OTHER RESIDENCE	7555
ALLEY	7066
PARK/PLAYGROUND	6653
CLOTHING STORE	6206
YARD (RESIDENTIAL/BUSINESS)	6202
GAS STATION	5742

Figure 4.87 Value Counts for the Premis Desc Column

We're only showing the top 20 values here because there are over 300 unique values for this column. The natural cutoff is after VEHICLE, PASSENGER/TRUCK. This reduces the total to eight unique values, including the Other group we'll create, while keeping all values above 20,000 crimes.

Again, remember to think in the broader context of your use case. Any time you're able to identify a few unique values that can be removed, you're removing columns that the model will need to evaluate.

As shown in Listing 4.94, we'll leverage a similar approach to the crime code descriptions, where we use the DataFrame to extract the values we want to keep rather than typing them out into their own list.

```
#Create list of values based on the data frame we already created
top_7_crime_locations = list(df_review.head(15)['Premis Desc'].unique())

#Create the new column to consolidate the values
df['crime_premises'] = np.where(
  df['Premis Desc'].isin(top_7_crime_locations),
  df['Premis Desc'],
  'Other'
  )
```

Listing 4.94 Creating New Consolidated Crime Premises Field

Onto our last column to evaluate! The Weapon Desc (weapon description) field was another graph that was mostly unreadable when we attempted to visualize it. Listing 4.95 shows the code to create this and Figure 4.88 shows the resulting values counts.

Weapon Desc	DR_NO
STRONG-ARM (HANDS, FIST, FEET OR BODILY FORCE)	174721
UNKNOWN WEAPON/OTHER WEAPON	36382
VERBAL THREAT	23840
HAND GUN	20184
SEMI-AUTOMATIC PISTOL	7267
KNIFE WITH BLADE 6INCHES OR LESS	6837
UNKNOWN FIREARM	6581
OTHER KNIFE	5880
MACE/PEPPER SPRAY	3730
VEHICLE	3260
ROCK/THROWN OBJECT	2739
PIPE/METAL PIPE	2469
BOTTLE	2414
FOLDING KNIFE	2266
STICK	2249
CLUB/BAT	2088
KITCHEN KNIFE	1921
AIR PISTOL/REVOLVER/RIFLE/BB GUN	1864
KNIFE WITH BLADE OVER 6 INCHES IN LENGTH	1697
BLUNT INSTRUMENT	1387

Figure 4.88 Value Counts for the Weapon Desc Field

```
#count number of unique values in the column
print(df['Weapon Desc'].nunique())

#create new data frame that counts the number of crimes for each value
#sorts the values from largest to smallest
df_review = df.groupby(
  'Weapon Desc')['DR_NO']
  .count()
  .reset_index()
  .sort_values('DR_NO', ascending = False)

#displays top 20 values of the data frame
df_review.head(20)
```

Listing 4.95 Unique Values and Counts of the Weapon Desc Field

There are 79 unique values, all of which will need to be consolidated. Of all the fields, this one has one of the more obvious cutoff points to start grouping values in Other. After the HAND GUN value, there is quite the drop-off. We'll use this as the point to consolidate all other values, as shown in Listing 4.96.

```
#create list of the top 4 weapons by volume
top_4_weapons = list(df_review.head(4)['Weapon Desc'].unique())

#create new consolidated column that groups weapons values
df['crime_weapon'] = np.where(
  df['Weapon Desc'].isin(top_4_weapons),
  df['Weapon Desc'],
  'Other'
  )
```

Listing 4.96 Create Consolidated Weapon Value Column

Now that we've created these new columns, we can execute the dummy coding (see Listing 4.97).

```
#list of columns to dummy code
cols_to_dummy = [
  'AREA NAME', 'crime_description', 'victim_gender',
  'victim_race', 'crime_premises', 'crime_weapon'
  ]

#execute dummy coding
df_dummied = pd.get_dummies(df, columns = cols_to_dummy)
```

Listing 4.97 Dummy Code Data

This created a DataFrame with 94 columns. While we expected this, we should now take a moment to drop all the columns we don't need anymore (see Listing 4.98). Spoiler: This includes a *lot* of columns. It's helpful to use df_dummied.columns to print out all of the column names, especially when you're dealing with such a large number. You can then copy and paste the column names and minimize your risk of a typo.

```
#drop columns no longer needed
df_dummied = df_dummied.drop(
  columns = [
    'AREA', 'Rpt Dist No', 'Part 1-2', 'Crm Cd', 'Crm Cd Desc',
    'Mocodes','Vict Sex', 'Vict Descent','Premis Cd', 'Premis Desc',
    'Weapon Used Cd', 'Weapon Desc', 'Status','Status Desc', 'Crm Cd 1',
    'Crm Cd 2', 'Crm Cd 3', 'Crm Cd 4','LOCATION', 'Cross Street', 'LAT', 'LON'
    ]
  )
```

Listing 4.98 Drop Columns We No Longer Need

This brings us down to 72 columns in the overall dataset. As a final step in this section, we'll save this DataFrame to a *.csv* file so we don't need to rerun all of this code (see Listing 4.99). We'll pick up the use case in the dimensionality reduction section to see if there are any opportunities to further reduce the number of columns.

```
#save data to csv file
df_dummied.to_csv('crime_data_prepped.csv', index=False)
```

Listing 4.99 Save DataFrame to .csv File

4.3.6   Dimensionality Reduction

The goal of dimensionality reduction is to reduce the number of columns you're giving your machine learning model while minimizing the information loss. The math behind this can be quite complex, but the application is quite simple. It becomes another step in your data preparation process.

Dimensionality reduction can be a confusing topic. Why would you want to limit the information the model is given? Shouldn't these powerful machine learning algorithms be able to sift through the data? While in some cases they can, they still have limitations. In Chapter 5, we'll discuss machine learning models in more detail, but it's necessary to understand why dimensionality reduction is a valuable approach.

It's important to note that with dimensionality reduction, you're usually only applying one technique per set of columns. You'll learn about principal component analysis (PCA) and linear discriminant analysis (LDA) in the following sections. The best approach is to test both techniques, but ultimately, you'll only use one in your model. This choice will be represented in the use cases, where we'll apply each dimensionality reduction technique to each dataset. The resulting data will then serve as an input for testing within the machine learning process.

Why Perform Dimensionality Reduction?

Let's use a simple decision tree model to explain the value of dimensionality reduction. For example, consider a use case to predict the winner of a spelling bee. Your data contains variables on each contestant, including their previous performance in other spelling bee competitions, their education, their country of birth, and which school they're from. There are 10,000 entrants into this spelling bee. As you know from Section 4.3.5 on dummy coding, the result of the categorical data (even after limiting to the top n values) can create hundreds of columns. In its simplest form, a decision tree decides which column will explain the data the best. But here's what happens when a decision tree makes a decision over hundreds of columns:

- The processing time required is significant, and the code may not even run, depending on how robust your environment is.
- The quality of the decisions will be poor given the number of decisions required.

Consider this from the perspective of the spelling bee example. What would happen if you were to group students' country of origin into 5 categories, where each country has had a similar performance in spelling bee competitions to date? You could take the same approach with schools, grouping them into 25 categories instead. By reducing the number of categories, we're also reducing the number of columns that need to be dummy coded. The fewer columns going into the model, the fewer decisions it needs to make, which simplifies the process for the machine learning algorithm because fewer calculations are required.

As you'll discover in Chapter 5, a decision tree is a relatively simple model. So, it would be fair to argue that this is a strawman argument for a lower-complexity machine learning model. Decision trees are the foundation for other more progressively sophisticated models, such as random forest models or Gradient Boosting Machine (GBM) models. While these models have other mechanisms that reduce the impact of large datasets, at their core they're still decision trees. This means that the core mechanics—like a large number of columns and the need to reduce them—apply across both simple and complex algorithms!

Drawbacks of Dimensionality Reduction

The most significant drawback of dimensionality reduction is the lack of interpretability. Machine learning models are often flagged as *black box* models and are criticized because it's difficult to interpret how they're making decisions.

Unfair Bias Against Machine Learning?

Machine learning models often get criticized in the media and the corporate world. The concept of a black box model is often referenced because it's not always clear why the model is making the decision that it is. When we get into the traditional techniques in the next sections, you'll see how hard it can be to translate these techniques back to the real world.

What I find interesting is that these techniques actually are rooted in traditional statistical and regression-based models, which are usually seen as easier to interpret than machine learning models. Even if you can show the coefficients for a model, if those coefficients translate back to columns that can't necessarily be interpreted, are your results actually interpretable?

As you'll see in Chapter 6, there are ways to make the model's decision more interpretable. However, what happens if you're using dimensionality reduction?

Issues with interpretability for dimensionality reduction arise when you're trying to tie a column back to what the business user understands. Most dimensionality reduction approaches produce an output that helps the machine learning model by creating an index of sorts—but deciphering this as a human is often quite difficult.

To briefly return to the spelling bee example, most of the techniques we'll discuss in this chapter go beyond simply grouping or clustering each country or school. Instead, they involve creating a set of n new columns, with each country or school potentially belonging to more than one. If your use case indexes heavily on interpretability, this can be a significant challenge that limits your ability to use traditional dimensionality reduction techniques.

Real-World Story: When Dimensionality Reduction Can't Be Used

I was asked to create a model that helps identify the workload of open positions for recruiters at a company. When working with the stakeholders (and having worked with them in the past), I knew that transparency was key for two reasons. The first is the historical context around this request and the skepticism that previous attempts at measuring capacity had created. The second ties to how the stakeholders wanted to use the model. If a recruiter was over or under capacity, the stakeholders needed to know which actions would get the recruiter back to the desired capacity range.

Inputs into the model included the types of roles a recruiter works as well as their associated attributes (department, job experience, skillset, etc.). After dummy coding, there were hundreds of columns. When thinking about dimensionality reduction, I knew the interpretability had to be maintained, so I couldn't use many of the traditional techniques. Ultimately, I had to be more creative in my approach by creating multiple models. Using a regression model, I predicted capacity with each of the variables independently to retrieve their coefficients (relationship to target variable). I then grouped these into smaller categories that could be used in the final model.

Principal Component Analysis

PCA is the most common dimensionality reduction technique. At its core, PCA reduces the number of columns in your data while minimizing the loss of information. (For the purposes of this book, we won't get into the math behind how PCA executes this task.)

In its simplest form, PCA creates new principal components (columns) that are as uncorrelated with each other as possible. It only uses your input data, so it doesn't function like a model that predicts which columns are related to the target variable. This falls under a category in machine learning called *unsupervised learning*. PCA is finding the relationship within the *n* columns themselves. This is actually quite simple to execute, as shown in Listing 4.100.

```
from sklearn.decomposition import PCA
pca = PCA(n_components=5)
pca.fit(x)
pca_of_x = pca.transform(x)
```

Listing 4.100 PCA Example

This is our first real introduction into the concept of *fit and transform*. It's essential in the machine learning space, but it can be a bit confusing. It'll be a larger focus in Chapter 5, but the following mental model is helpful to follow:

► `pca = PCA(n_components=5)`
  Think about this step as setting the necessary parameters. You're creating an object called pca that can take *any* dataset from *n* columns into 5 columns.

► `pca.fit(x)`
  This step is where you're training the pca object for your specific dataset, which in this case is x. If x has 100 columns, this step makes pca specific to x and identifies how it will go from 100 columns to 5. At this point, your data hasn't been changed.

► `pca_of_x = pca.transform(x)`
  This step changes your data. Because you're creating a new dataset called pca_of_x, you aren't overwriting your original data. Now, pca_of_x represents your data as 5 columns.

A good question would be: how do I know how well this is being done? There are built-in metrics you can use from the PCA object, for example:

```
pca.explained_variance_ratio_
```

This metric shows what percentage of your original data is explained by the new columns. The higher, the better! You can use this metric to find the right number of components required for your data. We'll go through these examples with our use cases, but this often involves starting with a lower number and trying to reach at least 80%, if possible (*http://s-prs.co/v617001*).

Linear Discriminant Analysis

The key difference between LDA and PCA is that LDA operates like a predictive model. This is called *supervised learning* in the machine learning space. It requires the columns you'd use to predict the data as well as the column you're trying to predict. Essentially, LDA is trying to find which columns have a similar correlation to your target variable, with the goal of reducing your total numbers of columns. PCA ultimately only finds trends and frequencies within the columns to which you're applying the dimensionality reduction.

An example of LDA is shown in Listing 4.101.

```
from sklearn.model_selection import train_test_split
from sklearn.discriminant_analysis import LinearDiscriminantAnalysis
from sklearn.metrics import accuracy_score

x = columns_for_model
y = target_column

X_train, X_test, y_train, y_test = train_test_split(
    X,
    y,
    test_size=0.25,
    random_state=17)

lda = LinearDiscriminantAnalysis(n_components=5)
lda.fit(X_train, y_train)

y_pred = lda.predict(X_test)
accuracy = accuracy_score(y_test, y_pred)
print(accuracy)
```

Listing 4.101 LDA Example

Spoiler alert: This is the workflow for how machine learning models are put together. We're doing a test and train split, creating the object for the model, fitting the model, applying the results onto our data, and then testing it. As you explore the machine learning process in depth in Chapter 5, you'll hopefully find some comfort in the consistency of how the code is run. Your learning process then becomes more about understanding the specific parameters and theory behind each technique.

As with PCA, having a metric to test LDA allows us to optimize the results and get a better score. This will be shown in practice for each of the use cases.

Manual Business Context

There isn't a specific methodology or approach for using business context; however, it's worth mentioning as an option. Sometimes, there is business context that doesn't show up in the data and can't be identified with an analytical technique. Locations are a good example. If you have the specific location down to the city level, but your company has a predetermined set of regions, grouping your location by those regions is likely the best option to go from hundreds or thousands of options to a more manageable number.

The caution with this is the potential for biases in the decision-making process, which is why machine learning in practice is as much of an art as it is a science.

Real-World Story: Using Business Context to Reduce the Number of Columns

Returning to my recruiter capacity use case, I also had to wrangle with another challenge. An important input into calculating recruiter capacity is the skillsets of the roles they're hiring for. Some skillsets are very easy to hire for, while others are more competitive. In our data, we had something called *job families*, an internal categorization of skillsets that roles can be assigned to. Unfortunately, there were over 150 different job families.

Luckily, there was already an available hierarchy that grouped the job families together for us. I was then able to use the data because it had less than 30 unique values. Not only did this reduce how many columns I had, but it also aligned with what internal stakeholders were already familiar with and trusted!

Use Case 1: Dimensionality Reduction

As a refresher from the dummy coding done in Section 4.3.5, we have three DataFrames for our first use case that have been dummy coded from our three main categorical columns: Description, Country, and Customer ID. For the purposes of this example, and to avoid any computation limit issues, we'll apply the dimensionality reduction techniques on each column independently. With each iteration, we'll identify the optimal explained variance ratio to see the optimal number of components.

Listing 4.102 displays the code and Figure 4.89 shows the results. In Listing 4.102, we're creating a list of components to try and calling it versions_of_n. We're then iterating on this list within our for loop to execute the PCA process. The act of PCA is quite simple, as we create an object using the PCA function and provide it with the number of components (in this case, we call it n since we're using a for loop). We then use the fit function to build the PCA model and see how it works.

```
#create list of values for components to try
versions_of_n = [5, 10, 15, 20, 25]

#loop to try each iteration of components
for n in versions_of_n:
  #create pca object
  pca = PCA(n_components=n)

  #fit pca to df_country
  pca.fit(df_country)

  #print explained ratios of the components
  print("Explained variance ratio:", pca.explained_variance_ratio_)
```

Listing 4.102 PCA for Country

```
Explained variance ratio: [0.53948868 0.1072148  0.09590434 0.05104672 0.0299692 ]
Explained variance ratio: [0.53948868 0.1072148  0.09590434 0.05104672 0.0299692  0.02294434
 0.02004789 0.01735775 0.01385662 0.01132005]
Explained variance ratio: [0.53948868 0.1072148  0.09590434 0.05104672 0.0299692  0.02294434
 0.02004789 0.01735775 0.0138567  0.01132023 0.01007923 0.00957213
 0.00893617 0.00792954 0.00706204]
Explained variance ratio: [0.53948868 0.1072148  0.09590434 0.05104672 0.0299692  0.02294434
 0.02004789 0.01735775 0.0138567  0.01132023 0.01007924 0.00957213
 0.00893619 0.00792956 0.00706209 0.00629088 0.00546283 0.00454529
 0.00376609 0.00340661]
Explained variance ratio: [0.53948868 0.1072148  0.09590434 0.05104672 0.0299692  0.02294434
 0.02004789 0.01735775 0.0138567  0.01132023 0.01007924 0.00957213
 0.00893619 0.00792956 0.00706209 0.00629088 0.00546283 0.00454529
 0.00376609 0.00340661 0.003254   0.00281161 0.00258454 0.0023111
 0.00216038]
```

Figure 4.89 PCA Explained Variance Output for Country

Notice the repetition in each component! This is distinction is important for understanding how PCA works. Each component it's identifying is consistent, so regardless of creating 5 components or 10, the first 5 will all have the same explained variance. Now it's our job to identify the proper number of components to use. Let's stop at 11, as each additional component adds less than 1%. As shown in Listing 4.103, we're using the 11 components to fit a PCA model. The result of the code returns 0.919, which is interpreted as 92% of the signal in the data is retained.

```
#pick the 11 components
pca_country = PCA(n_components=11)

#fit the pca model to df_country
pca_country.fit(df_country)
```

```
#print the explained variance ratio
print("Explained variance ratio:", sum(pca_country.explained_variance_ratio_))
```

Listing 4.103 PCA on Country with 11 Components

Why does this matter? We've gone from 41 columns to 11, which will enable the predictive models we create to run more effectively and minimize the information loss.

Let's move onto the description data by running the code in Listing 4.104. Figure 4.90 shows the results.

```
#using 50 components for the description field
pca_description = PCA(n_components=50)

#fit the pca model to df_description
pca_description.fit(df_description)

#print the explained variance ratio
print("Explained variance ratio:", pca_description.explained_variance_ratio_)
```

Listing 4.104 PCA on Description

```
Explained variance ratio: [0.0586984  0.0389234  0.03156677 0.03069755 0.02437006 0.02415793
 0.02403801 0.02210171 0.02196503 0.02186512 0.02161366 0.02140195
 0.02124718 0.02119584 0.02109496 0.02101431 0.02084483 0.02012485
 0.02005993 0.01987263 0.01968657 0.01958428 0.0192483  0.01886669
 0.01871777 0.01812862 0.01785278 0.01780634 0.01771951 0.01764159
 0.01747844 0.01711189 0.01696205 0.0166877  0.01649591 0.01635185
 0.01623273 0.01574097 0.01530771 0.01525098 0.01512538 0.01497118
 0.0145831  0.01447635 0.0144469  0.01438101 0.0142796  0.01413733
 0.01400023 0.01387212]
```

Figure 4.90 PCA on Description

Notice anything different? Each of the columns are still adding a reasonable amount of value. In a case like this—and *at this stage in the modeling process*—it's best to keep the description column the way it is. Take a moment to consider why we were able to reduce country down from 41 columns to 11 but each of the columns in the description data add value.

This relates back to the preprocessing step, before we dummy coded the data. We've already identified which values of the description column are the most frequent in the data, so it isn't surprising that we'd see an insignificant drop-off in the explained variance. When you run PCA, you'll often see the first two to three components have more than 90% of the explained variance. However, we don't see this here, which is likely related to how many unique values this column has.

So, we'd expect to see the same thing for customer ID in Figure 4.91 when we run the code in Listing 4.105, right?

```
#use 50 for the components on customer
pca_customer = PCA(n_components=50)

#fit the pca model to df_customers
pca_customer.fit(df_customers)

#print the explained variance ratio
print("Explained variance ratio:", pca_customer.explained_variance_ratio_)
```

Listing 4.105 PCA on Customer ID

```
Explained variance ratio: [0.44767076 0.03891113 0.03291038 0.02212475 0.01978241 0.01568597
 0.01368008 0.01252962 0.01179399 0.01045417 0.00999975 0.00894846
 0.00878155 0.00861002 0.00823137 0.00761056 0.00724325 0.0060076
 0.00596177 0.00587381 0.00577839 0.00561631 0.00550255 0.00545616
 0.00517505 0.00498668 0.00483008 0.00480758 0.00475552 0.00464433
 0.00452821 0.00443285 0.0043424  0.00431635 0.00422753 0.0041771
 0.00411937 0.0040648  0.00401546 0.00396998 0.00393563 0.00393077
 0.00387291 0.00380537 0.0037323  0.00369682 0.00363394 0.00360131
 0.00355133 0.00348697]
```

Figure 4.91 PCA on Customer ID Results

Nope! After the first 10 components, the drop-off dips below 1% in explained variance. If we rerun with 10 components, we get a total of 63% explained variance. While we're losing quite a bit of explanation in the data, we're also significantly reducing the size of the data that we feed into our model.

In summary, with our PCA selections, we'd go from nearly 200 columns down to 71 columns. This is quite favorable as a starting point.

Next, we'll work through the LDA process. Recall that LDA differs from PCA in that we're using the concept of supervised learning versus unsupervised learning. This requires us to bring in the price column in addition to our dummy coded data. Let's start with the country data, as shown in Listing 4.106. Here, we're following what you'll see in Chapter 5: a workflow that looks more like a model process (because it is). As with PCA, we create a list of components to iterate over in a for loop. However, now we need to define what we're trying to reduce, which we define as x. Then, we define what we're trying to ultimately predict as y. We use the train_test_split function to break this data up into 4 objects (X_train, X_test, y_train, and y_test). The next two steps are similar to PCA. We create the model using the LinearDiscriminantAnalysis function, passing it the number of components, and then fit the model to the data using the fit function. Because we're essentially building a model, we need to create the predictions using the predict function. Finally, we test the results using the accuracy_score function.

If at this point you're a bit uncertain about how many steps go into this, it will be like second nature after you work through Chapter 5!

```
#pick number of components
versions_of_n = [5, 10, 15]

#for loop to try various iterations of components
for n in versions_of_n:

  #create inputs for LDA
  x = df_country

  #create target variable for LDA
  y = df_cleaned['Price'].astype(int)

  #split into training and test data
  X_train, X_test, y_train, y_test = train_test_split(x, y, test_size=0.25)

  #create lda model with number of components from the loop
  lda = LinearDiscriminantAnalysis(n_components=n)

  #fit the LDA model
  lda.fit(X_train, y_train)

  #predict on the test dataset with the lda model
  y_pred = lda.predict(X_test)

  #create the accuracy score
  accuracy = accuracy_score(y_test, y_pred)

  #print the accuracy score
  print(accuracy)
```

Listing 4.106 LDA on Country

You'll notice something interesting from the results. Regardless of the components, the difference is minimal and the accuracy is quite low, generally around 27%. This suggests that LDA is not an effective mechanism for dimensionality reduction on our Country data. PCA's results are much better, so we'll want to leverage PCA.

Important Note: Comparing Metrics

When executing PCA we're using explained variance, while with LDA we're using accuracy. As you test different approaches, it's important to think through what the metrics actually mean. PCA itself cannot have an accuracy metric, but the explained variance tells us how much of the information from the data is kept. With LDA, the accuracy metric tells us how accurate the new components are when trying to identify the commonality in relation to a target variable.

Use Case 2: Dimensionality Reduction

This is one of the shortest sections in the book, because we don't need to consider dimensionality reduction for this use case's data. Because we were intentional about which columns we dummy coded, we were able to keep our number of columns to 40. This means we don't need to do any dimensionality reduction on the data!

4

Use Case 3: Dimensionality Reduction

When executing PCA in the first use case, our approach was to execute each column we dummy coded by itself. For this use case, we're going to attempt the PCA dimensionality reduction across the entire dataset. When we go through the modeling process in Chapter 5, we'll try both datasets to show you what dimensionality reduction looks like in practice.

When we feed *all* columns into the dimensionality reduction process, we also need to ensure we're feeding the *right* columns into the process. For example, we should remove the area names, as their explicit values are important for this specific use case. Other columns, like our dates, cannot be part of the dimensionality reduction process.

Manually selecting each column would be tedious in this case, but at the same time, explicitly dropping each column is just as tedious. We'll use an approach where we explicitly select each column, with some help from some additional Python code that checks whether a column begins with specific strings (see Listing 4.107). This uses the filter function, which allows us to select columns dynamically. We're using regex to automatically select columns that contain any of the strings we pass it. As you'll see in Figure 4.92, only the columns we want remain.

```
#use regex to select columns in the dataset
df_to_reduce = df.filter(
  regex='Vict Age|
  crime_description|
  victim_gender|
  victim_race|
  crime_premises|
  crime_weapon'
  )

#print columns in the new dataset
print(df_to_reduce.columns)
```

Listing 4.107 Select Columns Based on Their Name Containing a String

```
Index(['Vict Age',
       'crime_description_ASSAULT WITH DEADLY WEAPON, AGGRAVATED ASSAULT',
       'crime_description_BATTERY - SIMPLE ASSAULT',
       'crime_description_BURGLARY', 'crime_description_BURGLARY FROM VEHICLE',
       'crime_description_INTIMATE PARTNER - SIMPLE ASSAULT',
       'crime_description_Other', 'crime_description_ROBBERY',
       'crime_description_SHOPLIFTING - PETTY THEFT ($950 & UNDER)',
       'crime_description_THEFT FROM MOTOR VEHICLE - GRAND ($950.01 AND OVER)',
       'crime_description_THEFT FROM MOTOR VEHICLE - PETTY ($950 & UNDER)',
       'crime_description_THEFT OF IDENTITY',
       'crime_description_THEFT PLAIN - PETTY ($950 & UNDER)',
       'crime_description_THEFT-GRAND ($950.01 & OVER)EXCPT,GUNS,FOWL,LIVESTK,PROD',
       'crime_description_VANDALISM - FELONY ($400 & OVER, ALL CHURCH VANDALISMS)',
       'crime_description_VANDALISM - MISDEAMEANOR ($399 OR UNDER)',
       'crime_description_VEHICLE - STOLEN', 'victim_gender_F',
       'victim_gender_M', 'victim_gender_Other', 'victim_race_B',
       'victim_race_H', 'victim_race_O', 'victim_race_Other', 'victim_race_W',
       'victim_race_X', 'crime_premises_DEPARTMENT STORE',
       'crime_premises_DRIVEWAY', 'crime_premises_GARAGE/CARPORT',
       'crime_premises_MARKET',
       'crime_premises_MULTI-UNIT DWELLING (APARTMENT, DUPLEX, ETC)',
       'crime_premises_OTHER BUSINESS', 'crime_premises_OTHER PREMISE',
       'crime_premises_OTHER RESIDENCE', 'crime_premises_Other',
       'crime_premises_PARKING LOT',
       'crime_premises_PARKING UNDERGROUND/BUILDING',
       'crime_premises_RESTAURANT/FAST FOOD', 'crime_premises_SIDEWALK',
       'crime_premises_SINGLE FAMILY DWELLING', 'crime_premises_STREET',
       'crime_premises_VEHICLE, PASSENGER/TRUCK', 'crime_weapon_HAND GUN',
       'crime_weapon_Other',
       'crime_weapon_STRONG-ARM (HANDS, FIST, FEET OR BODILY FORCE)',
       'crime_weapon_UNKNOWN WEAPON/OTHER WEAPON',
       'crime_weapon_VERBAL THREAT'],
      dtype='object')
```

Figure 4.92 Remaining Columns After Selecting Based on Their Names Containing a String

Now that we've narrowed down our columns for dimensionality reduction, we can run PCA on the data. We'll take a similar approach as we did in for the first use case, where we loop through a number of combinations of components to select (see Listing 4.108). Figure 4.93 shows the results. Take a few moments to review the results before reading on and consider how you'd proceed. Which number of components would you pick? Are you able to glean any other insights from the results?

```
from sklearn.decomposition import PCA

#create list of component options to try
versions_of_n = [5, 10, 15, 20, 25]

#loop through each component and print out the results
for n in versions_of_n:
  pca = PCA(n_components=n)
  pca.fit(df_to_reduce)
  print("Number of Components: ", n)
```

```
print("Overall Explained Variance: ",sum(pca.explained_variance_ratio_))
print("Explained Variance Ratio: ", pca.explained_variance_ratio_)
print("")
```

Listing 4.108 Execute PCA Test on the Data

```
Number of Components:  5
Overall Explained Variance:  0.9958538517016797
Explained Variance Ratio:  [9.93110541e-01 8.95900485e-04 7.68773142e-04 6.04167914e-04
 4.74469213e-04]

Number of Components:  10
Overall Explained Variance:  0.997490121113856
Explained Variance Ratio:  [9.93110541e-01 8.95900538e-04 7.68773770e-04 6.04169553e-04
 4.74470778e-04 3.81216985e-04 3.56349914e-04 3.24717790e-04
 3.10867992e-04 2.63112845e-04]

Number of Components:  15
Overall Explained Variance:  0.9983891971184181
Explained Variance Ratio:  [9.93110541e-01 8.95900539e-04 7.68773773e-04 6.04169563e-04
 4.74470905e-04 3.81217454e-04 3.56351704e-04 3.24719983e-04
 3.10868410e-04 2.63139993e-04 2.52480713e-04 1.86387043e-04
 1.81601928e-04 1.40718745e-04 1.37855416e-04]

Number of Components:  20
Overall Explained Variance:  0.9989484787261598
Explained Variance Ratio:  [9.93110541e-01 8.95900539e-04 7.68773773e-04 6.04169564e-04
 4.74470929e-04 3.81217529e-04 3.56351888e-04 3.24720131e-04
 3.10868898e-04 2.63140767e-04 2.52481572e-04 1.86388769e-04
 1.81618394e-04 1.40812763e-04 1.37861539e-04 1.21287132e-04
 1.16556717e-04 1.14582192e-04 1.04338789e-04 1.02395893e-04]

Number of Components:  25
Overall Explained Variance:  0.9993754886117083
Explained Variance Ratio:  [9.93110541e-01 8.95900539e-04 7.68773773e-04 6.04169564e-04
 4.74470929e-04 3.81217532e-04 3.56351888e-04 3.24720142e-04
 3.10868909e-04 2.63140784e-04 2.52481578e-04 1.86389296e-04
 1.81618886e-04 1.40812926e-04 1.37870898e-04 1.21291886e-04
 1.16560721e-04 1.14623846e-04 1.04343179e-04 1.02444175e-04
 9.89260598e-05 9.39129060e-05 8.24764611e-05 8.00334357e-05
 7.15473495e-05]
```

Figure 4.93 Results of PCA Test

What conclusions did you come to with the results? You may have noticed that the first column captures almost all of the variance. This means that the underlying data could have a simple structure that can be captured with relative ease. However, now is the time to take a step back and consider the use case.

Recall that our goal is to predict the probability of crimes occurring on a given day in a given location so resources can be allocated accordingly. Could you do that with our current dataset? Unfortunately, we can't achieve this with our current data setup because we need to create a column that identifies crime in the future for a given location. This is a subtle nuance, and one we'll discuss more in Chapter 5. The best practice is to execute PCA on the dataset we'll use directly before the modeling step. That means we'll need to take a quick sidestep to properly set up our dataset.

For this type of data problem, we'll need to complete two primary steps:

1. Create a daily dataset to ensure that dates where crime didn't occur are still included in the dataset.

2. Summarize our dataset into our use case's definition of a row. This means we need one row to equal one day for one location. For example, if we had 10 locations for 10 days, we'll have 100 rows.

In order to execute the first step, we need a unique list of the areas and then we'll generate a list of dates using the range of the data. Unfortunately, we've already dummy coded the area names. This is often part of the modeling process where we need to go back and revisit our approach. The only change we'll need to make is remove the Area Name column from the dummy coding process. Once we've done this, we can rerun the dummy code, and we'll have the area name back in our code.

First, we'll create a DataFrame for all the unique area names (see Listing 4.109). The key column will be used for the join, which will be explained in more detail shortly.

```
#create areas data frame from unique list of area names
areas = pd.DataFrame(
  df['AREA NAME'].unique(),
  columns=['Area Name']
  )


#create key as join column
areas['key'] = 0
```

Listing 4.109 Create areas DataFrame

Next, we'll create the range of dates using the pandas function date_range (see Listing 4.110).

```
#create data frame of dates ranging from the first and last day of the dataset
dates = pd.DataFrame(
  pd.date_range(
    start=df['DATE'].min(),
    end=df['DATE'].max()
    ),
  columns=['Date']
  )


#create key as join column
dates['key'] = 0
```

Listing 4.110 Create date DataFrame

Now we need to join the two DataFrames, as shown in Listing 4.111. We want to join all of the combinations together, so we'll use the outer option. This is where the key columns come in, since we must specify a column to join by; the key column is essentially a fake column to execute the join by.

```
#join the areas and dates data frames using outer join
df_full = dates.merge(
  areas,
  how='outer',
  on='key'
  )


#drop the key column
df_full = df_full.drop(columns=['key'])
```

Listing 4.111 Join the areas and dates Columns to Create Full List of Dates and Areas

For those who like to check that the math works out: The areas DataFrame has 21 rows and the dates DataFrame has 1,826 rows. The df_full DataFrame has 38,346 rows, which is the result of multiplying 21 by 1,826. This created a unique combination of dates and areas names for the entirety of our dataset.

Now we need to summarize the crimes up to this daily area level. To do this, we'll need to settle on either an average or count of all our numeric variables. We'll use an average, as it's the most flexible measure and enables better comparisons over time, given that counts can fluctuate. Pandas has a nifty feature that allows you to take the mean of all numeric columns. We'll also take an overall count of crime to ensure it's preserved. Listing 4.112 shows that we're grouping the data by date and area and then taking the mean for all numeric columns and calling it df_summarized. We then repeat the process, but instead of taking the mean, we take a count of each area and date.

```
#summarize using mean for all numeric variables, grouping by date and area name
df_summarized = df.groupby(
  ['DATE', 'AREA NAME']
  ).mean(numeric_only=True).reset_index()


#summarize the count of crime for the date and area name
df_count = df.groupby(
  ['DATE', 'AREA NAME']
  )['DR_NO'].count().reset_index()


#rename the DR NO to crime count for clarity
df_count.columns = ['DATE', 'AREA NAME', 'crime_count']
```

Listing 4.112 Summarize Crime Dataset into Date and Area Name

After completing this, you can check the number of rows it produces. The result is 38,320. If you compare this to the 38,346 number in our full dataset, this means there are a few combinations of area name and date that didn't have a crime on a given day. We'll first join the df_summarized and df_count DataFrames together, and then we'll join the result back onto the full set of dates to create a holistic DataFrame that contains all combinations of dates and area names (see Listing 4.113).

```
#join together the summarized data with avergages with the count of crime data
df_summarized = df_summarized.merge(
  df_count,
  how = 'left',
  on = ['DATE', 'AREA NAME']
  )

#convert date to pandas date time to ensure join works
df_summarized['DATE'] = pd.to_datetime(df_summarized['DATE'])

#join all dates onto the summarized dataset to ensure all combinations occur
df_summarized = df_full.merge(
  df_summarized,
  how = 'left',
  left_on = ['Date', 'Area Name'],
  right_on = ['DATE', 'AREA NAME']
  )

#fill all blanks with 0
df_summarized = df_summarized.fillna(0)

#drop extra columns from join
df_summarized = df_summarized.drop(columns=['Date', 'Area Name', 'DR_NO'])
```

Listing 4.113 Join Datasets to Get a Single Unified Version of Summarized Data

We now have a summarized dataset after this brief detour. It's common to find yourself in these situations, so this gives you a realistic idea of how the modeling process works. Rarely is it linear—you'll often find yourself backtracking to fix an error you've discovered.

Now we can run PCA again to see if the results changed based on our new summarized dataset (see Listing 4.114 and Figure 4.94).

```
#remove columns we don't want in PCA
df_to_reduce = df_summarized.drop(columns=['AREA NAME', 'DATE', 'crime_count'])
```

```
#versions of the components to test
versions_of_n = [5, 10, 15, 20, 25]

#loop through each of the component options
for n in versions_of_n:
  pca = PCA(n_components=n)
  pca.fit(df_to_reduce)
  print("Number of Components: ", n)
  print("Overall Explained Variance: ",sum(pca.explained_variance_ratio_))
  print("Explained Variance Ratio: ", pca.explained_variance_ratio_)
  print("")
```

Listing 4.114 Retest PCA on the Summarized Data

```
Number of Components:  5
Overall Explained Variance:  0.997027826350221
Explained Variance Ratio:  [9.94235803e-01 1.10622242e-03 7.40585069e-04 5.42742625e-04
 4.02473730e-04]

Number of Components:  10
Overall Explained Variance:  0.9983379579159609
Explained Variance Ratio:  [9.94235803e-01 1.10622242e-03 7.40585072e-04 5.42742632e-04
 4.02474234e-04 3.53933111e-04 2.77398733e-04 2.68473380e-04
 2.13082454e-04 1.97243373e-04]

Number of Components:  15
Overall Explained Variance:  0.9989576862919035
Explained Variance Ratio:  [9.94235803e-01 1.10622242e-03 7.40585072e-04 5.42742633e-04
 4.02474248e-04 3.53933295e-04 2.77399544e-04 2.68474129e-04
 2.13102416e-04 1.97248162e-04 1.89372752e-04 1.28761889e-04
 1.20755252e-04 9.54177590e-05 8.53942138e-05]

Number of Components:  20
Overall Explained Variance:  0.9993366068192041
Explained Variance Ratio:  [9.94235803e-01 1.10622242e-03 7.40585072e-04 5.42742633e-04
 4.02474248e-04 3.53933300e-04 2.77399547e-04 2.68474147e-04
 2.13102431e-04 1.97248308e-04 1.89372818e-04 1.28770648e-04
 1.20757625e-04 9.54339110e-05 8.54486827e-05 8.30210150e-05
 8.00537695e-05 7.52964935e-05 7.26370673e-05 6.78301767e-05]

Number of Components:  25
Overall Explained Variance:  0.999609111372218
Explained Variance Ratio:  [9.94235803e-01 1.10622242e-03 7.40585072e-04 5.42742633e-04
 4.02474248e-04 3.53933300e-04 2.77399547e-04 2.68474147e-04
 2.13102436e-04 1.97248313e-04 1.89372820e-04 1.28770668e-04
 1.20758436e-04 9.54342541e-05 8.54499533e-05 8.30222907e-05
 8.00544587e-05 7.53093198e-05 7.26386605e-05 6.78381359e-05
 6.49133714e-05 5.68996857e-05 5.26668965e-05 5.04945826e-05
 4.75032172e-05]
```

Figure 4.94 Retesting PCA on Summarized Data

Interestingly, we get similar results! The first component includes the majority of the variance. We'll save our summarized DataFrame to a *.csv* file to preserve our steps (see Listing 4.115). We'll revisit the PCA process once we've made our way into modeling in Chapter 5.

```
#write to data to csv file
df_summarized.to_csv("df_summarized", index=False)
```

Listing 4.115 Save Summarized DataFrame to .csv File

4.4 Summary

This chapter demonstrated that data preparation is the foundation of every successful machine learning project, often consuming the majority of your time and effort. You discovered that raw data—whether it's from spreadsheets, databases, or APIs—rarely arrives in a model-ready format. Instead, it requires thoughtful exploration, cleaning, and transformation to become useful.

Let's quickly recap what we've covered in this chapter:

- You learned about the various types of data sources. We categorized them into manual files, such as CSV or Excel files, and automated approaches such as databases or APIs. While this book focuses on the manual files for ease of use, it will be important to also learn how to leverage automated connection approaches.

- You learned how to explore the data using various approaches and Python functions. This includes using generative AI tools to work smarter and faster. We also demonstrated how to visualize data in Python so that you can understand it quickly and thoroughly.

- You learned about the data cleaning process and how messy it can be. You'll often need to account for missing data, outliers, and other gremlins in the data.

- You learned how to use simple statistics to understand the data within your data, as well as how to use and interpret statistical measures. This only scratched the surface of the world of statistics, but it provided a foundation upon which you can build.

- You learned that you need to aggregate your data before it goes into a machine learning model. The concept of "what does one row equal" is critically important. You'll rarely put highly granular data (orders, claims, etc.) into your model to generate predictions. The data will almost always be aggregated up to another level.

- You learned about the dummy coding process. Ultimately, machine learning models can only accept numeric values, so the primary way to make a text column numeric is to dummy code it. We discussed the nuances associated with dummy coding; specifically, we talked about the need to be intentional about the number of columns that will be generated if you don't group unique values from a column (a concept called cardinality).

- Lastly, you learned about dimensionality reduction. This is a high-level concept that's often difficult to understand. The goal of dimensionality reduction is to make the model's job easier by grouping together columns while maintaining as much information as possible. We'll show this in action for our third use case in the next chapter.

As you proceed to the next chapter on models (likely the reason you've purchased this book), remember that your success in building a model depends on completing the steps in this chapter thoroughly and carefully.

Chapter 5
Picking Your Model

At long last, you've gotten through much of the hard work required to build an effective machine learning model. Take a pause and give yourself some credit! In this chapter, we'll dive into the foundational algorithms of machine learning.

We'll now shift our focus to the models themselves and explore strategies you can use to pick them. This chapter focuses on select algorithms and models and is not intended to be comprehensive. Instead, we focus on the most commonly used algorithms to help you understand how they build upon each other. Although knowing all these algorithms is useful, it can be unnecessary noise when you're just getting started. Speaking from experience, we discussed upwards of 20 different algorithms during my master's program, and I've only used a small subset of these in practice.

This chapter begins with a discussion on the model selection approach and a framework you can use for it. We'll then discuss a selection of algorithms, starting with regression models and moving into the traditional tree-based models typically associated with machine learning. We'll also talk about clustering, the misfit of the machine learning world.

A word of warning: Don't skip Section 5.5.1 on decision trees! The decision tree algorithm is the foundation of random forest and Gradient Boosting Machine (GBM) models. The chapter is deliberately structured to introduce these important topics using simple decision trees before you progress to more complex models. While you're unlikely to use the decision tree algorithm in practice, understanding it is critical to learn the random forest and GBM algorithms that are more common on the job.

5.1 The Simpler the Model, the Better

The longer you work in the technical space, the more you'll come to understand that the simpler solution is usually the better solution. If you need to add an additional 1,000 lines of code to make your model 1% better, is it actually worth it? If you work on incredibly high-stakes use cases where 1% represents millions or billions of dollars, then maybe it is. However, in most enterprise analytics scenarios, the risk and time associated with the additional 1,000 lines of code isn't worth the 1% gain.

When Complicated Becomes a Problem

I had just started a new role, and one of my responsibilities was to take over an existing set of models. The underlying tech stack was not ideal and the complexity of the models created several challenges for me. I spent many hours digging through the code only to realize that the stakeholders weren't actually using the model.

As I worked with the previous owner of the models, it became clear how complex the modeling process was from the various data sources and how they were brought into the process to the actual model itself. This resulted in a situation where my team wasn't adding the right value for the organization. It also impacted the former owner of the model because he received a number of additional questions he otherwise wouldn't have. If nothing else, heed this warning for your own self-interest!

That said, I've haven't always created simple processes in my own work. When transitioning out of one of my roles, I had a number of technical processes to hand over to another member of the team. While the nature of the work was complicated, it was my own code, so I found it easy to understand. This meant I had minimal documentation and hadn't gone back to see where I could have simplified the code and the process.

The takeaway here should be to simplify your approach as much as possible. It'll help you and others support your stakeholders now and in the future.

Think about this from the stakeholder perspective as well. A simpler approach is easier for others to explain, but it's also easier to explain yourself. When you're in the depths of building out your process, you'll be at the height of your understanding of the code and all its small nuances. Once you've deployed your model into production, time will pass, and you'll become less familiar with those nuances. You may revisit your code after a few months in a panic thinking you missed something or you're doing something in error.

You'll also experience this when conducting a knowledge share or cross-training session for others who are new on your team or need to understand the model. A simple process and a simple model make the knowledge share easier on yourself and other members of your team. Part of maturing in your machine learning journey is understanding that scalability and longevity are important, so you'll sometimes need to sacrifice incremental increases in your model's performance to ensure its longevity.

It's helpful to think about this from a business perspective. If the incremental increase in accuracy is 0.25%, does the added complexity of a change justify this increase? Depending on your business case, sometimes that answer is yes, and sometimes it's no. However, in a situation where the change is a connection to two additional data sources, complicated functions, and an overall increase in your code base, it's unlikely to be a favorable tradeoff.

There's also a courtesy component to this as well. It's unlikely you'll be the one who creates a model and owns it for its entire life, so thinking ahead is important! The ease of taking over a model varies significantly depending on how it was structured.

From a purely technical perspective, the more complex your model, the longer it takes to train and execute. Say you can train a simpler model in five minutes, but a more complex model trains in twenty minutes. This extra time slows down your ability to iterate on the model and add new features with additional value.

In practice, you'll likely need to be iterative about which data you bring into your model at which stage. Preparing your data for a model is the most time-consuming part of the modeling process, so you shouldn't attempt to bring in all the possible data at once. It's better to focus on a smaller subset of your data to improve your development speed.

Finally, consider how likely it is that your stakeholders will introduce additional complexity. If your data is generated by a human-dependent process, you'll probably have to account for nuances that make your model more complex.

5.2   Model Decision Framework

There are a number of resources available online that show you the full suite of options in the modeling space. You can find examples in the scikit-learn (sklearn) documentation at *https://scikit-learn.org/stable/machine_learning_map.html*.

Figure 5.1 shows the framework we'll use for this chapter, which consists of the general order of algorithms to consider as well as their associated complexity.

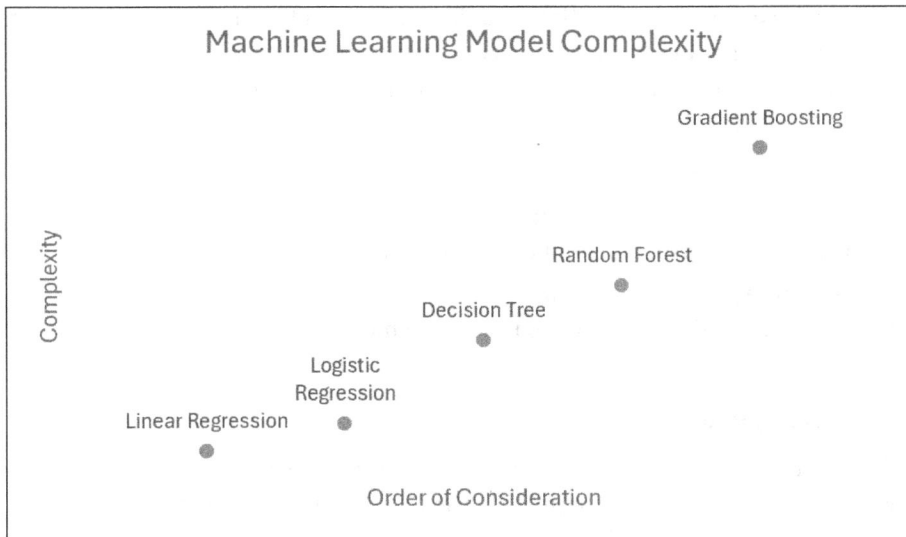

Figure 5.1 Model Complexity

You'll see that regression should usually be considered first, and it has relatively low complexity (more on this in Section 5.4). Then, we get into the tree-based models, which have an increasing level of complexity.

The trends and themes in this framework won't always hold true in every situation. The goal is to give you a general framework you can use to think about model selection, since it can be a confusing and intimidating task when you're getting started.

5.2.1   How Important Is Interpretability?

Interpretability is the ability to understand *why* your model is making the decisions it's making. We'll discuss this concept in more detail in Chapter 6, but for the purposes of picking our model, we need to take a moment to think about it. We've just established that regression is lower in complexity, but let's throw another curve ball into the mix. If interpretability isn't high on your priority list, you can skip regression. This is primarily due to the additional data cleaning and considerations required for regression, which are discussed in Section 5.4. The tradeoff in time it takes to train a regression model versus a decision tree is relatively minimal. Any minimal increase in time to train a decision tree offsets the added effort required to ensure your data is set up properly for a regression model.

The biggest drawback you'll see with tree-based models is that they're *black boxes* that are hard to interpret. However, you can use the techniques we'll discuss in Section 5.5 to interpret tree-based models, so this is by no means a binary decision where you're either getting a better model or a model with interpretability. However, if your stakeholder is hyperfocused on understanding the *why* behind the prediction, or the model is being used in a day-to-day operational setting, understanding how a model makes its predictions yields significant value. The coefficients generated by regression models provide a high degree of specificity in interpretation that is hard to replicate with a machine learning model.

Let's summarize our decision-making process:

- **If interpretability is very important**
  Stick with regression as the default and move to the next question.

- **If interpretability is not important**
  Skip the next question and adjust to tree-based models.

5.2.2   How Many Rows and Columns?

If you've identified that interpretability is important, the next step is identifying whether your data can support a regression use case. You should think about rows and columns together because of the limitations in how the backend math of regression works.

Here's a mental model for considering this is: The more rows you have, the more columns you can use. As the size of your data and number of records grows, it allows the math behind the model to look across more columns and find relationships between them. A dataset with only 100 rows and 30 columns won't work for regression.

As a rule of thumb, it's best to keep your number of columns at 30 or fewer if you have fewer than 100,000 rows of data to train your model on. If you have hundreds of thousands of rows, then 50 columns or fewer is generally acceptable. For larger datasets, the number of rows per column should be around 3,000–5,000 rows per column.

These guidelines allow the math behind the regression to operate correctly. They also ensure you're not breaking any of the assumptions in regression.

> **Keep Multicollinearity in Mind**
>
> As we'll discuss more in the next section, multicollinearity occurs when you have two columns that are highly correlated with each other. Regression assumes there is no multicollinearity in the dataset. As your column count grows, this becomes a more challenging dynamic to manage.

To summarize our rules:

- **If you have an appropriate column-to-row ratio**
  Stick with regression and move to the next question.

- **If you do not have an appropriate column-to-row ratio**
  Adjust to a tree-based model and move to the next question.

5.2.3   What Is Being Predicted?

At this point, you've identified the category of model you'll be using: either a regression approach or tree-based approach. Now, you'll need to understand what you're predicting and the associated next step based on the type of model you selected. The two categories of what we're predicting are called *regression* (if you're confused, keep reading) and *classification*.

Regression

Regression in this context translates to predicting a number. If you're predicting the number of sales, this is a regression or regressor prediction. Regression models naturally do this, and linear regression does this explicitly. When most people think about predictive models, they're likely thinking about a model that predicts a specific number.

> **Sometimes Regression Is Regression... Sometimes It's Not**
>
> I'm not entirely sure who thought it was a good idea to name the overall approach to predicting a number "regression" when this terminology is already reserved for linear and logistic regression, but it is what it is. This has confused me on a few occasions when onboarding onto a new project or team, so it's never a bad idea to clarify what someone means when they say "regression."

Classification

Classification is well-named. The objective of the classification model is to classify your data. In practice, it's still technically predicting a number. For example, if you're building a model to predict whether an employee will leave the company, your model will classify them either as someone who will stay or as someone who will leave. Someone who will leave is often coded in the data as a 1 and someone who will stay is coded as a 0.

Probabilities are at the core of classification. While it can depend on the use case, how valuable is it to provide a binary prediction? Psychologically, it creates a perception of confidence. However, using the employee turnover model example, what if your model predicts an employee will leave in the next six months, but they're still employed at the company on month seven?

Thinking probabilistically is often more valuable for stakeholders, but it also keeps your model from taking unnecessary heat for being wrong. All models are wrong—the good ones are just less wrong. As an alternative, what if your model predicted the probability someone would leave in the next six months? For a specific employee, the same binary prediction that they may leave could actually only be a 25% probability. Most decision-makers will interpret a 25% probability of an event occurring differently than just being told the event *will* occur.

So why the lecture about probabilistic thinking? Because all classification models start with probability and are converted into binary terms that minimize the frequency of false positives and false negatives. The difference in the code is relatively trivial, as the model is outputting both, so as always it goes back to the use case.

There is also a spectrum to consider as well. Going back to the employee turnover example, what if a probability of 15% is considered high in this business context? In scenarios where your target variable is considered imbalanced (one outcome you're predicting is more likely than another), it can be challenging for your stakeholders to understand the full context. For employee turnover, most companies will retain the majority of their employees in a six-month span rather than see them leave (I hope). This can lead to your model output recommending that the optimal binary cutoff point for turnover should be 15%. While the math may be optimal, a stakeholder is likely to question the value of your model's outputs. In this scenario, you can consider grouping your data together into logical categories. One approach could be to group anyone with a probability of

50% or greater as high risk, 15% to 49% as medium risk, and anything below that as low risk. While you've introduced subjectivity into the model's output, you've also met the stakeholder where they need to be to effectively consume your model's output.

Translating to Model Selection

Linear regression and logistic regression models have distinct use cases. If you're trying to predict a number, you'd use linear regression (should we call it "regression regression"?). If you're trying to classify data, you'd use logistic regression.

For tree-based models, the regression versus classification distinction is almost completely abstracted from our perspective. Each model has a regressor and classifier function that can be loaded in, and the inputs required are more or less the same (e.g., `DecisionTreeClassifier` and `DecisionTreeRegressor`). In practice, this is quite nice. It's easier to switch between approaches when you can just change the name of the function without having to change all your hyperparameters. However, it can muddy the waters from a learning perspective, because there isn't much of a distinction when you're applying it.

As we work through the use cases, we'll apply the various models to each one so you can see what we discussed in practice.

5.3   Train-Test Split

Regardless of what model you're using, you need to split your data into a test set and a train set. The training data is what you build your model with. The test data is withheld to understand the quality of your model. When we split the data, we're splitting the *rows* of the data, not the columns. This means our training and test data will have the same columns, but different rows.

You may have heard of the term *overfitting* in the context of modeling. If your model performs very well on your training data but not on your test data, it may be overfit (that is to say, your model isn't generalizable). Having a test dataset helps you combat this. The goal of building a machine learning model is to help predict the future. By definition, we don't know the future, so we need to make sure our model is generalizable with the data we already know about. Withholding some amount of data to test the model on is necessary to do this.

Conceptually, it's most helpful to think about this in the context of a time series dataset. Even though this book doesn't formally cover time series predictive modeling, the same underlying context applies. Say you have five years' worth of data for your company sales, and you need to predict sales for year six, which hasn't started yet. The recommended approach would be to build your model as if year five had yet to start. You'd build your model on data from the first four years and then test its performance on year five. The goal of time series is to predict future dates, which is why it's best practice to use a specific date cutoff instead of randomly splitting the data into the training and test sets.

This same concept applies to the general train_test_split function available in sklearn, as shown in Listing 5.1. If our data isn't specific to time series data (our second and third use cases are examples of this), we use the train_test_split function. In creating a training and testing dataset in this way, we're randomly holding out X% of the data rather than identifying a logical point in time to split the data. The purpose of this function is to split the rows from x and y into their respective train and test objects.

```
from sklearn.model_selection import train_test_split
X_train, X_test, y_train, y_test = train_test_split(
    X,
    y,
    test_size=0.2,
    random_state=42
    )
```

Listing 5.1 Train-Test Split Code

There are a few components to understand here, the first being the objects you're creating. The output of train_test_split() is four different objects:

1. Your input variables that are used to create the model.
2. Your input variables for the testing of your model.
3. What you're trying to predict so your model has the "answer" to train itself.
4. The "answer" to compare against the model's predictions for your testing data.

The inputs to the train_test_split() function are:

▶ **X**
All the columns used to create the prediction.

▶ **y**
The column you're trying to predict.

▶ **test_size**
What percentage of the data should be held out for testing.

▶ **random_state**
Any number that will enable you to create reproducible results as you iterate on your model.

The test_size parameter may require some explanation. What is the right amount of data to leave out for testing? In general, you'll likely select a number between 20% and 30%. The smaller and/or more nuanced your dataset, the more data you want to reserve for testing. This selection of your test_size ties back to the concept of overfitting and how to reduce the risk of overfitting going undetected. In general, the larger your test size, the lower the risk of you not detecting the overfitting in your modeling process.

The `random_state` parameter enables your results to be more reproducible—but how? The specific number you provide does not matter; it's the consistency in which you apply that number. If you use 17 and you run the `train_test_split()` function multiple times, you'll always get the same rows going into your train and test datasets. Especially when you're iterating on your model, this ensures your model performance results are driven by intentional changes, rather than the difference in how the function split your data into a training set and test set.

As we go through each model, you'll find that the process remains consistent. Applying the same X and y allows you to test multiple models and evaluate their performance. This increases your speed of iteration significantly!

5.4   Regression Models

Regression models are the backbone of analytics, and the underlying math is integral to how value is derived from data. At their core, regression models are a fancier version of correlations. As promised at the beginning of the book, this section won't include a complicated math lesson, but we'll revisit high school algebra class for a quick refresher of the underlying concepts. We'll then walk through linear and logistic regression in detail and apply them to our three use cases.

5.4.1   What Are Regression Models?

Regression models are based on the $y = mx + b$ equation. High school math may have been many years ago for some of you, so as a refresher, y is the value we're trying to predict, m is the slope of the line, x is the data point we know, and b is the y intercept. The biggest difference is that as you add additional variables, you create more variables like mx.

The regression model looks to optimize this equation to identify the line of best fit, which essentially means it's trying to create the version of the $y = mx + b$ equation that best matches the data by reducing the distance between the line the equation would make and the actual data. To further explain this, we'll walk through the simplified example shown in Figure 5.2.

You'll notice that we're only using two variables (student intelligence and exam scores), which is intentional for visualization purposes. It's easy for us to consume a two-dimensional line graph, but in practice you're not building a model with only one column to predict another column. When you have many columns in your data, the same concept is applied—you're just adding another variable to your equation. For example, if your data has three columns being used to predict a fourth column, $y = mx + b$ will become $y = m1x + m2x + m3x + b$. As you add new columns, the regression model adds another mx for each of them.

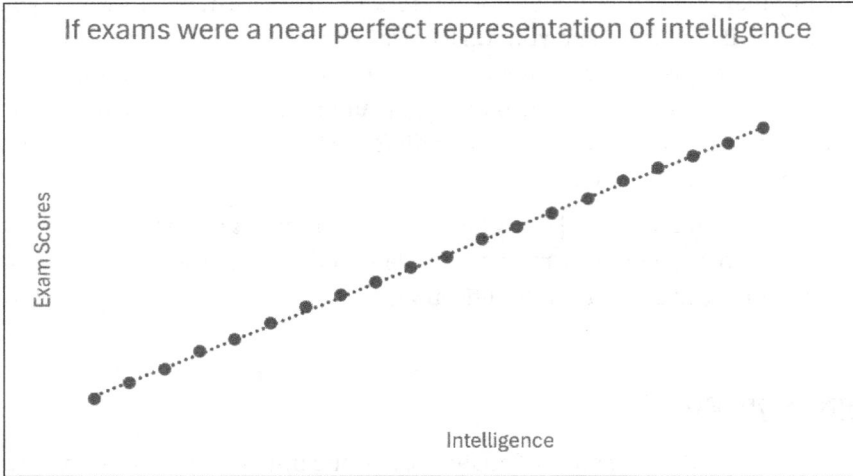

Figure 5.2 Near-Perfect Regression Line

For this example, the equation of this line is $y = 1.8256x + 1.7258$ (Excel has an option to show you any line graph line's linear equation). Since this is an equation, the predictive nature of regression involves inputting x when its value is still unknown. If you haven't already started making this connection, this is why regression is so much more interpretable than other machine learning algorithms! You have an equation where you can set all other variables to 0, which gives you the overall relationship to the column you're trying to predict. Regression models will output this for you as *coefficients*. In our example, the coefficient for is 1.8256, or the m in our linear equation.

Chapter 6 will cover evaluating your model, but the most common evaluation metric in regression is R^2. This metric evaluates the amount of variance your regression equation is account for, with 1 meaning it's a perfect match and 0 meaning it's capturing none of the variance. For this equation, the R^2 is 0.9995 (almost as if it was planned to be nearly perfect).

Now what happens when we take the same example, but the data is messier? Let's look at Figure 5.3.

With this messier data, we still see a distinct linear pattern. The R^2 for this line is still very high at 0.9553, meaning the model can pick up 96% of the variance in this dataset.

What happens when we take this same data and give it a trend where students at the upper end of the intelligence distribution score higher on their exams (see Figure 5.4)?

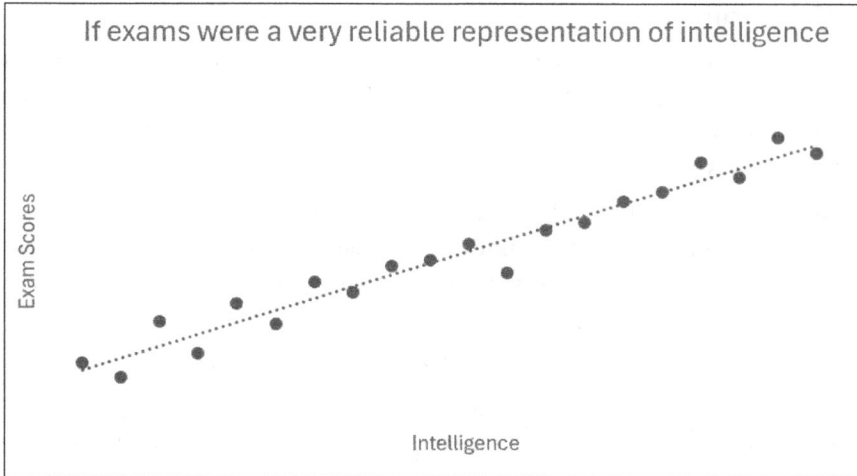

Figure 5.3 Messier Example of Data

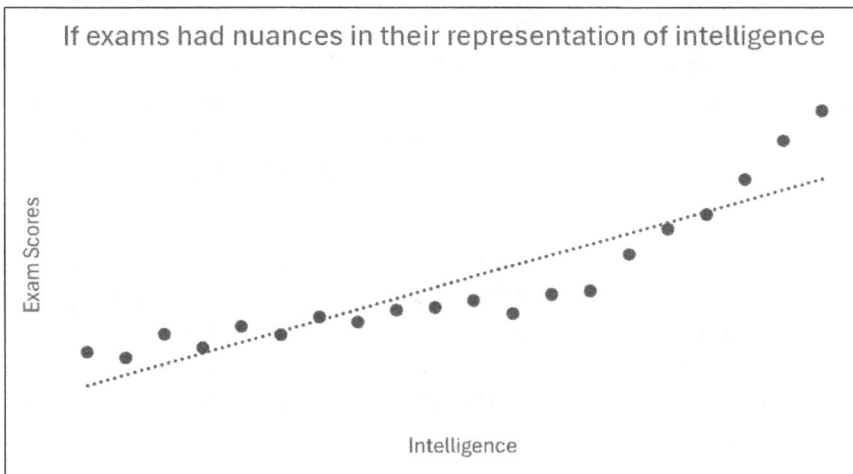

Figure 5.4 Higher Intelligence Students Score Higher on Exam

By making this change, the line is more than twice as steep, and the R^2 has dropped down to 0.8125! It's important to understand why this is and where there are limitations in linear regression. Linear regression is only able to map a straight line through the data. As you add additional variables, if those variables don't explain the trend of students with higher intelligence, the model has no way of knowing the top 25% of students have a different trend of exam scores. This also creates a dangerous dynamic where your model is better at predicting one group versus another. If you want to sound smart at a dinner party, this is called *Simpson's paradox*.

5.4.2   Multicollinearity

The math that regression is based on can be quite picky. A regression model has many assumptions. The most important assumption to consider when you're preparing your data is to make sure none of your columns are highly correlated with each other. This is called multicollinearity.

In general, you should address any correlation above 0.7 between two of the columns you're using to build a model. You can find the correlation of any two columns using the pandas function corr. When you provide the function with your columns, it outputs a correlation matrix.

There are varying approaches for addressing the high correlation between columns:

- The simplest approach is to remove one of the two correlated variables. One common cause for multicollinearity is bringing together different datasets where two of the columns have essentially the same data, just from a different source. In a case like this, it makes sense to just remove one of those columns. In other cases where it isn't as clear-cut, use your knowledge of the data and business process. If one column is more reliable than another, use the more reliable one!

- The other approach is dimensionality reduction, which we covered in Chapter 4, Section 4.3.6. The goal of dimensionality reduction is to reduce the number of columns while maintaining as much of the information as possible. Regardless of the technique, two highly correlated columns will be consolidated based on how that technique operates.

You may still be wondering why this actually matters. What harm can multicollinearity cause? The biggest issue is that your coefficients won't be calculated properly. One of the benefits of regression is that it isolates a variable's impact on what you're trying to predict. When you have two columns with the same impact, the underlying math cannot isolate properly. This results in small changes in your data leading to significantly different predictions.

5.4.3   Linear Regression

The demonstration shown in Section 5.4.1 was linear regression. In its simplest form, your model is a formula that generates a straight line through your data. One of the major challenges with linear regression is that straight line component. If there are different trends for different groups, the best approach is often to create two models instead of one. For example, if you're predicting the sales of two brands, their underlying sales drivers may be different, so a linear regression model may get confused by the independent trends of these distinct brands. The output is a bad model that compromises between both groups, and that isn't useful for anyone.

In practice, this creates three major limitations:

- Finding trends across your data can be time-consuming and difficult. Trends also change over time, so the effort required to identify the trends that will drive how you separate your data can lead to a time-intensive retraining process.
- As you split your data, each additional split reduces your sample size. Combine this with your required test train split, and you can run into sample size issues. You should have hundreds of rows at an absolute minimum to consider linear regression in this predictive modeling capacity.
- Creating separate models increases the effort and time required to maintain the models, because each one requires testing and validation.

The use case for linear regression is to predict a number, meaning it works best for numeric predictions, such as the score of a test, a sales number, or someone's IQ.

The code for creating a linear regression is quite straightforward. There aren't any hyperparameters to tune like there are with the machine learning models (which we'll cover in Section 5.5). As you can see in Listing 5.2, after you've created your training and testing data, it only takes five lines of code to build the model and identify its accuracy on the testing data. First, we use the `train_test_split` function to generate the four necessary components to train and test our model. Then, we instantiate our model (a fancy word for creating it) and fit it using the `fit` function. Finally, we generate our predictions using the `predict` function and generate the accuracy with the `accuracy_score` function.

```
from sklearn.linear_model import LinearRegression
from sklearn.metrics import accuracy_score

X_train, X_test, y_train, y_test = train_test_split(
  X,
  y,
  test_size=0.2,
  random_state=42
  )

model = LinearRegression()

model.fit(X_train, y_train)

y_pred = model.predict(X_test)

accuracy = accuracy_score(y_test, y_pred)
print(f"Accuracy: {accuracy}")
```

Listing 5.2 Linear Regression Code

Now that we've gone through the basics of linear regression, let's discuss it in the context of our use cases. You'll see that this is where many of the nuances and complexities of these particular use cases become reality.

Use Case 1: Linear Regression

Can we satisfy Chris with linear regression? Let's find out! As a reminder, Chris wants to know how many sales he can expect in the next 3 months.

We'll leverage some of our existing data prep, which we saved to a *.csv* file. This will be used as the data source. As a reminder, we're working with the columns listed in Table 5.1.

Column Name	Column Type
Invoice	Integer
StockCode	Object
Description	Object
Quantity	Float
InvoiceDate	Object
Price	Float
Customer ID	Float
Country	Object
New_invoicedate	Object
Invoicedate	Object
Description Grouped	Object
Customers Grouped	Float

Table 5.1 Column Names and Types

You may notice a handful of things. First, we're not bringing in our dummy column data, which is intentional for the purposes of linear regression—we're seeing how the model will perform without it. As a starting point, we'll see if we can build a model without this data. Listing 5.3 shows us loading in the data using the `pd.read_csv` function.

```
#read libraries
import pandas as pd
import numpy as np
```

```
#create in cleaned data from previous chapter
df_cleaned = pd.read_csv("df_cleaned.csv")
```

Listing 5.3 Loading In Data

It's important to think in the context of the model being used to actually predict the future. There is an entire category of predictive modeling specific to time series data. Time series models are designed to identify time patterns in the data, such as seasonality and other recurring trends. Our use case has elements of time series data, so we could consider leveraging a traditional time series model. While this isn't a book about time series models, this distinction can help you understand why it makes more sense to use a regression or machine learning model in the real world.

Time series models have many benefits and limitations. The challenge with these models is you're predicting an end result. For example, in this use case, we'd be predicting sales with no other variables as inputs. This means the prediction relies solely on the *historical* trends of sales data to predict the future. Feature-based models are the opposite; they're able to account for other information that may be influencing the number you're trying to predict, but they don't naturally identify the time-based trends in the data. That's why feature engineering is important—it enables a feature-based model to see trends in the data. Ultimately, it's best to use an ensemble approach, where you build both a time series model and a more feature-based model (such as regression or machine learning). Given that the scope of this book is specific to machine learning models, we won't go into building a time series model.

Lagging variables is one approach to creating a feature-based model. You perform lagging by joining your data onto itself after shifting it back *n* number of hours, days, weeks, or months. This is what enables a forward-looking prediction once you've trained your feature-based model for a time-based use case. For example, in this use case, we're predicting sales. If we only have features that are the last *n* number of months ago in our data, we can use it to predict future sales.

Before we start playing with this data, let's take a step back to make sure you understand how to set it up. Our data is currently a log of all transactions, but that isn't the data structure that will get us what we need. When building a model—especially a regression model—you need to summarize your data to an appropriate level for the model. How this looks goes back to your use case.

Chris wants to predict the next 3 months of sales. To do this, we'll aggregate our data up to the day level. From there, we're able to add the results up to monthly data. This strikes a balance between keeping enough data to build a model (which is why we don't aggregate up to a monthly level) and giving too much detail (like keeping it at the product level). We'll use the pandas `groupby()` function for this, as shown in Listing 5.4. This summarizes our data up to the daily level. The rationale for taking a distinct count of `Description`, `Customer ID`, and `Country` is that we're trying to capture the potential relationship of how products, customers, and countries may be impacting sales.

```
#create an aggregated version of the data
df_data_grouped = df_cleaned.groupby(
  'invoicedate'
  ).agg({
    'Description': 'nunique',
    'Customer ID': 'nunique',
    'Country': 'nunique',
    'Quantity': 'sum',
    'Price':'sum'
  }).reset_index()
```

Listing 5.4 Grouping Data by Day

To facilitate this variable lagging, we'll use the datetime library. We'll start by lagging our data by 3 months, as shown in Listing 5.5. This adds a new invoicedate column to our summarized dataset, which will then be used to join the data for the purposes of lagging our data.

```
#load in data specific libraries
from datetime import datetime
from dateutil.relativedelta import relativedelta

#create date from the invoice timestamp
df_data_grouped['invoicedate'] = pd.to_datetime(
  df_data_grouped['invoicedate']
  ).dt.date

#add 3 months to the date
df_data_grouped['invoicedate_minus_3'] =
  df_data_grouped['invoicedate'] + relativedelta(months=3)
```

Listing 5.5 Lagging by 3 Months

Next, we set up our data for the join. Lagging variables can get confusing; it's easiest to find a date to ground yourself on to be the focus of predicting future values. The following code contains the main date (invoicedate) and the target variable we'll use to create the model (Price):

```
#select only the date and price
df_for_model = df_data_grouped[['invoicedate', 'Price']]
```

To set ourselves up for success, we'll only select our desired columns. It's helpful to reduce the columns before joining data, which can be done using double brackets ([[]) to name each column you'd like to keep (see Listing 5.6). This is cleaner and reduces the risk of confusing you (or others) about which columns belong to the original dataset and which are lagged.

```
#select only needed columns for the model
df_minus_3 = df_data_grouped[[
  'invoicedate_minus_3',
  'Description',
  'Customer ID',
  'Country',
  'Quantity',
  'Price'
]]
```

Listing 5.6 Cleaning Up Data Before Join

We'll also want to do some feature engineering. As shown in Listing 5.7, feature engineering gives the model context about time. The first step is identifying in which month an activity occurred, with the goal of giving the model context about yearly trends by identifying the month within the year. We use the `dt.month` function to extract the month from the date. The second step gives the model context about the overall timeline. We're calculating how many days have passed since sales were tracked. We do this by subtracting the two dates and then converting the difference into the number of days using the `dt.days` function.

```
#create the month from the date
df_minus_3['month'] = pd.to_datetime(
  df_minus_3['invoicedate_minus_3']
  ).dt.month

#calculate number of days between the date and the first sale
df_minus_3['since_first_sale'] =
  pd.to_datetime(df_minus_3['invoicedate_minus_3']) - min(pd.to_datetime(df_for_
    model['invoicedate']))

#convert calculation to a numeric value
df_minus_3['since_first_sale'] = df_minus_3['since_first_sale'].dt.days
```

Listing 5.7 Time-Based Feature Engineering

Now we can bring the data together by using the `pd.merge` function, joining the date column from each DataFrame. The tradeoff with lagging your variables is that your oldest day now has to be reduced by your largest time gap. In this example, we'll lose out on the first 3 months of our dataset because those data elements don't have any data to reference 3 months prior. This blank data is filtered out in the last line of Listing 5.8.

```
#merge lagged data back onto main dataset
df_for_model = pd.merge(
  df_for_model,
```

```
df_minus_3,
how = 'left',
left_on = 'invoicedate',
right_on = 'invoicedate_minus_3',
suffixes = ('','_3months')
)
```

```
#remove records where lagged data is blank
df_for_model = df_for_model[pd.notna(df_for_model['invoicedate_minus_3'])]
```

Listing 5.8 Joining and Removing Blank Data That Can't Be Lagged

Our data is now ready for the actual modeling process! In most cases, we want to randomly split our data; however, this doesn't make sense when we're dealing with time-based data. If the goal is to predict the next 3 months of sales, it's better to first see if you can predict the most recent 3 months of sales. This replaces the train_test_split() you'd normally do to randomly split the dataset. You can also use shape to get a sense of how many rows are in the training set and test set.

As shown in Listing 5.9, we use the relativedelta function to calculate the date at which to split the data into training and test setsz. This date is then used to split our data into the training and test sets using the traditional Python subsetting approach.

```
#calculate the maximum data date
max_data_date = max(df_data_grouped['invoicedate']) - relativedelta(months=3)

#split into training and test data
training_data = df_for_model[df_for_model['invoicedate'] <= max_data_date]
test_data = df_for_model[df_for_model['invoicedate'] >= max_data_date]

#print number of rows and columns for each training and test datasets
print(training_data.shape)
print(test_data.shape)
```

Listing 5.9 Creating Training and Test Datasets

In this split, there are 372 rows in the training set and 67 in the test set. Only 15% is left for the test set, which is lower than ideal, but it's okay for our time-based use case. The goal of a larger test set is to ensure generalizability. In a time series example like this one, if the model can generalize for the next 3 months of test data, we can feel more confident that it will create reliable future predictions. We'll make sure only the numeric columns are included by removing our date columns, as shown in Listing 5.10.

```
#select only numeric columns for the training data
training_data = training_data[[
  'Description',
```

```
  'Customer ID',
  'Country',
  'month',
  'since_first_sale',
  'Price_3months',
  'Price'
]]

#select only numeric columns for the test data
test_data = test_data[[
  'Description',
  'Customer ID',
  'Country',
  'month',
  'since_first_sale',
  'Price_3months',
  'Price'
]]
```

Listing 5.10 Selecting Only the Necessary Columns

Since we're not using `train_test_split()` to split the day, we now need to split both the training and test datasets into the x and y datasets using the `iloc` function, which uses the position of a row or column to select data. In this code, we're selecting all columns, noted by the colon (:) before the comma. We then use `:-1` to select all but the last column for the X datasets and `-1` to select only the last column for the y datasets, as shown in Listing 5.11.

```
#create the predictors for the model's training and test data
X_train = training_data.iloc[:, :-1]
X_test = test_data.iloc[:, :-1]

#create the target for the training and test data
y_train = training_data.iloc[:, -1]
y_test = test_data.iloc[:, -1]
```

Listing 5.11 Split into Required x and y Columns for Model

Now we can train the model! Since we're predicting a number, we need to select a different metric to evaluate the model. A good one to start with is the mean absolute error (MEA) function, `mean_absolute_error()`.

> **Mean Absolute Error**
>
> Chapter 6 discusses the various metrics in more depth, but MAE may be the easiest to explain to stakeholders since it shows the average of the error. For example, by taking the absolute value, errors of +5 and -5 don't cancel out to 0.

In Listing 5.12, we've used the same model workflow where we instantiate the model, fit, predict, and then measure the accuracy on the test dataset.

```
#read in the linear regression and mae libraries
from sklearn.linear_model import LinearRegression
from sklearn.metrics import mean_absolute_error

#create model
model = LinearRegression()

#fit the model to the data
model.fit(X_train, y_train)

#create the predictions for the test data
y_pred = model.predict(X_test)

#use MAE to measure the accuracy
accuracy = mean_absolute_error(y_test, y_pred)

#print the accuracy results
print(f"Accuracy: {accuracy}")
```

Listing 5.12 Building Linear Regression Model with 3-Month Lag

The MAE result is 2,052, meaning for each day the model is off by roughly $2,000 (on an absolute basis). Is this a lot? Yes, given that that's a notable amount of each day's overall sales. Another way to look at this is by summing your predictions and the test data, as follows:

```
#print the sum of both the actual target for the test data and the model's pre-
dictions
print(sum(y_test))
print(sum(y_pred))
```

The result is $423,620 for the test data and $301,129 for the predictions, which means our prediction is only 71% of the actual sales for the period we're predicting. This is a pretty good indication that our model is underpredicting. If you recall what we found in our data exploration in Chapter 4, Section 4.2, sales increase around September through the end of the year due to the holiday season. Our model doesn't seem to have this context, so what can we do to get it into the model? The simplest way is to add a dummy variable that will identify this trend in the data. In the following code, we've created the same new column on both the X_train and X_test datasets, identifying any month after September as a holiday month using the np.where function, which behaves like an if-then statement:

```
#create new holiday season feature on both the training and test data
X_train['holiday_season'] = np.where(X_train['month'] >= 9, 1, 0)
X_test['holiday_season'] = np.where(X_test['month'] >= 9, 1, 0)
```

After making this update, we can now rerun the model using the same code we used to run the model in Listing 5.12. The results are a notable improvement! Our MAE is down to 1,676. Our model now predicts 83% of our actual sales data, so we've gotten much closer just by adding this new variable.

Let's see if we can improve it even further. We'll plot our predictions of sales compared to the test data over time to see if we can glean any insights using the code from Listing 5.13. First, we use the plot function, passing it our date and y_test data. Next, we pass the same date but with the y_pred data. This allows us to add multiple lines onto the same graph. Figure 5.5 displays the resulting graph.

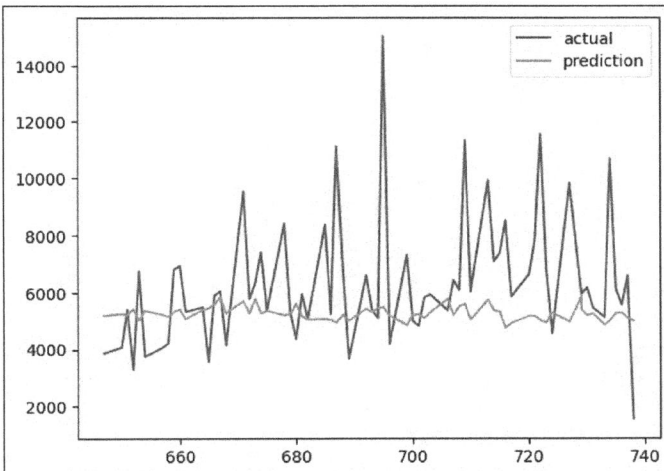

Figure 5.5 Plot of Actuals Versus Predictions

```
#read in plotting library
import matplotlib.pyplot as plt

#plot the actual data
plt.plot(test_data['since_first_sale'], y_test, label = "actual")

#plot the predictions
plt.plot(test_data['since_first_sale'], y_pred, label = "prediction")
```

```
#show the legend
plt.legend()
```

Listing 5.13 Plotting Actuals Versus Prediction

When we plot the data this way, it's clear there are spikes in sales on specific days that aren't being accounted for. There appears to be some type of trend. Let's see if the day of the week has any impact on sales. This presents a new question: What's the right way to bring in this data? Do we want to bring it in as a numeric variable (0-6), or should we bring it in as a categorical variable and dummy code? Let's try both!

To start, we'll go back and recreate our dataset, as shown in Listing 5.14.

```
#select only the date and price
df_for_model = df_data_grouped[['invoicedate', 'Price']]

#select only specific fields
df_minus_3 = df_data_grouped[[
  'invoicedate_minus_3',
  'Description',
  'Customer ID',
  'Country',
  'Quantity',
  'Price']]

#create month feature
df_minus_3['month'] = pd.to_datetime(
  df_minus_3['invoicedate_minus_3']
  ).dt.month

#create days since first sale feature
df_minus_3['since_first_sale'] =
  pd.to_datetime(df_minus_3['invoicedate_minus_3']) - min(pd.to_datetime(df_for_
    model['invoicedate']))

#convert days since first sale to numeric value
df_minus_3['since_first_sale'] = df_minus_3['since_first_sale'].dt.days

#join onto original data
df_for_model = pd.merge(
  df_for_model,
  df_minus_3,
  how = 'left',
  left_on = 'invoicedate',
```

```
  right_on = 'invoicedate_minus_3',
  suffixes = ('','_3months'))
```

Listing 5.14 Recreate the Data

Now we add in the new weekday data and recreate our training and test datasets, as shown in Listing 5.15.

```
#add day of the week feature
df_for_model['day_of_week'] = pd.to_datetime(
  df_for_model['invoicedate']).dt.weekday

#remove values that the lagged data isn't populated for
df_for_model = df_for_model[pd.notna(df_for_model['invoicedate_minus_3'])]

#calculate the max data date for splitting our model into train and test
max_data_date = max(df_data_grouped['invoicedate']) - relativedelta(months=3)

#split into training and test data
training_data = df_for_model[df_for_model['invoicedate'] <= max_data_date]
test_data = df_for_model[df_for_model['invoicedate'] >= max_data_date]

#select training data fields
training_data = training_data[[
  'Description',
  'Customer ID',
  'Country',
  'month',
  'since_first_sale',
  'Price_3months',
  'day_of_week',
  'Price'
  ]]

#select test data fields
test_data = test_data[[
  'Description',
  'Customer ID',
  'Country',
  'month',
  'since_first_sale',
  'Price_3months',
  'day_of_week',
  'Price'
  ]]
```

```
#split into X training and X test
X_train = training_data.iloc[:, :-1]
X_test = test_data.iloc[:, :-1]

#split into y training and y test
y_train = training_data.iloc[:, -1]
y_test = test_data.iloc[:, -1]

#create the holiday season feature on both the training and test data
X_train['holiday_season'] = np.where(X_train['month'] >= 9, 1, 0)
X_test['holiday_season'] = np.where(X_test['month'] >= 9, 1, 0)
```

Listing 5.15 Recreate Our Training and Test Datasets

Then, we rerun our model and the code for the graph, as shown in Listing 5.16.

```
#load in linear regression and MAE libraries
from sklearn.linear_model import LinearRegression
from sklearn.metrics import mean_absolute_error

#create linear regression model
model = LinearRegression()

#fit model to the data
model.fit(X_train, y_train)

#create prediction on the test data
y_pred = model.predict(X_test)

#calculate and print the MAE of the test data results
accuracy = mean_absolute_error(y_test, y_pred)
print(f"Accuracy: {accuracy}")

#load in plotting library
import matplotlib.pyplot as plt

#plot both the actual and prediction data on the graph
plt.plot(test_data['since_first_sale'], y_test, label = "actual")
plt.plot(test_data['since_first_sale'], y_pred, label = "prediction")
plt.legend()
```

Listing 5.16 Rerun Data with the Weekday Numeric Variable

The outcome is essentially unchanged, meaning we didn't get any value in adding the weekday data as a number. The graph in Figure 5.6 shows how our model still doesn't account for the spikes in sales.

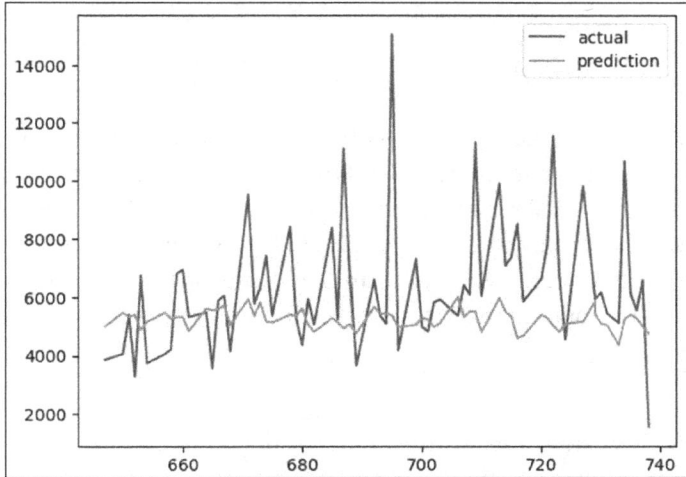

Figure 5.6 Model with Numeric Weekday

When happens when we bring in the data as categorical variables? We can do this using the get_dummies function we used in Chapter 4, as shown in Listing 5.17. We provide the function with the data of the week column. It then converts it into 7 columns, and we use the pd.concat function to add this back to the X_train DataFrame. We repeat these steps for the X_test data to ensure both DataFrames have the same transformations executed on them. If we don't do this, the model will not operate as intended.

```
#dummy code the day of the week on the training data
train_weekday_dummies = pd.get_dummies(
  X_train['day_of_week'],
  prefix='weekday'
  )


#add dummy coded day of the week to the training data
X_train = pd.concat(
  [X_train, train_weekday_dummies],
  axis = 1
  )


#dummy code the day fo the week on the test data
test_weekday_dummies = pd.get_dummies(
  X_test['day_of_week'],
  prefix='weekday'
  )
```

```
#add dummy coded day of the week to the test data
X_test = pd.concat(
  [X_test, test_weekday_dummies],
  axis = 1
  )
```

Listing 5.17 Dummy Code the Weekday Variable

Now we can rerun the model with the same code we've been using. Our error gets lower: It's now 1,560, about 100 lower than our last iteration. What's interesting now is that if you sum the predictions and the actual prices for the testing day, you'll find the gap has grown to about 82%. So, what's happening? The model is predicting a lower value than the last iteration, but if you look at Figure 5.7, you can see it's starting to account for some of the spikes in price—just not to the degree it needs to.

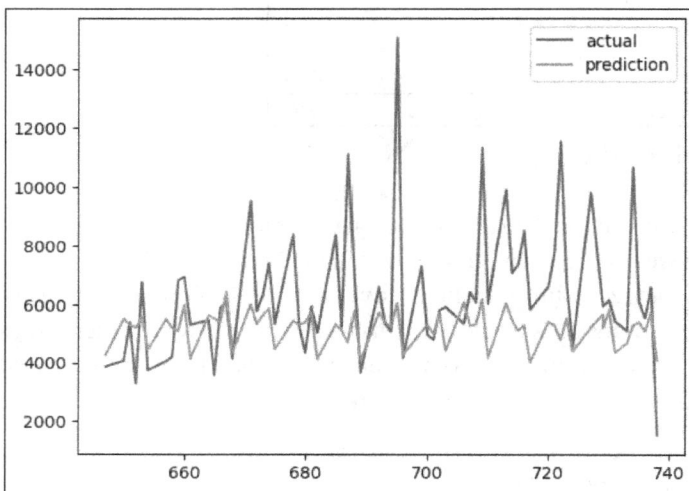

Figure 5.7 Weekday as Categorical Variables

Where do we go from here? This is a good example of a situation where you would likely want to evaluate other models that are better at identifying the trends in the data. This example was written without the benefit of hindsight, and we can hypothesize that the interaction of month and day of week from a tree-based model is more likely to yield the results we're looking for to better account for the spikes. This is covered in Section 5.5.1, where you'll see that a primary benefit to the tree-based models is their ability to find these nonlinear interactions within the data.

Something to remember—especially when Chris is your stakeholder—is that the past can't always predict the future. This is a common aspect to consider when educating your stakeholders. If they don't have an analytics background, they may think you can build a model for anything. Sometimes that just isn't the case!

Use Case 2: Linear Regression

The objective for our second use case is to predict the chances that someone will repeat an order within the next 7 days. Any time you hear any word related to the probability of an outcome (chance, probability, etc.), this is the sure sign that linear regression won't be the model you're going to use. Instead, you'll use logistic regression, which we'll discuss in Section 5.4.4.

Use Case 3: Linear Regression

This use case is about predicting the probability of an area having crime on a given day. You might be thinking that linear regression isn't the best approach here because the stakeholder is requesting a probability, not a number. However, think about how the data has presented itself. When we summarized it by area and day, almost every single area had crime on that given day. So why would we try to predict a probability? The model will ultimately be used to inform stakeholders on where to allocate resources, and it can achieve this by knowing which area on a given day will have more or less crime.

We'll explore how linear regression can achieve this. As always, the first step is to read our data, as shown in Listing 5.18. We'll use the summarized dataset we created in Chapter 4, Section 4.3.6.

```
#read in pandas
import pandas as pd

#read in data from previous chapter
df = pd.read_csv("df_summarized")
```

Listing 5.18 Import Summarized Dataset

We'll run a simple model right away to find our starting point with the final model. For now, we'll intentionally exclude the area name as a feature in the model to see what the results look like without it. We'll also drop the DATE column, which can't be included because it's not a numeric column, as well as the crime_count column, which is our target variable. (We're already pushing the limit of linear regression with the nearly 50 columns we're attempting to use as predictors.) Listing 5.19 shows the code for this. We follow the same modeling workflow as other use cases, which entails splitting our data into train and test sets, creating the model, fitting it, and then evaluating it. Our evaluation metric in this case is a combination of MAE on its own as well as MAE divided by the mean of y_test. We do this to normalize the MAE metric to the context of the dataset.

```
#read in sklearn libraries for the modeling process
from sklearn.linear_model import LinearRegression
from sklearn.model_selection import train_test_split
from sklearn.metrics import mean_absolute_error
```

```
#split data into what will be used to predict and what is being predicted
X = df.drop(columns=['DATE', 'AREA NAME', 'crime_count'])
y = df['crime_count']

#perform standard train test split
X_train, X_test, y_train, y_test = train_test_split(
  X, y, test_size=0.2, random_state=42)

#create and fit the regression model
model = LinearRegression()
model.fit(X_train, y_train)

#predict with the modeling using the test data
y_pred = model.predict(X_test)

#calculate the mean absolute error
mae = mean_absolute_error(y_test, y_pred)

#calculate the mean absolute error divided by average of the target
mae_by_avg_target = mae / y_test.mean()

#print both results
print(mae)
print(mae_by_avg_target)
```

Listing 5.19 Create Linear Regression Model

The result is relatively promising, given the context. The MAE comes back at 6.5, which is roughly 25% of the average target value (for those familiar, this is a similar calculation to mean absolute percentage error [MAPE]). The MAE number is harder to interpret on its own, but the result of 25% can loosely be used as an accuracy percent: The model has an error rate of 25%. Not bad for an initial model where we've done nothing to account for the trends in the data with date specific features!

Given this good starting point, let's take a look at the coefficients to discover what the model finds valuable. In Listing 5.20, we've used the coef_ function to pull out the coefficients from the model into a DataFrame. The results are shown in Figure 5.8.

```
#extract the coefficents from the model
coefficients = model.coef_

#get the feature names from the model
feature_names = model.feature_names_in
```

```
#add the these into a series to display the coefficients alongside their feature
names
pd.Series(data=coefficients, index=feature_names)
```

Listing 5.20 View Coefficients of the Model

```
Vict Age                                                             -0.082311
crime_description_ASSAULT WITH DEADLY WEAPON, AGGRAVATED ASSAULT      17.965487
crime_description_BATTERY - SIMPLE ASSAULT                           11.685328
crime_description_BURGLARY                                           21.373779
crime_description_BURGLARY FROM VEHICLE                               2.213384
crime_description_INTIMATE PARTNER - SIMPLE ASSAULT                  -4.621731
crime_description_Other                                              12.631013
crime_description_ROBBERY                                            10.403193
crime_description_SHOPLIFTING - PETTY THEFT ($950 & UNDER)          -14.095029
crime_description_THEFT FROM MOTOR VEHICLE - GRAND ($950.01 AND OVER) 0.086295
crime_description_THEFT FROM MOTOR VEHICLE - PETTY ($950 & UNDER)   -16.550620
crime_description_THEFT OF IDENTITY                                  27.568402
crime_description_THEFT PLAIN - PETTY ($950 & UNDER)                 -7.034606
crime_description_THEFT-GRAND ($950.01 & OVER)EXCPT,GUNS,FOWL,LIVESTK,PROD -1.684420
crime_description_VANDALISM - FELONY ($400 & OVER, ALL CHURCH VANDALISMS) -2.449467
crime_description_VANDALISM - MISDEAMEANOR ($399 OR UNDER)           -1.689151
crime_description_VEHICLE - STOLEN                                  -14.324866
victim_gender_F                                                     14.944671
victim_gender_M                                                     12.896172
victim_gender_Other                                                 13.636149
victim_race_B                                                       24.288182
victim_race_H                                                        0.693986
victim_race_O                                                        7.012393
victim_race_Other                                                    3.854551
victim_race_W                                                        3.260104
victim_race_X                                                        2.367777
crime_premises_DEPARTMENT STORE                                     10.114845
crime_premises_DRIVEWAY                                             -7.053405
crime_premises_GARAGE/CARPORT                                       13.705114
crime_premises_MARKET                                               -4.317378
crime_premises_MULTI-UNIT DWELLING (APARTMENT, DUPLEX, ETC)          3.061521
crime_premises_OTHER BUSINESS                                        5.280746
crime_premises_OTHER PREMISE                                         4.523803
crime_premises_OTHER RESIDENCE                                     -14.245634
crime_premises_Other                                                4.543209
crime_premises_PARKING LOT                                           3.744716
crime_premises_PARKING UNDERGROUND/BUILDING                         11.733309
crime_premises_RESTAURANT/FAST FOOD                                  2.175149
crime_premises_SIDEWALK                                             14.644929
crime_premises_SINGLE FAMILY DWELLING                               -5.715358
crime_premises_STREET                                                0.531290
crime_premises_VEHICLE, PASSENGER/TRUCK                             -1.249864
crime_weapon_HAND GUN                                                5.589493
crime_weapon_Other                                                   1.690407
crime_weapon_STRONG-ARM (HANDS, FIST, FEET OR BODILY FORCE)         13.973588
crime_weapon_UNKNOWN WEAPON/OTHER WEAPON                            13.333885
crime_weapon_VERBAL THREAT                                           6.889619
```

Figure 5.8 Coefficient Results

The values in Figure 5.8 suggest that the coefficient shows how much the predicted value changes when the variable increases by one unit. If the coefficient is negative, the result goes down, if the coefficient is positive, the result goes up. Think of this as a correlation between the variable and the result you're trying to predict. We see a number of high and low values, which tells us there may be specific columns that will become important as we continue to build out our model.

Given how many columns we already have, we'll explore this further as we progress into tree-based models in Section 5.5.1.

In Chapter 4, Section 4.3.6, we promised to revisit principal component analysis (PCA)—and now is the time to do so. We ultimately won't use it for the final model, because as you'll see when we get to the tree-based models, our lower number of columns doesn't require us to apply PCA. However, for learning purposes, we'll apply PCA to the dataset and run through the linear regression model to give you an example of how you can apply PCA in practice to reduce your columns.

Training Principal Component Analysis

You should always train a PCA model on the training dataset, not your test dataset. This ensures that you're not at risk of leaking the answer to your model via the PCA process. Think about your model in practice. After you've built it and start using it to predict future events, you won't be able to run PCA on data that hasn't happened yet.

As shown in Listing 5.21, we're creating the PCA model with five components, fitting it to our data, using the transform function to actually apply the PCA to our training and test sets, and then printing out the explained variance of the PCA model.

```
from sklearn.decomposition import PCA

#create the pca object using 5 components
pca = PCA(n_components=5)

#fit the pca model to the training data
pca.fit(X_train)

#apply the pca model to the X_train data
X_train = pca.transform(X_train)

#apply the pca model to the X_test data
X_test = pca.transform(X_test)

#check the explained variance ratio
print("Overall Explained Variance: ",sum(pca.explained_variance_ratio_))
print("Explained Variance Ratio: ", pca.explained_variance_ratio_)
```

Listing 5.21 Apply PCA to the Data

The explained variance looks good—it's essentially 100% with only five components. Let's see how it performs when we run it through the model in Listing 5.22.

```
#create and fit the regression model
model = LinearRegression()
model.fit(X_train, y_train)

#predict with the modeling using the test data
y_pred = model.predict(X_test)

#calculate the mean absolute error
mae = mean_absolute_error(y_test, y_pred)

#calculate the mean absolute error divided by average of the target
mae_by_avg_target = mae / y_test.mean()

#print both results
print(mae)
print(mae_by_avg_target)
```

Listing 5.22 Run Linear Regression on Data That Has Gone Through PCA

The resulting MAE is 6.9, which is 26% of the average target value. This tells us that the PCA process worked extremely efficiently, with only a slight reduction in model performance while reducing our number of columns from almost 50 down to 5.

We'll end up skipping logistic regression for this use case since we're predicting a numeric value. We'll pick it back up in Section 5.5.1.

5.4.4 Logistic Regression

Logistic regression is used for classification. As discussed in Section 5.2.3, classification models predict probabilities, which then are translated to a binary outcome. Logistic regression is no exception to this. Just like linear regression, logistic regression models are formulas. The logistic regression formula is more complicated than linear regression, so we won't dive into the details. However, the overall concept is the same. The underlying math is what differs.

The use cases for logistic regression usually revolve around binary outcomes. Examples include:

- Customer retention
- Employee turnover
- Customer conversion

As you familiarize yourself with the various models at your disposal, you may start to see a trend. The vast majority of machine learning problems *could* be adjusted to be classification. For example, instead of predicting a sales number, what if you build a model that predicts the probability of reaching a sales goal? This type of approach can be more valuable for stakeholders. For example, if the stakeholder is expected to reach $10 million in sales for the next quarter, predicting a 75% probability of reaching that goal could have more value than predicting that sales over that period of time will reach $9 million.

The code for logistic regression should look very similar to linear regression. As you can see in Listing 5.23, the only change is that we're using `LogisticRegression()` instead of `LinearRegression()`. All the other code remains the same!

```
from sklearn.linear_model import LogisticRegression
from sklearn.metrics import accuracy_score

X_train, X_test, y_train, y_test = train_test_split(
    X,
    y,
    test_size=0.2,
    random_state=42
    )

model = LogisticRegression()

model.fit(X_train, y_train)

y_pred = model.predict(X_test)

accuracy = accuracy_score(y_test, y_pred)
print(f"Accuracy: {accuracy}")
```

Listing 5.23 Logistic Regression Code

Let's dive into logistic regression for our use cases.

Use Case 1: Logistic Regression

Logistic regression isn't the right tool for this use case. Each row we're predicting is by day, so what would the probability or classification be? While we could try to predict a sales number by each day to identify whether a specific target could be met, we'd be using logistic regression for the sake of using logistic regression. Our next use case will be a better demonstration of the value of logistic regression.

Use Case 2: Logistic Regression

Our second use case is an ideal example of logistic regression. Our goal is to identify the probability that a customer will order again in the next 7 days. Logistic regression outputs a value from 0 to 1 identify the probability an event will occur. Logistic regression is the right model to start with for this use case.

Since this is our first model for this use case, we'll start with our data. What is the target variable for this use case? We've already stated the objective is identifying if someone will order within the next 7 days, but how does that translate to our data? Which column should we be using? Take a few moments to think about this before moving to the next paragraph.

This can be a tricky concept to consider when you're getting started with predictive models. The presence of another row indicates there is another order and therefore indicates our target variable. However, our algorithms can't magically pull from multiple rows; all the data required to train the model for each observation must be on the same row. Table 5.2 shows an abbreviated sample of data to illustrate this point.

Customer ID	Order Date	Ordered Again in 7 Days?
12345	1/1	Yes
98765	1/3	No
56789	1/8	Yes

Table 5.2 Example of How Target Variable Needs to Be Set Up

This example illustrates that we're adding *future* values to *historical* values. This is important! Think about it this way: If we're using the model to predict the future, we need to set up our data to look forward into the future, not backwards. In order to build the model, we need to provide it with snapshots in time that translate to what we ultimately want to predict.

For the model's purpose, the yes and no values become a 1 and 0, respectively. This is how the model can then predict the probability of the outcomes, since it's identifying the value between 0 and 1.

So how do we do this? Let's get started!

The first step is the easiest. We'll load in the data that we previously saved from our data cleaning and dummy coding in Chapter 4, Section 4.3. We'll then create a new DataFrame with only a handful of columns that we'll need (who ordered and when) using double brackets ([[]) to select the columns. Listing 5.24 shows this code and Figure 5.9 shows a preview of the new DataFrame.

```
#read in pandas
import pandas as pd

#read in data from previous chapter
df = pd.read_csv("use_case_2_cleaned.csv")

#create dataset for lagging
df_lagging = df[['Customer ID', 'order_date']]

#preview data
df_lagging.head()
```

Listing 5.24 Creating Target Variable–Specific DataFrame

Customer ID	order_date
5d6c2b96db963098bc69768bea504c8bf46106a8a5178e...	2024-09-10
0781815deb4a10a574e9fee4fa0b86b074d4a0b36175d5...	2024-09-10
f93362f5ce5382657482d164e368186bcec9c6225fd93d...	2024-09-10
1ed226d1b8a5f7acee12fc1d6676558330a3b2b742af5d...	2024-09-10
d21a2ac6ea06b31cc3288ab20c4ef2f292066c096f2c5f...	2024-09-10

Figure 5.9 Preview of Target Variable DataFrame

There's nothing fancy going on here; we're just creating a log of each order. This Data-Frame can now be joined onto our main DataFrame to get the future dates onto our data. This will be a simple task for anyone who's used SQL before, because SQL allows you to use conditional logic to join date (e.g., date column x is greater than or equal to date column y). Unfortunately, this is a bit harder with Python. The merge function in pandas doesn't support joins on values like "less than or equal to" or "greater than or equal to." You can only join values with an equal comparison between two columns.

Regardless, the first step is to create the start (or floor) of the time window to which the future hours should be allowed to join (see Listing 5.25 and Figure 5.10). We're adjusting by 8 days using the timedelta function, so we don't need to be inclusive of the bottom range, which will help us when we join the data.

```
#read in date specific libraries
from datetime import datetime, timedelta

#create new date with difference of 8 days
df_lagging['order_minus_8'] = pd.to_datetime(
  df_lagging['order_date']) - timedelta(days=8)
```

```
#preview the data
df_lagging.head()
```

Listing 5.25 Create Floor Date for Which a Future Order Can Be Associated with a Previous Order

	Customer ID	order_date	order_minus_8
	5d6c2b96db963098bc69768bea504c8bf46106a8a5178e...	2024-09-10	2024-09-02
	0781815deb4a10a574e9fee4fa0b86b074d4a0b36175d5...	2024-09-10	2024-09-02
	f93362f5ce5382657482d164e368186bcec9c6225fd93d...	2024-09-10	2024-09-02
	1ed226d1b8a5f7acee12fc1d6676558330a3b2b742af5d...	2024-09-10	2024-09-02
	d21a2ac6ea06b31cc3288ab20c4ef2f292066c096f2c5f...	2024-09-10	2024-09-02

Figure 5.10 Preview of Customer Order with Date Window

Now that we have our date window, we need to bring the data together. There really isn't a great way to execute this in pandas, so we'll leverage a cross-join scenario where we join each customer ID transaction first and then we filter the dataset using our date window.

Listing 5.26 shows how we clean up the new DataFrame by using the drop function to remove the original order_date column and then merge the data. We're using an inner join in this case because we only want to return rows where there is a match between each dataset. If there isn't a match, it's too early in the dataset to create the lag. Since this is specific to the lagging, we're only selecting Customer ID and order_date from df. This will simplify future steps—when lagging, we only need the unique columns required for joining and the lagged variable, because we'll ultimately join this data back to the original dataset. This join will result in too many duplicate rows, since we're only joining on the Customer ID column.

```
#identify create new column called max_date
df_lagging['max_date'] = df_lagging['order_date']

#drop the order_date column
df_lagging = df_lagging.drop(columns=['order_date'])

#join lagged data onto the original data
df_lag_7 = pd.merge(
  df[['Customer ID', 'order_date']],
  df_lagging,
  how = 'inner',
  on = 'Customer ID'
  )
```

```
#print number of columns and rows for both df and df_lag_7
print(df.shape)
print(df_lag_7.shape)
```

Listing 5.26 Join All Customer Orders Together

The resulting DataFrame now has duplicates that we'll need to carefully address. We started with over 21,000 rows and are now up to almost 87,000 rows. When filtering by date, we also need to keep in mind that some orders won't have any other orders that fall within our date window. This is why we didn't overwrite our original DataFrame. Ultimately, we'll join all this data back onto that DataFrame to get the proper target variable column. As you may be realizing, the complexity and nuance of properly lagging this data can be quite tedious.

Next, we'll filter the dates by selecting rows where the order date is greater than that of the last 8 days and less than the max date column we've created, as shown in Listing 5.27. This reduces the data to give each row a single lag value since our initial join could only be done using Customer ID.

```
#filter the data to ensure only the necessary rows are matched
df_lag_7 = df_lag_7[
   (df_lag_7['order_date'] > df_lag_7['order_minus_8']) &
   (df_lag_7['order_date'] < df_lag_7['max_date'])
   ]

#show the number of columns and rows in the data
df_lag_7.shape
```

Listing 5.27 Filter the Cross-Joined Dataset to Be Within the Date Window

The results show we're back down to 4,524 rows. This indicates we're seeing 4,524 instances of customers ordering again within 7 days.

We can now bring this dataset back together with the main dataset, as shown in Listing 5.28. As you'll see, we do some additional cleaning to the target dataset to make the join easier. This includes adding a column of 1s to represent our target variables of 1 and 0. Since not all rows will have a match, and we need to ensure all rows have either a 1 or 0, we're filling in any values with a 0 using the fillna function.

Dropping duplicates is extremely important! As we only care if the customer orders at least one time in the next week, we want to ensure we don't create duplicate records for the instances where someone orders more than once in that given week.

```
#select only the customer id and order date fields
df_lag_7 = df_lag_7[['Customer ID', 'order_date']]
```

```
#drop duplicates
df_lag_7 = df_lag_7.drop_duplicates()

#create dummy column of all 1 values
df_lag_7['target_7'] = 1

#join the data to the original dataset after correction to the lagged dataset
df = pd.merge(
  df,
  df_lag_7,
  how = 'left',
  on = ['Customer ID', 'order_date']
  )

#fill in any blanks with 0
df['target_7'] = df['target_7'].fillna(0)
```

Listing 5.28 Merge Target Variable Data Back onto Main Dataset

Before we actually build the model, we should take a step back and think about our use case. Our stakeholder wants to view the last 7 days; however, now is the easiest time to add another interval. Given the number of customer orders, it's worth adding a 2-week or 14-day interval as well. Even if the stakeholder doesn't end up using it, it could be a natural next question. Having this already ready to review, shows you're well prepared and goes over favorably with stakeholders. It also presents an opportunity for you to identify additional insights between these different approaches that can bring value to your stakeholder.

The code for this is shown in Listing 5.29, where we've executed all the same steps we did for the 1-week lag but this time for 2 weeks. If you apply this same methodology more than twice, you should convert the code into a function that you can input the date interval. This will prevent the need to copy and paste the same code over and over again!

```
#select only the customer id and order date fields
df_lagging = df[['Customer ID', 'order_date']]

#create lag for 2 weeks column
df_lagging['order_minus_15'] = pd.to_datetime(
  df_lagging['order_date']) - timedelta(days=15)

#create max_date column from order_date
  df_lagging['max_date'] = df_lagging['order_date']
```

```
#drop the originally named order_date column
df_lagging = df_lagging.drop(columns=['order_date'])

#join the 14 day lag with the lagged dataset
df_lag_14 = pd.merge(
  df[['Customer ID', 'order_date']],
  df_lagging,
  how = 'inner',
  on = 'Customer ID'
  )

#select only rows that fit the desired criteria
df_lag_14 = df_lag_14[
  (df_lag_14['order_date'] > df_lag_14['order_minus_15']) &
  (df_lag_14['order_date'] < df_lag_14['max_date'])
  ]

#select only the customer ID and order date
df_lag_14 = df_lag_14[['Customer ID', 'order_date']]

#drop the duplicates
df_lag_14 = df_lag_14.drop_duplicates()

#create dummy column of all 1's
df_lag_14['target_14'] = 1

#join onto original dataset
df = pd.merge(
  df,
  df_lag_14,
  how = 'left',
  on = ['Customer ID', 'order_date']
  )

#fill the target field with 0's
df['target_14'] = df['target_14'].fillna(0)
```

Listing 5.29 Create 14-Day Target Variable

We're now ready to build the model! We'll build it in stages, since a full 40 columns going into a logistic regression model isn't the best idea. As a starting point, we'll want to evaluate the columns we think will have the most value. In Listing 5.30, we're adding the columns that are likely to be the most valuable using double brackets ([[) and assigning them to X. Then we're noting our target variable and assigning it to y.

```
#select only desired columns
X = df[[
  'Total', 'KPT duration (minutes)', 'Rider wait time (minutes)',
  'DistanceNumeric', 'Restaurant name_Masala Junction',
  'Restaurant name_Swaad', 'Restaurant name_Tandoori Junction',
  'Restaurant name_The Chicken Junction', 'Subzone_Chittaranjan Park',
  'Subzone_DLF Phase 1', 'Subzone_Greater Kailash 2 (GK2)',
  'Subzone_Sector 135', 'Subzone_Sector 4', 'Subzone_Shahdara',
  'Subzone_Sikandarpur', 'Subzone_Vasant Kunj',
  'Cancellation / Rejection reason_Cancelled by Customer',
  'Cancellation / Rejection reason_Cancelled by Zomato',
  'Cancellation / Rejection reason_Items out of stock',
  'Cancellation / Rejection reason_Kitchen is full',
  'Cancellation / Rejection reason_Merchant device issue'
  ]]

#select target_7 as the target variable
y = df['target_7']
```

Listing 5.30 Select the Columns Most Likely to Be Predictive

Now that we've assigned these columns to X and y, we can move them into a standard model, as shown in Listing 5.31. One difference you'll see is that there are now parameter options when instantiating our model (solver and random_state). Other than that, it's the same process we've seen before, just using the LogisticRegression function.

```
#read in necessary libraries for the modeling process
from sklearn.linear_model import LogisticRegression
from sklearn.model_selection import train_test_split
from sklearn.metrics import accuracy_score
from sklearn.metrics import roc_auc_score

#split the data into train and test
X_train, X_test, y_train, y_test = train_test_split(
  X,
  y,
  test_size=0.3,
  random_state=42
  )

#create logistic regression model
model = LogisticRegression(solver='liblinear', random_state=42)
```

```
#fit the model to the data
model.fit(X_train, y_train)
```

Listing 5.31 Split Data into Test and Train Sets, Then Create the Model

Now that we have a model, we'll create the predictions on the testing data and then evaluate how the model performs. In Listing 5.32, we're calculating two different approaches to measuring the model's value. The first is the accuracy score using the accuracy_score function. The second is the receiver operating characteristic (ROC) area under the curve (AUC)—say that 10 times fast!—using the roc_auc_score function.

```
#use the model to predict on the test data
y_pred = model.predict(X_test)

#calculate and print the accuracy score
accuracy = accuracy_score(y_test, y_pred)
print(f"Accuracy: {accuracy}")

#calculate and print out the ROC AUC score
y_pred_proba = model.predict_proba(X_test)[:, 1]
auc_score = roc_auc_score(y_test, y_pred_proba)
print(f"ROC AUC Score: {auc_score:.4f}")
```

Listing 5.32 Create and Test Initial Logistic Regression Model

The results from each metric are different. We'll discuss metrics in more detail in Chapter 6, but here, accuracy represents how many times the model guesses either 1 or 0 correctly. The ROC AUC is more complex (and we'll get there in Chapter 6, Section 6.2.6), but anything below 0.7 is not a great model. The accuracy score we get when running the model is 85%, while the ROC AUC score is 0.6. This may seem confusing initially, but think about the data. Most people don't order again within a week, meaning a model that simply says that no one will order again in the next week would return what appears to be a good accuracy score. That's why the ROC AUC score is a better metric to leverage, given its ability to handle how our data is distributed.

Since we also calculated the target for repeat orders within 14 days, let's see what the model looks like when we use that for the target variable in Listing 5.33.

```
#select y as the 2 week target
y = df['target_14']

#execute the train test split
X_train, X_test, y_train, y_test = train_test_split(
    X,
    y,
    test_size=0.3,
```

```
   random_state=42)

#create logistic regression model
model = LogisticRegression(solver='liblinear', random_state=42)

#fit the model to the data
model.fit(X_train, y_train)

#use the model to predict on the test data
y_pred = model.predict(X_test)

#calculate and print the accuracy metric
accuracy = accuracy_score(y_test, y_pred)
print(f"Accuracy: {accuracy}")

#calculate and print the roc auc metric
y_pred_proba = model.predict_proba(X_test)[:, 1]
auc_score = roc_auc_score(y_test, y_pred_proba)
print(f"ROC AUC Score: {auc_score:.4f}")
```

Listing 5.33 Rerun Using the 14-Day Target Variable Column

The results are even worse. The accuracy is 76%, and the ROC AUC score is 0.58. Neither of these models are likely to be valuable in practice. You may have noticed how quickly the model runs, which can be a significant benefit. However, it appears that the data may be more complicated than a logistic regression model can handle. Instead, we'll try out tree-based machine learning models to see how those algorithms handle this dataset.

Use Case 3: Logistic Regression

Since this use case calls for predicting a number of crimes rather than a probability, logistic regression isn't the proper model.

5.5   Machine Learning Models

This section covers what is more traditionally considered machine learning. We'll cover tree-based models in increasing complexity. We'll start with the simpler decision tree algorithm and then move to the random forest algorithm, which is a collection of decision trees. Lastly, we'll discuss GBM, a more complex algorithm in which decision trees learn from each other.

A Word of Caution

One of the best—and also most dangerous—aspects of these algorithms is their ease of use. They require few assumptions about your data and can cover up poor data preparation. This allows you to learn about and create models quickly, but it can also encourage bad habits. One of the most common is the tendency to just throw all your data into the model and see what sticks. This is more computationally intensive, but if you're working in an enterprise-grade environment, it may have only a minimal impact on runtimes. From a use case perspective, this can make the model hard to interpret and may even raise ethical concerns. It's best practice to have a reasonable hypothesis for why you include data in the model.

5.5.1   Decision Tree

Decision trees are like flowcharts. The model evaluates the data given to it and creates a quantitative decision. The model will continue to do this until it's instructed to stop. The mechanism used to determine how it splits the data can vary, but in general, it's looking for the split that will best explain the data. If the model can only make one split, what will yield the best result?

Decision trees have a high explainability, meaning you can easily identify each decision being made. The decision tree algorithm is rarely used in practice because of its simplicity and inability to leverage multiple smaller models. However, it's foundational for understanding other machine learning models, and you'll benefit from taking the time to learn about it. The following sections introduce many new concepts and approaches. While they're not specific to decision trees, it's easier to understand them using a simpler algorithm.

We'll start with the basic construction of a decision tree and explore core concepts and capabilities before applying decision trees to our use cases.

Decision Tree Modeling

Listing 5.34 is an example of decision tree code using the Iris dataset, a publicly available dataset commonly used in machine learning examples. The Iris dataset is included in sklearn, so we can simply import the dataset rather than needing to download any files. As you'll notice, it's not notably different from how we create the regression models (except for our usage of DecisionTreeClassifier). As we progress through the various algorithms in this chapter, you'll learn this exact model isn't one you'd want to implement. For now, observe the structure and steps of the code.

```
from sklearn.datasets import load_iris
from sklearn.model_selection import train_test_split
from sklearn.tree import DecisionTreeClassifier
```

```
from sklearn.metrics import accuracy_score
from sklearn import tree
import matplotlib.pyplot as plt

iris = load_iris()
X = iris.data
y = iris.target

X_train, X_test, y_train, y_test = train_test_split(
    X,
    y,
    test_size=0.3,
    random_state=42
    )

clf = DecisionTreeClassifier(random_state=42)

clf.fit(X_train, y_train)

y_pred = clf.predict(X_test)

accuracy = accuracy_score(y_test, y_pred)
print(accuracy)
```

Listing 5.34 Decision Tree Code Example

The output of this code is interesting. The accuracy is 100%—and if you have a model with 100% accuracy, you should be skeptical. It's important to go back and consider your use case. In this case, the model is predicting a type of flower, so it may be possible to build a 100% accurate model. However, in most real-world applications, 100% accuracy is not achievable.

As we discussed, decision trees can be interpretable. Listing 5.35 shows how you can visualize the decisions being made by the model, with the results shown in Figure 5.11. We're using the plot_tree function to create a visual representation of how the model is coming to the predictions. We pass this function the model (clf), the names of our features (iris.feature_names), and what we're trying to predict (iris.target_names) to have this plot generated.

```
fig = plt.figure(figsize=(10, 7))
tree.plot_tree(
    clf,
    feature_names=iris.feature_names,
    class_names=list(iris.target_names),
```

```
   filled=True
   )
plt.show()
```

Listing 5.35 Visualizing Decision Tree Model

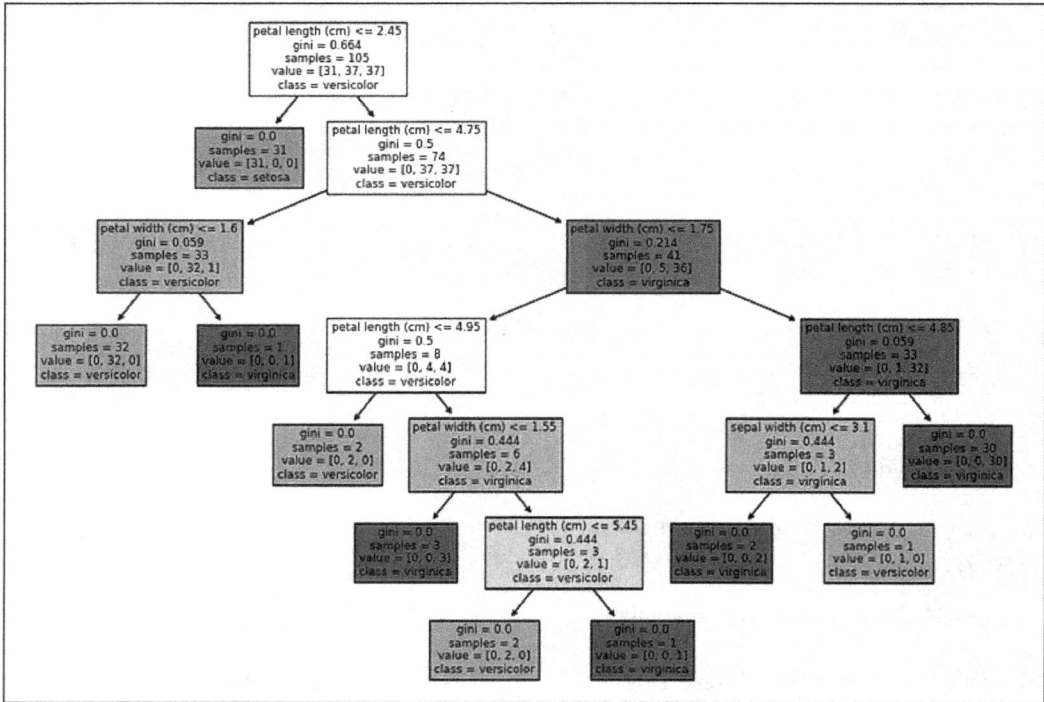

Figure 5.11 Visualization of Iris Dataset in Decision Tree Model

As you look at Figure 5.11, you'll see some new terms (like Gini, which we'll get to later in this chapter). The components in each of the boxes should give you insight into the model and also raise some questions. Particularly of note are the first line, which explains the decision being made, and the last line, which tells you what the prediction would be if the model stopped in that box. These are both helpful, and in theory, they would allow you to recreate the model by hand.

You should be skeptical when reviewing the samples. The sample number shows the amount of data being considered at that decision. As you move towards the bottom of the tree, you start seeing smaller sample sizes. In tree-based models, this almost always leads to overfitting (meaning your model is not generalizable to new data). To demonstrate this, let's move to a dataset that isn't as clean. Listing 5.36 uses the Digits dataset from sklearn as an example.

```
from sklearn.datasets import load_digits
from sklearn.model_selection import train_test_split
from sklearn.tree import DecisionTreeClassifier
from sklearn.metrics import accuracy_score
from sklearn import tree
import matplotlib.pyplot as plt

digits = load_digits()

X_train, X_test, y_train, y_test = train_test_split(
  digits.data,
  digits.target,
  test_size=0.2,
  random_state=42
  )

clf = DecisionTreeClassifier(random_state=42)

clf.fit(X_train, y_train)

y_pred = clf.predict(X_test)

accuracy = accuracy_score(y_test, y_pred)
print(accuracy)
```

Listing 5.36 Decision Tree Model Code on Messier Data

Our accuracy score comes back at 84%. However, we want to see what that accuracy was when the model was created. In Listing 5.37, we're creating the predictions for the training set and checking the accuracy.

```
y_train_pred = clf.predict(X_train)
print(accuracy_score(y_train, y_train_pred))
```

Listing 5.37 Predicting on Training Dataset

The result is 100%. If the accuracy is only 84% for the testing data but 100% for the training data, it indicates that we're overfitting. The model perfectly fits the training data but not the testing data. So, how do we prevent this from happening? We use hyperparameter tuning, which we'll discuss in the next section.

Beyond our accuracy metric, Figure 5.12 also shows how overfit the model is. This generates the same decision tree we saw for the Iris dataset, but now with the Digits dataset (see Listing 5.38). This image is chaos, as each box represents a decision from the model!

```
target_names = list(
  " ".join(
    map(
      str, list(digits.target_names)
      )
    )
  )

fig = plt.figure(figsize=(10, 7))
tree.plot_tree(
  clf,
  feature_names=digits.feature_names,
  class_names=target_names,
  filled=True
  )
plt.show()
```

Listing 5.38 Decision Tree Model Visual for Digits Dataset

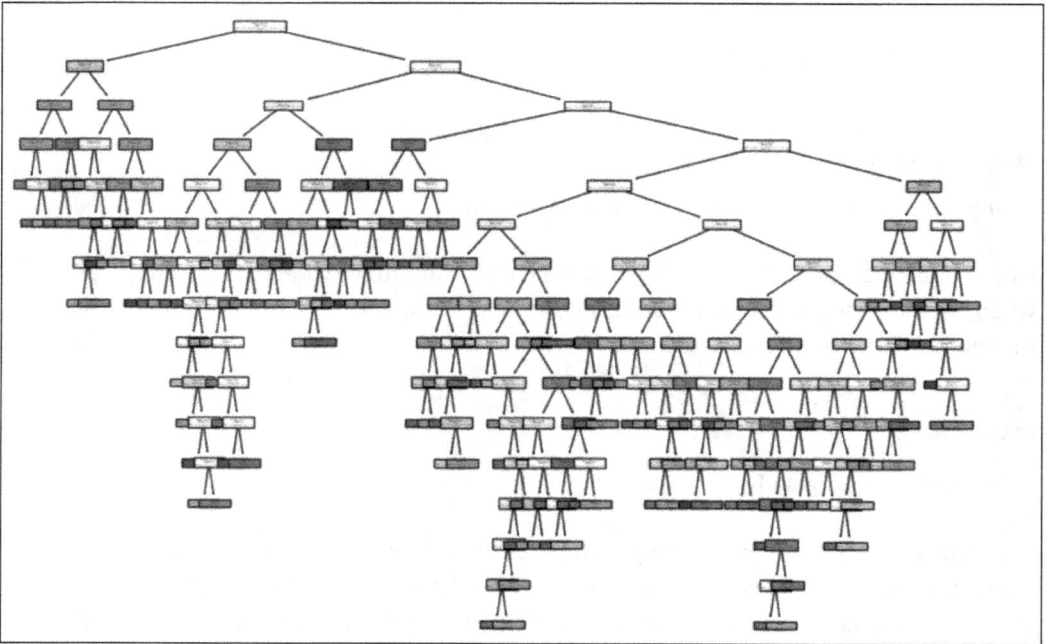

Figure 5.12 Overfit Decision Tree Model Visual for Digits Dataset

Decision Tree Hyperparameters

The concept of *hyperparameters* is core to machine learning algorithms. These are the human-defined inputs into the model. They're where your understanding of machine learning algorithms starts making a notable impact on your model's results. There are many hyperparameters, but we'll focus on only the most important ones.

The first is `criterion`. This is the approach the model uses to determine how it creates its next decision. There are three options, with Gini as the default, which you saw in Figure 5.11:

- **Gini**
  The Gini `criterion` is a measure of impurity. When constructing a decision, the goal is to reduce the impurity as much as possible. The lower Gini value, the less impurity in the data. A commonly used example is thinking about your data as a jar of marbles. If you only have blue marbles, the impurity is low. However, as you add red marbles to the jar, the impurity increases.

- **Entropy**
  Entropy is a measure of uncertainty in the data. In the context of decision trees, it's used to create the next decision, choosing the one that most reduces uncertainty in the data. This sounds pretty similar to Gini, right? The math on the backend is different. As a result, entropy tends to penalize uncertainty more than Gini does.

- **Log loss**
  Log loss looks at the data from a probabilistic perspective, not a pure classification perspective. This is the metric used by logistic regression. A key difference between log loss and entropy or Gini is that it penalizes incorrect highly confident predictions more heavily.

It's best to select the `criterion` as part of the grid search process, as we'll discuss in the next section. You don't necessarily need to know which `criterion` to pick right away; it's part of the testing process. If you're not using grid search, you can keep the default of Gini.

The next hyperparameter is `max_depth`. This represents the maximum *level* of decisions a decision tree can go down to. An important nuance here is that it's not limiting how many decisions a model can make; rather, it specifies the most decisions on any given path that a model can make. This helps combat overfitting.

The last hyperparameter we'll discuss is `min_samples_split`. This hyperparameter also combats overfitting. The default value is 2, which in practice is rarely used. It's best to avoid setting this value below 10 for smaller datasets, and keep it closer to 25 for larger datasets. If you take the step back and think about how you'd make decisions, how many examples of something occurring would you want before you felt comfortable with the results?

Let's see what happens to the model's results when we implement some of these hyperparameters. Using the Digits dataset, we rerun creation of the model with these hyperparameters adjusted in Listing 5.39. Here, we're using `criterion`, `max_depth`, `min_samples_split`, and `random_state` to adjust what the model can do.

```python
digits = load_digits()

X_train, X_test, y_train, y_test = train_test_split(
    digits.data,
    digits.target,
    test_size=0.2,
    random_state=42)

clf = DecisionTreeClassifier(
    criterion = 'entropy',
    max_depth = 4,
    min_samples_split = 10,
    random_state=42
    )

clf.fit(X_train, y_train)

y_pred = clf.predict(X_test)

accuracy = accuracy_score(y_test, y_pred)
print(f" Test Accuracy: {accuracy}")

y_train_pred = clf.predict(X_train)
accuracy = accuracy_score(y_train, y_train_pred)
print(f" Training Accuracy: {accuracy}")
```

Listing 5.39 Decision Tree with Hyperparameters

The resulting model produces a 74% accuracy on the test data, with 73% accuracy on the training data. Looking more closely at these accuracy numbers, it's clear that we're not overfitting the model, which was our goal. However, now the accuracy is much lower. To help you conceptualize what impact this had on how the model is created, look at Figure 5.13, which was generated from this model. As you did with Figure 5.12, focus on the number of boxes rather than trying to read the specific decisions.

Let's keep playing with the hyperparameters. What happens if we increase the max_depth from 4 to 5 in Listing 5.40?

```python
clf = DecisionTreeClassifier(
    criterion = 'entropy',
    max_depth = 5,
    min_samples_split = 10,
    random_state=42
    )
```

Listing 5.40 Decision Tree with max_depth of 5

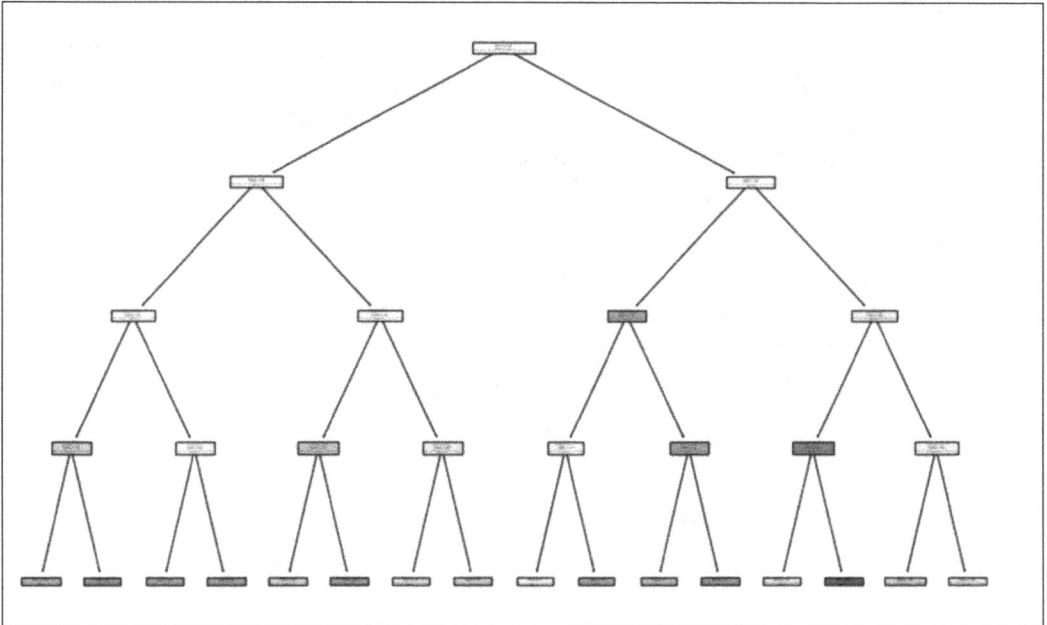

Figure 5.13 Decision Tree Visual with Some Hyperparameters

The result is 81% accuracy on the test data, with 84% on the training data. There is a bit of overfitting, but now we're getting much closer to the 84% accuracy from the model with no hyperparameters. What happens if we adjust max_depth up to 6 while also doubling min_samples_split to 20 in Listing 5.41?

```
clf = DecisionTreeClassifier(
  criterion = 'entropy',
  max_depth = 6,
  min_samples_split = 20,
  random_state=42
  )
```

Listing 5.41 Increasing min_samples_split While Increasing max_depth

Our results are incredible! We're now at 85% accuracy on the test data and 89% on the training data. This means we've created a model that is more accurate but also not dangerously overfit.

These steps might feel like we're guessing and checking, which isn't efficient, and that's partly true. In practice, you could run smaller models like this through another algorithm to identify the optimal solution by iterating through all possible options. This is expensive and impractical with larger datasets, as it can take hours to train a model. However, you can use the grid search capability to search over a smaller subset of options, as we'll discuss next.

Grid Search

The solution to testing a bunch of hyperparameter combinations is the GridSearchCV() function. You'll speed up your model development process considerably by using this workflow. Listing 5.42 shows what this looks like in full. We'll then break down each of the newly added components and discuss them in more detail. As you'll see, the majority of the process remains the same—we're just adding a new layer to it.

```python
from sklearn.model_selection import GridSearchCV
from sklearn.datasets import load_digits
from sklearn.model_selection import train_test_split
from sklearn.tree import DecisionTreeClassifier
from sklearn.metrics import accuracy_score

digits = load_digits()

X_train, X_test, y_train, y_test = train_test_split(
    digits.data,
    digits.target,
    test_size=0.2,
    random_state=42
)

param_grid = {
    'criterion': ['gini', 'entropy'],
    'max_depth': [5, 10, 15],
    'min_samples_split': [5, 10, 20]
}

clf = DecisionTreeClassifier(random_state=42)

grid_search = GridSearchCV(
    estimator=clf,
    param_grid=param_grid,
    cv=3,
    scoring='accuracy'
)

grid_search.fit(X_train, y_train)

print("Best parameters:", grid_search.best_params_)

print("Best score:", grid_search.best_score_)

best_dtree = grid_search.best_estimator_
```

```
accuracy = best_dtree.score(X_test, y_test)
print("Test accuracy:", accuracy)
```

Listing 5.42 Grid Search Example

When we take this approach, we're getting the best parameters to a `max_depth` of 10 and a `min_samples_split` of 5. Note that this is different from the previous section's results because we didn't allow the model to go as deep as 10 levels. The accuracy on the test data is 88% and 84% on the training data with the best model parameters. This was also much faster than if we tried running each of these combinations one by one.

Now let's run through the various chunks of code to explain what they're doing. We'll start with Listing 5.43, which is creating the parameters in the form of a dictionary for the grid search function to use. Here, we've added the `criterion`, `max_depth`, and `min_samples_split`. Each of these have at least two options that the grid search process will bring together to create unique combinations for testing. The goal is to test which model generates the best results.

```
param_grid = {
  'criterion': ['gini', 'entropy'],
  'max_depth': [5, 10, 15],
  'min_samples_split': [5, 10, 20]
  }
```

Listing 5.43 Grid Search Dictionary

We can use the results of the grid search process to identify which combinations to test.

Selective Testing

It's important to note that the `GridSearchCV()` function will test each unique combination. This means you need to be selective to ensure the grid search can run in a reasonable amount of time, especially if you have a larger dataset. If you have a lot of hyperparameters to search over, the `RandomizedSearchCV()` function will randomly select a subset of the total unique combinations.

The `GridSearchCV()` function takes only a few hyperparameters, as shown in Listing 5.44. The `estimator` is the model, the `param_grid` is the dictionary of values to try, and `scoring` is which metric should be used to identify the performance of the model. For now, know that the `cv` hyperparameter is a mechanism to prevent overfitting by creating n copies of the data. It's added to the combinations created as part of our grid search dictionary. Using this example, our current grid search has 18 unique combinations (2 × 3 × 3). Now that we have the `cv` hyperparameter, we multiply it by 3, resulting in 54 unique combinations.

```
grid_search = GridSearchCV(
  estimator=clf,
  param_grid=param_grid,
  cv=3,
  scoring='accuracy'
  )
```

Listing 5.44 GridSearchCV Function

By creating `grid_search`, we're not actually running the model iterations yet, which can be confusing. Think about this step as setting up all the right inputs and staging them but not actually executing on them.

By fitting `grid_search` in the following code, we now check each of the iterations of the model we created from different options of hyperparameter values:

```
grid_search.fit(X_train, y_train)
```

One notable difference is that we're not fitting the model itself; we're fitting the `grid_search` object. Once we fit the `grid_search` object, we have other outputs available to look at:

```
print("Best parameters:", grid_search.best_params_)
print("Best score:", grid_search.best_score_)
```

Specifically, we want to see the best parameters (the entire point of a grid search). Once we get those, we can then rerun the decision tree function using these best parameters to create our winning model (see Listing 5.45), which we can then use create predictions.

```
clf = DecisionTreeClassifier(
  criterion = 'entropy',
  max_depth = 10,
  min_samples_split = 5,
  random_state=42
  )
```

Listing 5.45 DecisionTreeClassifer with the Winning Hyperparameter Options

Grid Search in Practice

Grid search is a huge time saver when you're building your models. In practice, the number of options will vary significantly. If you're running this code as you read, you'll see it can execute the grid search for combinations I've provided within seconds. When you're building models with real data, the time to execute will be significantly longer. Grid search isn't magic; it's more or less a loop over the combinations you provide. The grid search is still training a model for each combination you provide it.

When using grid search in the real world, start with a wide range for a few key parameters of the algorithm you're using. Aim for 4–6 combinations in total. This initial round of testing gives you an idea of which parameters perform best without creating so many options that it takes a long time for the code to run. You can then get more granular with your hyperparameter selections based on what you've learned from your initial grid search. Listing 5.46 gives you an example of what this looks like in practice with four iterations.

```
first_search = {
  'criterion': ['entropy'],
  'max_depth': [5, 15],
  'min_samples_split': [5, 20]
  }

second_search = {
  'criterion': ['entropy'],
  'max_depth': [10, 15],
  'min_samples_split': [5, 10]
  }

third_search = {
  'criterion': ['entropy'],
  'max_depth': [10, 12],
  'min_samples_split': [5, 7]
  }

fourth_search = {
  'criterion': ['gini', 'entropy'],
  'max_depth': [10, 12],
  'min_samples_split': [5, 7]
  }
```

Listing 5.46 Iterating Through Grid Search Options

Free Feature Engineering

We'll discuss feature engineering in more detail in Chapter 6, Section 6.6.1, but because it's a foundational topic, we'll provide a brief introduction here. At a high level, feature engineering is the act of creating new columns from your data to provide to your model. In regression models, feature engineering is exclusively done by the individual creating the model. Regression, by its nature, only isolates the variables in your data to determine their impact on the target variable. However, in tree-based models, the model does some of this feature engineering for you! While this is a great feature (pun intended) of tree-based models, it has its limitations.

As we've discussed, we limit our decision tree from going too deep with the `max_depth` hyperparameter. If we haven't already done any feature engineering, it takes a minimum of two levels for the decision tree to perform feature engineering for us. For example, say we're a company trying to predict when someone will retire. Our company had a traditional pension plan that was discontinued 10 years ago, so some employees were grandfathered into the plan and others weren't. Other data elements like age play an important role in predicting when someone will retire. Without any feature engineering, our model would likely first split on age, then split on if an individual is part of the since-discontinued pension plan. This requires two splits. However, if we create a new set of features that puts individuals into the following groups, we've executed the same type of information gain for the model in a single split:

1. Over 67 and part of the pension plan
2. Over 67 and not part of the pension plan
3. Over 62 and part of the pension plan
4. Over 62 and not part of the pension plan
5. Over 55 and part of the pension plan
6. All other employees

This introduces some subjectivity into the process. To start, it's best to use your knowledge of the use case. If you're predicting retirement of employees, you'll want to understand the various ages and factors that go into retirement. After you have a starting point, you'll then identify which breakouts are important to the model and iterate from there.

Decision Tree Regressor

If you've been reading closely, you likely noticed the decision tree model we've been using has been classifying data, not generating a numeric prediction. How is this done? Sklearn makes it easy with the `DecisionTreeRegressor()` model. As you can see in Listing 5.47, it follows an almost identical process to our previous models. The function has a new name, and the model evaluation metric is different, but otherwise the process is mostly the same!

```
from sklearn.tree import DecisionTreeRegressor
from sklearn.model_selection import train_test_split
from sklearn.metrics import mean_squared_error, r2_score

X_train, X_test, y_train, y_test = train_test_split(
  X,
  y,
  test_size=0.2,
  random_state=42
  )
```

```
regressor = DecisionTreeRegressor()
regressor.fit(X_train, y_train)

# Make predictions
y_pred = regressor.predict(X_test)

# Evaluate the model
mse = mean_squared_error(y_test, y_pred)
r2 = r2_score(y_test, y_pred)

print("Mean Squared Error:", mse)
print("R-squared:", r2)
```

Listing 5.47 Decision Tree Regressor Example

The underlying mechanics of the decision tree model are the same—the only difference is the output. For example, if you're trying to predict the lifetime value of a prospective customer, the output of the model will be a value. This value is calculated by taking the average of all samples remaining in the split. This same concept applies to the subsequent tree-based models as well. The underlying mechanisms between a classification and regression model are the same, but the output differs based on the use case.

This concludes our exploration of the core concepts that underpin decision trees. In all honesty, I've used the decision tree algorithm once in my corporate jobs, and only as a means to explain a more sophisticated model to a nontechnical audience. Decision tree models aren't commonly used in practice because they're a single model that doesn't predict as well. As you'll see in Section 5.5.2, a random forest algorithm can encompass hundreds of different decision trees, resulting in a stronger and more accurate model.

However, decision trees are a great way to learn about tree-based models and other concepts such as hyperparameters and grid search. Taking the time to understand the content we covered in this section will make it easier for you to learn the complex models we cover in Section 5.5.2, Section 5.5.3, and Section 5.6.

Use Case 1: Decision Tree

Let's pick up where we left off with our first use case: performing linear regression in Section 5.4.3. As a refresher, we created a few additional columns to represent the month and day of week, but the linear regression model struggled to pick up on all the spikes in the data. Now, we'll use this same data to build the model. This is where you'll see the modeling process really accelerate. You can reuse the features and data you build here in each of your subsequent models. This emphasizes that setting up your data in the right way provides significant benefits for model development.

First, we'll run the decision tree regressor with no hyperparameter tuning to find our starting point (see Listing 5.48).

```
#read in decision tree libraries
from sklearn.tree import DecisionTreeRegressor
from sklearn.model_selection import train_test_split
from sklearn.metrics import mean_squared_error, r2_score

#create the decision tree model
regressor = DecisionTreeRegressor()

#fit the decision tree model to our training data
regressor.fit(X_train, y_train)

#use the model to predict on our test data
y_pred = regressor.predict(X_test)

#calculate MAE
mae = mean_absolute_error(y_test, y_pred)

#print MAE
print("Mean Absolute Error:", mae)

#sum the actuals and predictions for the test data
print(sum(y_test))
print(sum(y_pred))
```

Listing 5.48 Decision Tree with Default Parameters

The results are horrible. The MAE is 5,306, and our prediction on the test data is only 25% of the actual target variable for the test data. Let's implement a grid search to go through the data, as shown in Listing 5.49. We'll apply it using the approach we've learned, using the max_depth and min_samples_split hyperparameters to search over the data. We'll also specify our scoring hyperparameter when executing the grid search. We've been using MAE for this use case, so we tell the grid search process use the neg_mean_absolute_error option.

```
#read in library for grid search
from sklearn.tree import DecisionTreeRegressor
from sklearn.model_selection import GridSearchCV

#create decision tree model
regressor = DecisionTreeRegressor()
```

```
#establish the parameter grid to search over
param_grid = {
  'max_depth': [5, 10, 15],
  'min_samples_split': [5, 10, 20]
  }

#fun the grid search function
grid_search = GridSearchCV(
  estimator=regressor,
  param_grid=param_grid,
  cv=3,
  scoring='neg_mean_absolute_error'
  )

#fit the model to the best results of the grid search
grid_search.fit(X_train, y_train)

#use the model to predict on the test data
y_pred = grid_search.predict(X_test)

#print out the parameters of the best model
print("Best parameters:", grid_search.best_params_)

#print out the best score from the grid search
print("Best score:", grid_search.best_score_)

#calculate MAE
mae = mean_absolute_error(y_test, y_pred)

#print MAE
print("Mean Absolute Error:", mae)

#sum both the actuals and predictions of the test data
print(sum(y_test))
print(sum(y_pred))
```

Listing 5.49 First Iteration of Decision Tree Grid Search

Our results are better, but still not as good as the linear regression model. The MAE on the test data is 3,187 and the prediction of the test data is 51% of the actual sales for the test data (as a reminder, we were in the 2,000s for MAE with the linear regression model).

Before we move to another set of grid parameters, let's pause to compare the actual data and predictions for the test data using the code from Listing 5.50.We're using the same

approach as before to compare multiple lines on the same graph using the `plot` function. By passing both our test data (`y_test`) and prediction data (`y_pred`) as separate lines
with `plt.plot`, we can easily assess how well our model is (or isn't) doing. The results are
shown in Figure 5.14.

```
#read in plotting library
import matplotlib.pyplot as plt

#plot the actual data
plt.plot(test_data['since_first_sale'], y_test, label = "actual")

#plot the prediction data
plt.plot(test_data['since_first_sale'], y_pred, label = "prediction")

#show the legend
plt.legend()
```

Listing 5.50 Decision Tree After Grid Search

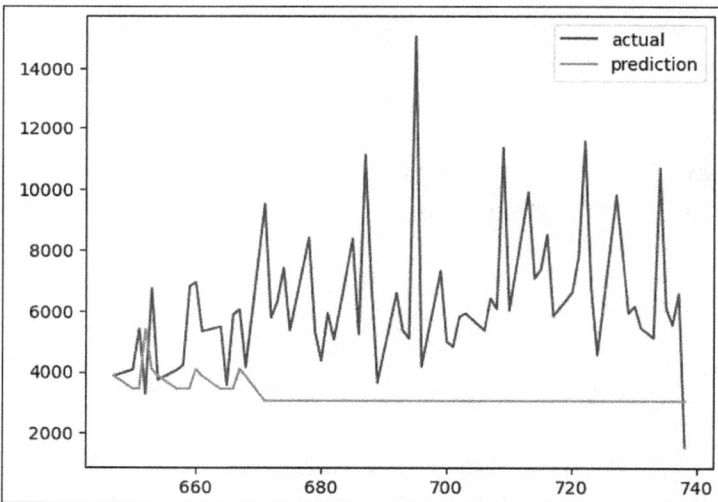

Figure 5.14 Decision Tree After First Grid Search

The flat line in Figure 5.14 tells us the prediction is the same across each of these days,
which isn't ideal, since we know there should be variation there. The next logical step is
to see whether we can give the model additional freedom to search the data by only
allowing it to test higher `max_depth` values and lower `min_samples_split` values, as shown
in Listing 5.51.

```
#updated parameter grid
param_grid = {
  'max_depth': [10, 15],
  'min_samples_split': [5, 10]
  }
```

Listing 5.51 New Parameter Grid for More Freedom in the Model

Unfortunately, this made it worse! The MAE went down to 4,200, and the prediction of the test data went down to 34%. This feels like it might be a losing battle at this point. Rather than keep trying to refine the decision tree model, we'll move on to the random forest model in Section 5.5.2 to see if it yields better results.

Use Case 2: Decision Tree

The logistic regression model struggled to yield meaningful results for the data in this use case. Now that we're using a tree-based model, our hope is that it will perform better. While a decision tree may not be the best model for this use case, it will help you understand how tree-based models work and how they become increasingly sophisticated.

When using logistic regression, we removed some columns to ensure we didn't overload the model. We won't need to do that with tree-based algorithms. First, let's get our data set up and start running it through the model (see Listing 5.52). Here, we've only dropped the columns that cannot go into the model because they're either our target variable columns (target_7 and target_14) or non-numeric columns (Customer ID, timestamp, and order_date).

```
#read in packages to prep data and calculate accuracy
from sklearn.model_selection import train_test_split
from sklearn.metrics import accuracy_score
from sklearn.metrics import roc_auc_score

#drop columns to define X
X = df.drop(
  columns=[
    'target_7',
    'target_14',
    'Customer ID',
    'timestamp',
    'order_date'
    ]
  )

#define y as the one week lag
y = df['target_7']
```

```
#split into train and test data
X_train, X_test, y_train, y_test = train_test_split(
  X,
  y,
  test_size=0.3,
  random_state=42
  )
```

Listing 5.52 Set Up Data for 7-Day Target with All Columns as Predictors

Now that we have our data in the necessary training and test split, we can run it through our decision tree model and see the results. Listing 5.53 shows the code for creating the model.

Classifier or Regressor?

Note that when we're using this model, it's called a classifier. This differs from the regressor model we discussed for the first use case. One of the best features of the sklearn environment is that the code and syntax are the same regardless of whether you use a classifier or regressor, which helps simplify the development process for these models. Personally, I've always found this helpful because I don't need to remember two distinct ways of modeling.

```
#read in decision tree classifier library
from sklearn.tree import DecisionTreeClassifier

#create model
dt_classifier = DecisionTreeClassifier(random_state=42)

#fit the model to the data
dt_classifier.fit(X_train, y_train)
```

Listing 5.53 Build Initial Decision Tree

Once we have our model, we can apply the same approach we used when testing logistic regression: comparing both the true accuracy metric and the ROC AUC metric (see Listing 5.54).

```
#create both the probability and binary predictions
y_pred_proba = dt_classifier.predict_proba(X_test)[:, 1]
y_pred = dt_classifier.predict(X_test)

#calculate both the ROC AUC and accuracy scores
auc_score = roc_auc_score(y_test, y_pred_proba)
accuracy = accuracy_score(y_test, y_pred)
```

```
#print both the ROC AUC and accuracy scores
print(accuracy)
print(auc_score)
```

Listing 5.54 Initial Decision Tree Testing Results

The results are unfortunately quite bad. The accuracy is 75%, and the ROC AUC is only 0.53. We'll discuss ROC AUC more in Chapter 6; however, for context, 0.5 means the model is adding no value.

We won't overengineer the decision tree, but we will make some adjustments to its inputs. First, let's try limiting how deep the model can go with the `max_depth` hyperparameter by keeping it at a modest value of 4 (see Listing 5.55).

```
#create the model with a max_depth of 4
dt_classifier = DecisionTreeClassifier(max_depth = 4, random_state=42)

#fit the model
dt_classifier.fit(X_train, y_train)

#calculate the binary and probability prediction
y_pred_proba = dt_classifier.predict_proba(X_test)[:, 1]
y_pred = dt_classifier.predict(X_test)

#calculate the ROC AUC and accuracy scores
auc_score = roc_auc_score(y_test, y_pred_proba)
accuracy = accuracy_score(y_test, y_pred)

#print the ROC AUC and accuracy scores
print(accuracy)
print(auc_score)
```

Listing 5.55 Limiting the Max Depth Hyperparameter

The result is 85% accuracy and 0.60 for the ROC AUC metric, so simply limiting how far into the data the model can go creates a notable improvement. This presents an interesting observation: When a decision tree isn't limited by how deep it can go, it ends up making back decisions and overfitting.

Let's pause to look at how the decision tree views the data and then visualize the tree, as shown in Listing 5.56. Figure 5.15 shows the results.

```
#import plotting libraries
import matplotlib.pyplot as plt
from sklearn import tree
```

5

```
#plot the decision tree
plt.figure(figsize=(20, 15))
_ = tree.plot_tree(dt_classifier,
          feature_names=list(X_train.columns),
          class_names=['No', 'Yes'],
          filled=True,
          rounded=True,
          fontsize=10)
plt.show()
```

Listing 5.56 Visualize the Decision Tree with Max Depth of 4

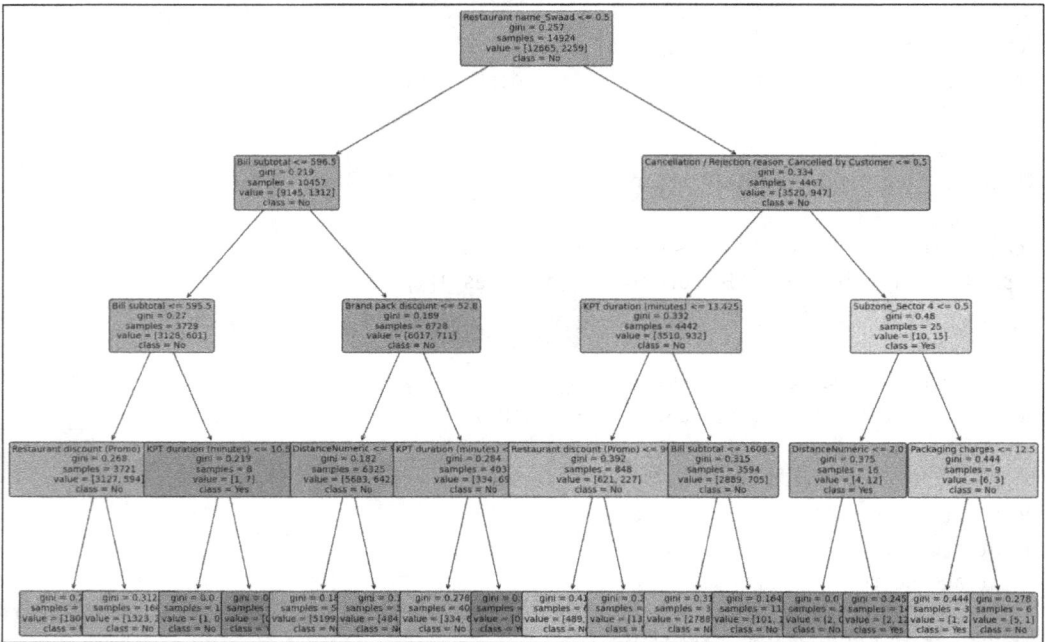

Figure 5.15 Decision Tree with Max Depth of 4

Blue boxes indicate predictions that the customer will place a repeat order within the next week. We'll discuss methods of evaluating the importance of variables in detail in Chapter 6, but this type of visualization helps you identify what the model is focusing on before the more complicated models make interpretation harder.

Now, what happens if we reduce max_depth down to 3 instead of 4? Take a look at Listing 5.57.

```
#create the model with a max_depth of 3
dt_classifier = DecisionTreeClassifier(max_depth = 3, random_state=42)
```

```
#fit the model to the data
dt_classifier.fit(X_train, y_train)

#calculate the binary and probability predictions
y_pred_proba = dt_classifier.predict_proba(X_test)[:, 1]
y_pred = dt_classifier.predict(X_test)

#calculate the ROC AUC and accuracy scores
auc_score = roc_auc_score(y_test, y_pred_proba)
accuracy = accuracy_score(y_test, y_pred)

#print the ROC AUC and accuracy scores
print(accuracy)
print(auc_score)
```

Listing 5.57 Reducing Max Depth to 3

We get essentially the same results: Accuracy is 85%, and the ROC AUC is 0.6. This is one of those situations where less is more. While the results are still not sufficient in practice, there's value in visualizing this simpler version of the decision tree to understand what the model is focusing on (see Figure 5.16).

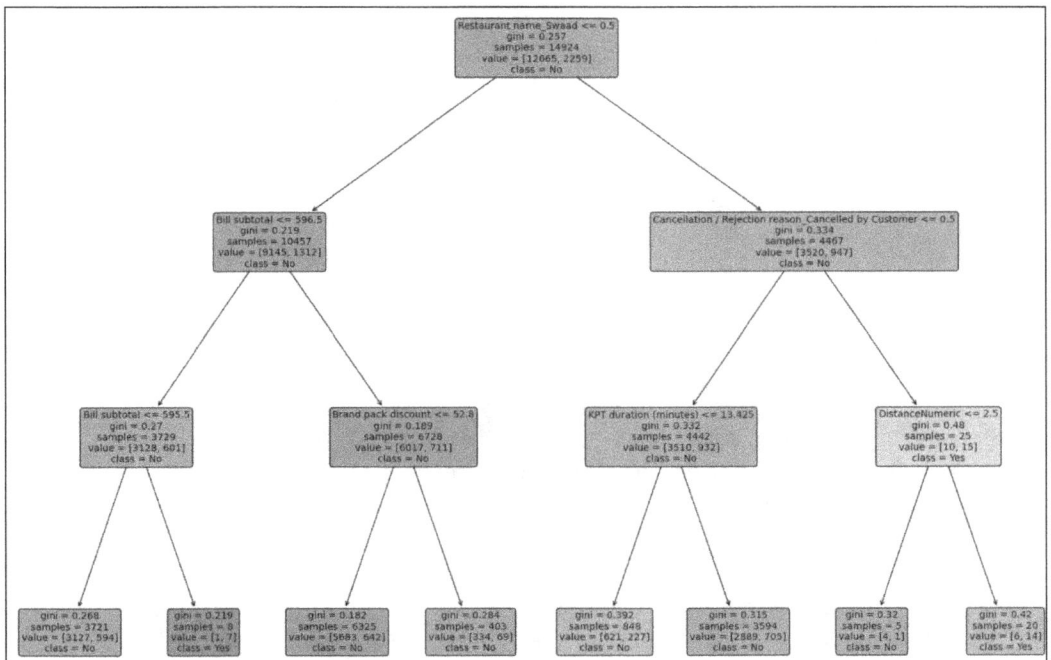

Figure 5.16 Decision Tree with Max Depth of 3

When you look at it from this digestible view, several interesting observations emerge, which also give insight into why decision trees are rarely used in practice:

- The model's first split is on a specific restaurant. While this can be useful, from a generalization perspective, it's unlikely that starting the entire model off from one restaurant will produce accurate predictions.
- Look at the left side of the first split and then follow it to the left side of the second split. The Bill subtotal column is used twice. While technically this provides the most information gain, it's unlikely that the difference between 596.5 and 595.5 will be a significant factor in determining whether someone will reorder within the next week.
- The right side of the first split seems to be more promising. The distance for the order is less than 2.5, which suggests individuals who are closer to the restaurant will be more likely to repeat their purchase within the next week.

All these can be valuable insights to keep in mind as you move into more complex tree-based models.

Before we move on, though, let's look at the 14-day target to see if it produces different results (see Listing 5.58).

```
#create the dataset for predictors for the model
X = df.drop(
  columns=[
    'target_7',
    'target_14',
    'Customer ID',
    'timestamp',
    'order_date'
    ]
  )

#create y as the 2 week target
y = df['target_14']

#execute the train test split
X_train, X_test, y_train, y_test = train_test_split(
  X,
  y,
  test_size=0.3,
  random_state=42
  )

#create model with max_depth of 3
dt_classifier = DecisionTreeClassifier(max_depth = 3, random_state=42)
```

```
#fit the model
dt_classifier.fit(X_train, y_train)

#calculate both the probability and binary predictions
y_pred_proba = dt_classifier.predict_proba(X_test)[:, 1]
y_pred = dt_classifier.predict(X_test)

#calculate the ROC AUC and accuracy scores
auc_score = roc_auc_score(y_test, y_pred_proba)
accuracy = accuracy_score(y_test, y_pred)

#print both the ROC AUC and accuracy scores
print(accuracy)
print(auc_score)
```

Listing 5.58 Build Model Using 14-Day Target

The results are similar to those of the logistic regression model. Let's see how the decision tree visualization compares to the 7-day target (see Listing 5.59 and Figure 5.17).

```
#plot the decision tree
plt.figure(figsize=(20, 15))
_ = tree.plot_tree(dt_classifier,
        feature_names=list(X_train.columns),
        class_names=['No', 'Yes'],
        filled=True,
        rounded=True,
        fontsize=10)
plt.show()
```

Listing 5.59 Create Decision Tree Visualization for the 14-Day Target Model

There are many similarities between this visualization and the 7-day decision tree visualization. The main exception appears to be that this decision tree doesn't consider the distance feature.

So, what's our takeaways about decision trees for this use case? Decision trees models in practice aren't great for generating accurate predictions. As we've discussed many times, the value of decision trees is more about understanding your data than being a practical model. With this use case, the next step will be exploring the random forest algorithm in Section 5.5.2 to see if we can add value with many decision trees working together.

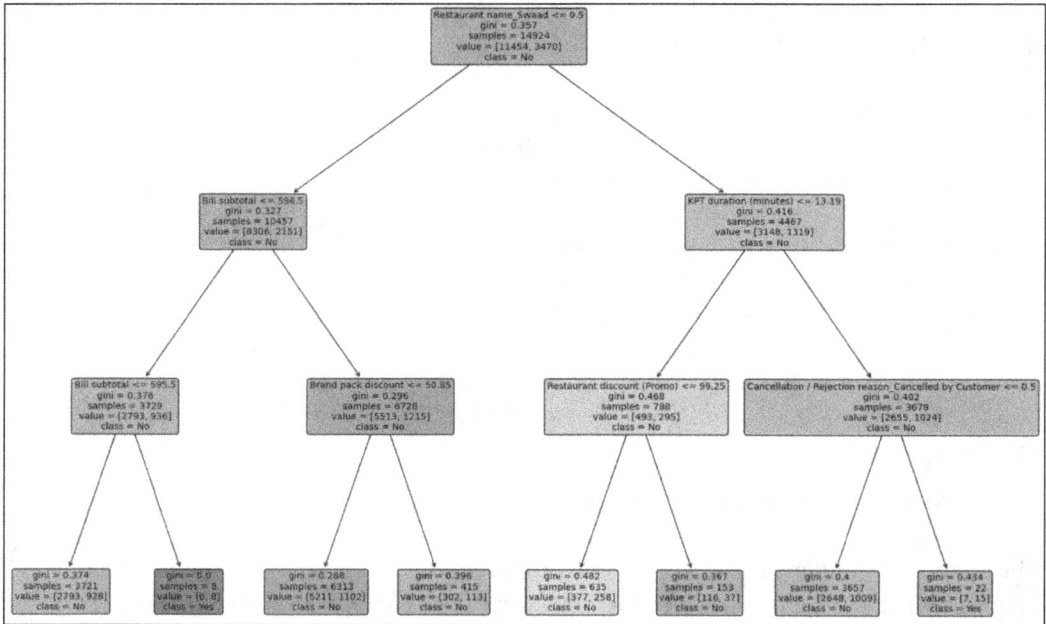

Figure 5.17 Decision Tree for 14-Day Target with Max Depth of 3

Use Case 3: Decision Tree

We'll follow a similar approach with the decision tree as we did for the first two use cases. It's ultimately not a practical model, as you've heard and observed so far, but it's foundational and provides insights into what features and splits may be important to consider in the future.

Let's start by running our nearly 50 columns through the model, as shown in Listing 5.60. We're starting from where we left off in the regression section, with our df Data-Frame. We won't leverage PCA moving forward because the number of columns is not a problem and we want to maintain an enhanced degree of interpretability with our model.

```
#read in libraries
from sklearn.tree import DecisionTreeRegressor
from sklearn.model_selection import train_test_split
from sklearn.metrics import mean_absolute_error

#split data into what will be used to predict and what is being predicted
X = df.drop(columns=['DATE', 'AREA NAME', 'crime_count'])
y = df['crime_count']

#perform standard train test split
X_train, X_test, y_train, y_test = train_test_split(
```

```
    X,
    y,
    test_size=0.2,
    random_state=42
    )

#create and fit the decision tree model
model = DecisionTreeRegressor(max_depth=3)
model.fit(X_train, y_train)

#predict with the modeling using the test data
y_pred = model.predict(X_test)

#calculate the mean absolute error
mae = mean_absolute_error(y_test, y_pred)

#calculate the mean absolute error divided by average of the target
mae_by_avg_target = mae / y_test.mean()

#print both results
print(mae)
print(mae_by_avg_target)
```

Listing 5.60 Initial Decision Tree Model

Our results improved a bit. If you recall from the linear regression model in Section 5.4.3, our MAE was 6.5, but with this model it's down to 6.1. The MAE divided by the average target is also down to 23% from 25%. The max_depth hyperparameter of 3 is relatively restrictive; however, it helps simplify the tree plot (see Listing 5.61 and Figure 5.18).

```
#read in plotting libraries
import matplotlib.pyplot as plt
from sklearn import tree

plt.figure(figsize=(20, 15))
_ = tree.plot_tree(model,
            feature_names=list(X_train.columns),
            class_names=['No', 'Yes'],
            filled=True,
            rounded=True,
            fontsize=10)
plt.show()
```

Listing 5.61 Create Tree Plot for Max Depth of 3 Decision Tree

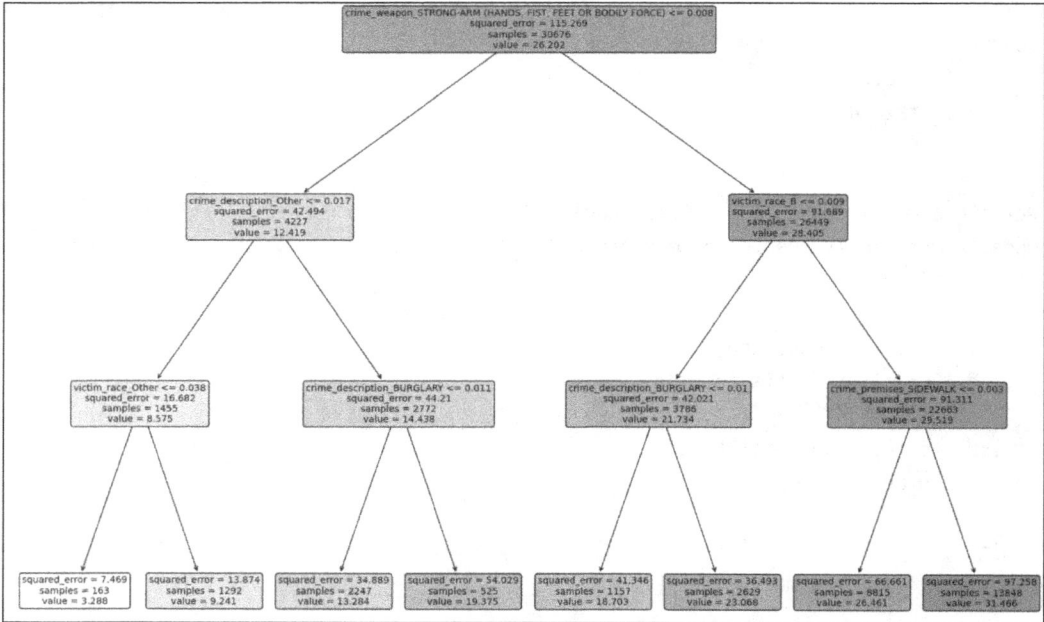

Figure 5.18 Tree Plot with Max Depth of 3

This is an interesting result. We have almost 50 columns in the model, but there are only a handful of groups of data, since most of the columns came from the dummy coding process. Many of these groups are represented in the data. The weapon, description, victim race, and premises are all represented at least once in this relatively simple decision tree. Even outside of our use cases, you'll often see a model (especially one like a decision tree) focus on one specific context of data. However, in this case, the model finds value in multiple pieces of information. If a simple model like a decision tree is already finding value in these various components, it suggests we can uncover additional value with more sophisticated models.

Next, let's increase the `max_depth` hyperparameter to 4, as shown in Listing 5.62, and see what happens.

```
#split data into what will be used to predict and what is being predicted
X = df.drop(columns=['DATE', 'AREA NAME', 'crime_count'])
y = df['crime_count']

#perform standard train test split
X_train, X_test, y_train, y_test = train_test_split(
  X,
  y,
  test_size=0.2,
  random_state=42
  )
```

```
#create and fit the decision tree model
model = DecisionTreeRegressor(max_depth=4)
model.fit(X_train, y_train)

#predict with the modeling using the test data
y_pred = model.predict(X_test)

#calculate the mean absolute error
mae = mean_absolute_error(y_test, y_pred)

#calculate the mean absolute error divided by average of the target
mae_by_avg_target = mae / y_test.mean()

#print both results
print(mae)
print(mae_by_avg_target)
```

Listing 5.62 Increase the Max Depth Hyperparameter to 4

The result is the MAE decreases to 5.7, and the MAE divided by the average target decreases to 22%. We're seeing continued progress by allowing the model to go deeper into the data. At this point, unfortunately, the tree plots become unreadable, so we'll skip right to a max_depth of 5 in Listing 5.63.

```
#split data into what will be used to predict and what is being predicted
X = df.drop(columns=['DATE', 'AREA NAME', 'crime_count'])
y = df['crime_count']

#perform standard train test split
X_train, X_test, y_train, y_test = train_test_split(
  X,
  y,
  test_size=0.2,
  random_state=42
  )

#create and fit the decision tree model
model = DecisionTreeRegressor(max_depth=5)
model.fit(X_train, y_train)

#predict with the modeling using the test data
y_pred = model.predict(X_test)

#calculate the mean absolute error
mae = mean_absolute_error(y_test, y_pred)
```

```
#calculate the mean absolute error divided by average of the target
mae_by_avg_target = mae / y_test.mean()

#print both results
print(mae)
print(mae_by_avg_target)
```

Listing 5.63 Increase the Max Depth to 5

We see yet another improvement! The MAE is down to 5.5, and the MAE divided by the average target decreases to 21%. Let's see what happens if we increase the max depth even further to 6, as shown in Listing 5.64.

```
#split data into what will be used to predict and what is being predicted
X = df.drop(columns=['DATE', 'AREA NAME', 'crime_count'])
y = df['crime_count']

#perform standard train test split
X_train, X_test, y_train, y_test = train_test_split(
  X,
  y,
  test_size=0.2,
  random_state=42
  )

#create and fit the decision tree model
model = DecisionTreeRegressor(max_depth=6)
model.fit(X_train, y_train)

#predict with the modeling using the test data
y_pred = model.predict(X_test)

#calculate the mean absolute error
mae = mean_absolute_error(y_test, y_pred)

#calculate the mean absolute error divided by average of the target
mae_by_avg_target = mae / y_test.mean()

#print both results
print(mae)
print(mae_by_avg_target)
```

Listing 5.64 Increase Max Depth to 6

We continue to see improvement! The MAE is down to 5.3, and the MAE divided by the average target is down to 20%. It would've been reasonable to expect that we'd hit a threshold where the model starts to overfit, resulting in decreased accuracy. In the spirit of experimentation, let's try a max_depth of 7 (see Listing 5.65).

```
#split data into what will be used to predict and what is being predicted
X = df.drop(columns=['DATE', 'AREA NAME', 'crime_count'])
y = df['crime_count']

#perform standard train test split
X_train, X_test, y_train, y_test = train_test_split(
  X,
  y,
  test_size=0.2,
  random_state=42
  )

#create and fit the decision tree model
model = DecisionTreeRegressor(max_depth=7)
model.fit(X_train, y_train)

#predict with the modeling using the test data
y_pred = model.predict(X_test)

#calculate the mean absolute error
mae = mean_absolute_error(y_test, y_pred)

#calculate the mean absolute error divided by average of the target
mae_by_avg_target = mae / y_test.mean()

#print both results
print(mae)
print(mae_by_avg_target)
```

Listing 5.65 Increase Max Depth to 7

Somehow, this is still improving performance! The MAE is down to 5, and the MAE divided by the average target is now 19%. You probably have an idea of how this iterative process works by now, so instead of walking through each incremental addition of max_depth, let's skip to our findings.

We'll continue to see improvements in performance until a max_depth of 16. The MAE went all the way down to 2.9, while the MAE divided by the average target went down to 11%. Is this a shocking finding, or should it be a warning sign? It's highly unusual for a tree-based model to be this deep and not overfit. Findings like this should raise a flag in your mind to be wary of what you're seeing. Although each dataset is different, if your results are far from what you expected, there's often an error or issue behind them.

5.5.2   Random Forest

The next algorithm we'll discuss is the random forest, which builds upon the decision tree algorithm. Here's a good analogy for thinking about decision trees compared to random forests: A decision tree is like trusting a single expert, while a random forest is like trusting a collection of experts. As with any topic, consulting a single expert puts you at risk of a narrow view. By asking multiple experts, you've removed the single point of failure and can feel more confident in your results. You'll often see this theory summarized as "multiple weak models are better than a single strong model." This concept is called *ensembling*.

In general, you'll see a noticeable increase in the quality of your model when you go from a decision tree to a random forest. Random forest algorithms do a better job at handling overfitting because they have multiple decision trees that calculate independently. The accuracy of the model tends to be higher because its multiple models work together.

You may be wondering exactly *how* the random forest creates decision trees. If a random forest algorithm builds multiple decision trees on the same data, why aren't they all the same? The answer resides in sampling and randomization.

For each decision tree in your random forest model, the algorithm takes a bootstrapped sample of the data. Sampling with replacement may seem odd, but the randomization and multiple decision trees help create diversity while keeping the data's distribution similar to the original dataset.

Bootstrapping

Randomization and sampling are powerful statistical tools that can help you build more effective models. If you're not familiar with *bootstrapping*, it's a fancy term for sampling with replacement. In other words, after a value is randomly selected, it's added back into the data so that it can be chosen again. Sampling with replacement is the secret sauce that lets each decision tree have a different underlying dataset.

Random forest algorithms also randomly select the columns that each decision tree can use. This further ensures diversity in each of your trees, as each one has a different subset of columns it can pick from. This also ties directly back to the theory that multiple weak models are better than a single strong one. By only allowing a model to consider a subset of columns, we're making it weak by not giving it all the context we have available. The benefit to this is that we're able to create decision trees that find patterns others couldn't before. For example, what if we have three columns that stand out as the most correlated with our prediction? Those three columns are the most likely to keep showing up in our single decision trees. However, if we create models in our random forest where only one or two of those columns are available, our decision trees will consider other columns it hadn't previously.

This approach of randomly selecting a subset of columns to build a model from ties back to feature engineering, which we discussed in Section 5.5.1. When randomly selecting columns to build each decision tree, doing more feature engineering up front can lead to better results. By doing the feature engineering yourself, you're grouping information in these columns so that each subset of your models can leverage those relationships.

Now, let's dive into working with the random forest algorithm and see it in action for our three use cases.

Random Forest Modeling

Listing 5.66 shows a random forest algorithm in practice, using the RandomForestClassifier() function. Apart from the specific name of the model, the primary difference is that we now need to input how many decision trees we want the model to build via the n_estimators hyperparameter (100 in this example).

```
from sklearn.ensemble import RandomForestClassifier
from sklearn.model_selection import train_test_split
from sklearn.metrics import accuracy_score
from sklearn.datasets import load_digits

digits = load_digits()

X_train, X_test, y_train, y_test = train_test_split(
    digits.data,
    digits.target,
    test_size=0.2,
    random_state=42
    )

model = RandomForestClassifier(n_estimators=100, random_state=42)

model.fit(X_train, y_train)

y_pred = model.predict(X_test)

y_train_pred = model.predict(X_train)
accuracy = accuracy_score(y_train, y_train_pred)
print(f" Training Accuracy: {accuracy}")
```

```
accuracy = accuracy_score(y_test, y_pred)
print(f"Accuracy: {accuracy}")
```

Listing 5.66 Random Forest Code Example

The result of this is interesting, especially if you compare it to almost 85% accuracy from the decision tree model. On the training data, we get 100% accuracy. This is a clear sign of overfitting, but it shouldn't come as a surprise since we're not limiting the model's `max_depth` nor `min_samples_split` hyperparameters. What's surprising, though, is that our test data accuracy is 97%. Yes, we're overfitting, but not by much, and our accuracy is more than 10 percentage points higher with the random forest than with the decision tree. A hundred weaker models appear to be better than a single strong one!

Now, let's start playing around with the hyperparameters by limiting how deep the model can go. In practice, it's unlikely you'll get to this level of accuracy, but for the sake of the example we're going to see if we can reduce the overfitting. First, let's try limiting `max_depth` and `min_samples_split` to 7 and 20, respectively, in Listing 5.67.

Before you get critical and ask why we're not using grid search for this, know that we're intentionally excluding it to avoid complicating these examples. The focus here is on foundational learning (which is the goal of this book).

```
from sklearn.ensemble import RandomForestClassifier
from sklearn.model_selection import train_test_split
from sklearn.metrics import accuracy_score
from sklearn.datasets import load_digits

digits = load_digits()

X_train, X_test, y_train, y_test = train_test_split(
  digits.data,
  digits.target,
  test_size=0.2,
  random_state=42
  )

model = RandomForestClassifier(
  n_estimators=100,
  m  ax_depth = 7,
  min_samples_split = 20,
  random_state=42
  )

model.fit(X_train, y_train)
```

```
y_pred = model.predict(X_test)

y_train_pred = model.predict(X_train)
accuracy = accuracy_score(y_train, y_train_pred)
print(f" Training Accuracy: {accuracy}")

accuracy = accuracy_score(y_test, y_pred)
print(f"Accuracy: {accuracy}")
```

Listing 5.67 Random Forest with Hyperparameter Restrictions

Our result shows a similar amount of overfitting, but with lower accuracy. The training data has an accuracy of 98%, and the test data has an accuracy of 95.5%. What do you think is the next logical step?

It makes sense to increase both the max_depth and min_samples_split hyperparameters, as shown in Listing 5.68. By letting the model go deeper, we'll hopefully enable it to uncover more of the relationships it found when it wasn't limited. We're also counteracting this depth by making sure the results have a larger sample size to execute the split.

```
from sklearn.ensemble import RandomForestClassifier
from sklearn.model_selection import train_test_split
from sklearn.metrics import accuracy_score
from sklearn.datasets import load_digits

digits = load_digits()

X_train, X_test, y_train, y_test = train_test_split(
    digits.data,
    digits.target,
    test_size=0.2,
    random_state=42
    )

model = RandomForestClassifier(
    n_estimators=100,
    max_depth = 10,
    min_samples_split = 40,
    random_state=42
    )

model.fit(X_train, y_train)

y_pred = model.predict(X_test)
```

```
y_train_pred = model.predict(X_train)
accuracy = accuracy_score(y_train, y_train_pred)
print(f" Training Accuracy: {accuracy}")

accuracy = accuracy_score(y_test, y_pred)
print(f"Accuracy: {accuracy}")
```

Listing 5.68 Random Forest with Wide Hyperparameters

Our results are even worse! The accuracy on our training data is 98%, but it decreased to 95% on our test data. It looks like we might be stuck—now is the time to do a grid search to see if there's a better alternative for our hyperparameters. As shown in Listing 5.69, the code is similar to that of our decision tree, with the main change being that we use RandomForestClassifier as our model function. The grid search options don't differ at this point, but they can and will in future iterations of this code.

```
from sklearn.model_selection import GridSearchCV
from sklearn.datasets import load_digits
from sklearn.model_selection import train_test_split
from sklearn.ensemble import RandomForestClassifier
from sklearn.metrics import accuracy_score

digits = load_digits()

X_train, X_test, y_train, y_test = train_test_split(
  digits.data,
  digits.target,
  test_size=0.2,
  random_state=42
  )

param_grid = {
  'max_depth': [5, 10, 15],
  'min_samples_split': [5, 10, 20]
  }

rfc = RandomForestClassifier(random_state=42)

grid_search = GridSearchCV(
  estimator=rfc,
  param_grid=param_grid,
  cv=3,
  scoring='accuracy'
  )
```

```
grid_search.fit(X_train, y_train)

print("Best parameters:", grid_search.best_params_)

print("Best score:", grid_search.best_score_)

best_random_forest = grid_search.best_estimator_

accuracy = best_random_forest.score(X_test, y_test)
print("Test accuracy:", accuracy)
```

Listing 5.69 Grid Search for Random Forest

The best model parameters came back with a max_depth of 15 and a min_samples_split of 5 (the combination with the most freedom for the model). The training accuracy for this combination was 96.5%, and the test accuracy was 98%. It's still a similar result, but we did start to address the overfitting problem!

Let's try one final combination using the grid shown in Listing 5.70. We're enabling the model to go much deeper with our high max_depth values. We also lower the restriction on how many records it needs to create a split with the lower min_samples_split options.

```
param_grid = {
  'max_depth': [15, 20, 25],
  'min_samples_split': [4, 5, 6]
  }
```

Listing 5.70 Grid Search Options for Random Forest

The grid search returned the best parameters as a max_depth of 15 and a min_samples_split of 4. The accuracy of the best parameters is 97% on the training data, which is the same (minus the rounding difference) on the test data. We've essentially eliminated the overfitting on our model! What's interesting about this is that you'd expect more overfitting, not less, with a smaller value in min_samples_split. Generally, the fewer records you let a model split on, the more likely the model is to be specific to the training data—but that isn't the case here. This is why testing your model and iterating on it is important!

Another lesson to take from this is the importance of random_state when testing your model. When we're testing different iterations, the randomness of the sample and column selection could cause differences in our results (especially when we're talking about a few percentage points, like this example). Since we've specified the random_state and kept it the same, we can feel confident that our changes to the hyperparameters are driving those differences instead of the random forest algorithm's randomized nature.

Hopefully, this small example highlighted an important point about the practical aspect of building models. If you felt like the example was splitting hairs, you wouldn't be wrong. The overall accuracy of the model hardly changed from the first iteration to the last. However, it's always important to test your model and not simply go with the first result. How much checking you should do depends on your role and the use case. If you're a dedicated data scientist working on a model that is critical to your business, you should spend a lot of time testing your model's hyperparameters. If you're a data analyst working on a model as a side project, you likely won't be doing as much iterating.

You've now seen how random forest models work and how their algorithms aren't much more complicated than a decision tree. Random forests are often the true entry-level algorithm in practice, so congratulations on reaching this threshold of under-standing real-world machine learning!

Use Case 1: Random Forest

As we did with the decision tree, we're going to build models from the same dataset we created for linear regression in Section 5.4.3. Let's first run the model using the default random forest hyperparameters to create a starting point, as shown in Listing 5.71.

```
#load in packages
from sklearn.ensemble import RandomForestRegressor
from sklearn.metrics import mean_absolute_error

#create the model
model = RandomForestRegressor(n_estimators=100, random_state=42)

#fit the model to the data
model.fit(X_train, y_train)

#use the model to create predictions
y_pred = model.predict(X_test)

#use the model to predict on the training data
y_train_pred = model.predict(X_train)

#calculate the accuracy on the training data
accuracy = mean_absolute_error(y_train, y_train_pred)

#print the accuracy of the training data
print(f" Training Accuracy: {accuracy}")

#calculate the accuracy on the test data
accuracy = mean_absolute_error(y_test, y_pred)
```

```
#print the accuracy on the test data
print(f"Accuracy: {accuracy}")
```

Listing 5.71 Random Forest with Default Hyperparameters

Our training data shows a great MAE! We're down to 405. However, if we check on the test data, it's over 4,000, so we're definitely overfitting. Let's build a graph that compares the actual test results to the predictions for the test data (see Listing 5.72 and Figure 5.19).

```
#load plotting library
import matplotlib.pyplot as plt

#plot both the actual and prediction data
plt.plot(test_data['since_first_sale'], y_test, label = "actual")
plt.plot(test_data['since_first_sale'], y_pred, label = "prediction")
plt.legend()
```

Listing 5.72 Random Forest Predictions Versus Actuals with Default Hyperparameters

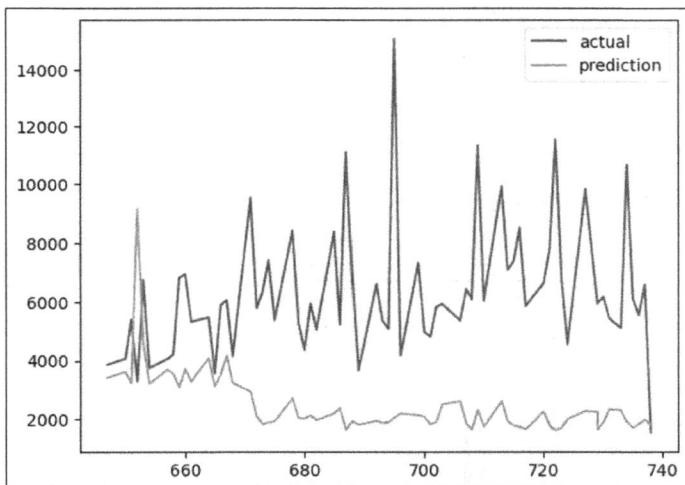

Figure 5.19 Random Forest with Default Hyperparameters

The model continues to underpredict the values in the test dataset and appears to get worse as time progresses. Let's see if adjusting the hyperparameters makes a difference by allowing it to go deeper with a higher max_depth of 10, as shown in Listing 5.73. Figure 5.20 shows the results.

```
#load in the necessary packages
from sklearn.ensemble import RandomForestRegressor
from sklearn.metrics import mean_absolute_error
```

```
#create the model with additional hyperparameters
model = RandomForestRegressor(
  n_estimators=300,
  max_depth = 10,
  min_samples_split = 15,
  random_state=42
  )

#fit the model to the data
model.fit(X_train, y_train)

#create prediction on test data
y_pred = model.predict(X_test)

#create prediction on the training data
y_train_pred = model.predict(X_train)

#calculate and print MAE on the training data
accuracy = mean_absolute_error(y_train, y_train_pred)
print(f" Training Accuracy: {accuracy}")

#calculate and print MAE on the test data
accuracy = mean_absolute_error(y_test, y_pred)
print(f"Accuracy: {accuracy}")
```

Listing 5.73 Random Forest with More Hyperparameters

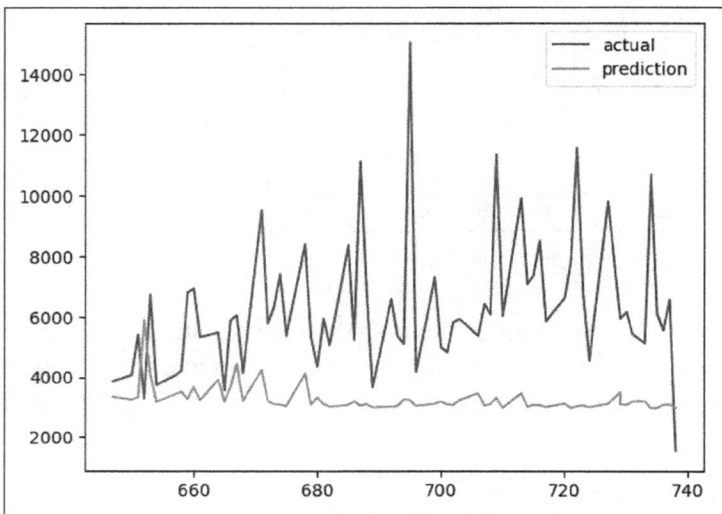

Figure 5.20 Random Forest with More Hyperparameters

The resulting MAE shows less overfitting at 3,165, but that's still not close to where our regression model was. The graph shows the model picks up some of the variance, but it still isn't accounting for the higher sales values, resulting in notable underpredictions.

What's our takeaway from this use case so far? Sometimes, the more complicated model isn't always the model you should use! That's why regression has its place. We'll continue to progress this use case in Section 5.5.3 to see if the GBM model can better account for these trends in the data.

Use Case 2: Random Forest

Back to our food delivery company use case! I'm excited to see what value a random forest algorithm can add to the modeling process for this data. From what we've seen with the decision tree, it's clear the data is nonlinear, so the ensembling approach of a random forest algorithm should be a good test to see whether our dataset can support a model as it's currently constructed.

As always, our first step is to set up our data into x and y, and then run it through so it splits into train and test sets, as shown in Listing 5.74.

```
#load in the train test split library
from sklearn.model_selection import train_test_split

#create the predictors for the model
X = df.drop(
  columns=[
    'target_7',
    'target_14',
    'Customer ID',
    'timestamp',
    'order_date'
    ]
  )

#use the 1 week variable lag as the target
y = df['target_7']

#execute train test split
X_train, X_test, y_train, y_test = train_test_split(
  X,
  y,
  test_size=0.3,
  random_state=42
  )
```

Listing 5.74 Train-Test Split for Random Forest

Next, we'll run the model with the default inputs to see what our accuracy and ROC AUC results yield, as shown in Listing 5.75.

```
#load in libraries for the modeling process
from sklearn.ensemble import RandomForestClassifier
from sklearn.metrics import accuracy_score
from sklearn.metrics import roc_auc_score

#create the random forest model
model = RandomForestClassifier(random_state=42)

#fit the model to the data
model.fit(X_train, y_train)

#create both the binary and probability predictions
y_pred_proba = model.predict_proba(X_test)[:, 1]
y_pred = model.predict(X_test)

#calculate both the ROC AUC and accuracy scores
auc_score = roc_auc_score(y_test, y_pred_proba)
accuracy = accuracy_score(y_test, y_pred)

#print both the ROC AUC and accuracy scores
print(auc_score)
print(accuracy)
```

Listing 5.75 Random Forest with Default Hyperparameters

For using default hyperparameters, our results are actually quite promising. The model has 85% accuracy and a 0.61 ROC AUC score. This means we should further explore our hyperparameter options. Before moving to a grid search, we'll try a few options ourselves to get a feel for what performs well in the model.

To start, we'll use more restrictive hyperparameter options, enforcing a max_depth of 3 and a min_samples_split of 20, as shown in Listing 5.76.

```
#create the model with specific hyperparameter
model = RandomForestClassifier(
  max_depth = 3,
  min_samples_split = 20,
  random_state=42
  )

#fit the model to the data
model.fit(X_train, y_train)
```

```
#create both the binary and probability prediction
y_pred_proba = model.predict_proba(X_test)[:, 1]
y_pred = model.predict(X_test)

#calculate both the ROC AUC and accuracy score
auc_score = roc_auc_score(y_test, y_pred_proba)
accuracy = accuracy_score(y_test, y_pred)

#print both the ROC AUC and accuracy score
print(auc_score)
print(accuracy)
```

Listing 5.76 Random Forest with Restrictive Hyperparameters

The resulting scores are 85% again for accuracy and 0.60 for ROC AUC. This is essentially the same, so it tells us that less restrictive hyperparameters are worth trying next. We'll bump up the max_depth to 5 and bump down the min_samples_split to 10 (see Listing 5.77).

```
#adjust the hyperparameters to let the model go deeper into the data
model = RandomForestClassifier(
  max_depth = 5,
  min_samples_split = 10,
  random_state=42
  )

#fit the model to the data
model.fit(X_train, y_train)

#calculate both the binary and probability predictions
y_pred_proba = model.predict_proba(X_test)[:, 1]
y_pred = model.predict(X_test)

#calculate both the ROC AUC and accuracy scores
auc_score = roc_auc_score(y_test, y_pred_proba)
accuracy = accuracy_score(y_test, y_pred)

#print both the ROC AUC and accuracy scores
print(auc_score)
print(accuracy)
```

Listing 5.77 Random Forest with Less Restrictive Hyperparameters

This results in the same 85% accuracy and 0.61 ROC AUC that we had with the default hyperparameters. Before we apply grid search, let's try using the same hyperparameters for `max_depth` and `min_samples_split` but increasing the `n_estimators` to 500. Listing 5.78 shows the code for this.

```
#adjust the hyperparameters
model = RandomForestClassifier(
  n_estimators = 500,
  max_depth = 5,
  min_samples_split = 10,
  random_state=42
  )

#fit the model to the data
model.fit(X_train, y_train)

#calculate both the probability and binary predictions
y_pred_proba = model.predict_proba(X_test)[:, 1]
y_pred = model.predict(X_test)

#calculate both the ROC AUC and accuracy scores
auc_score = roc_auc_score(y_test, y_pred_proba)
accuracy = accuracy_score(y_test, y_pred)

#print both the ROC AUC and accuracy scores
print(auc_score)
print(accuracy)
```

Listing 5.78 Random Forest with Less Restrictive Hyperparameters and 500 Trees

The results are still the same, with 85% accuracy and a 0.61 ROC AUC.

Let's take a quick pause. When you try out these hyperparameters and see such consistent results, you should start considering the possibility that your existing dataset may not support a usable model. In Chapter 6, we'll discuss additional steps you can take from here; however, you'll naturally develop this kind of intuition as you build more models.

Now we'll execute the grid search, using a modest grid of hyperparameters to see what model results we can yield (see Listing 5.79). In our previous examples, we only used the `max_depth` and `min_samples_split` options; we're now including the `n_estimators` as well. Our hyperparameter options take a relatively wide approach, and each option has a varying range of values. This gives us a starting point to understand which values of each hyperparameter may be most favorable for our data.

```
#load in the grid search library
from sklearn.model_selection import GridSearchCV
```

```
#create a paramater grid to search over
param_grid = {
  'n_estimators': [100, 300, 500],
  'max_depth': [None, 5, 10],
  'min_samples_split': [2, 5, 10]
  }

#create the model
model = RandomForestClassifier(random_state=42)

#execute the grid search
grid_search = GridSearchCV(
  estimator=model,
  param_grid=param_grid,
  cv=5,
  scoring='roc_auc',
  n_jobs=-1,
  verbose=1
  )

#fit the model to the best parameters
grid_search.fit(X_train, y_train)

#identify the best model's parameters and print them
best_model = grid_search.best_estimator_
print("Best Parameters:", grid_search.best_params_)
```

Listing 5.79 Random Forest Grid Search

You'll notice this code takes some time to run, which is expected. A nice feature of sklearn's grid search is that it will tell you how many combinations it's running, which can give you an idea of how long the code will take to run. These inputs have 135 combinations: 3 options for each hyperparameter and 5 cross folds ($3 \times 3 \times 3 \times 5 = 135$).

Our grid search says we should increase our max_depth to 10, reduce our min_samples_split to 5, and keep our n_estimator at 500. This is very telling, as the results encourage us to be even less restrictive. The model wants to go deeper and be less restricted on how many records it needs to perform another split. Let's try out these hyperparameters in the code, as shown in Listing 5.80.

```
#use the best parameters from the grid search to create model
model = RandomForestClassifier(
  n_estimators = 500,
  max_depth = 10,
  min_samples_split = 5,
  random_state=42
  )
```

265

```
#fit the model to the data
model.fit(X_train, y_train)

#calculate both the probability and binary predictions
y_pred_proba = model.predict_proba(X_test)[:, 1]
y_pred = model.predict(X_test)

#create both the ROC AUC and accuracy scores
auc_score = roc_auc_score(y_test, y_pred_proba)
accuracy = accuracy_score(y_test, y_pred)

#print both the ROC AUC and accuracy scores
print(auc_score)
print(accuracy)
```

Listing 5.80 Random Forest Best Hyperparameters

The resulting code gives us the same 85% accuracy, but the ROC AUC jumped up to 0.63! While this is still not a good model, it's shown improvement.

Your takeaway from the exploration of the random forest algorithm on this data should be how much the data can influence the results. The model kept telling us to let it dive deep into the data to find the trends. Even when we did so, it didn't create a usable model. However, we're moving in the right direction.

Before moving on, we'll run the 14-day target data through the model with the same hyperparameters to see how the results look (see Listing 5.81).

```
#create predictors for the model
X = df.drop(
  columns=[
    'target_7',
    'target_14',
    'Customer ID',
    'timestamp',
    'order_date'
    ]
  )

#use the 2 week lag as the target variable
y = df['target_14']

#execute the train test split
X_train, X_test, y_train, y_test = train_test_split(
  X,
  y,
```

```
  test_size=0.3,
  random_state=42
  )

#create the model
model = RandomForestClassifier(
  n_estimators = 500,
  max_depth = 10,
  min_samples_split = 5,
  random_state=42
  )

#fit the model to the data
model.fit(X_train, y_train)

#calculate both the probability and binary predictions
y_pred_proba = model.predict_proba(X_test)[:, 1]
y_pred = model.predict(X_test)

#calculate both the ROC AUC and accuracy scores
auc_score = roc_auc_score(y_test, y_pred_proba)
accuracy = accuracy_score(y_test, y_pred)

#print both the ROC AUC and accuracy scores
print(auc_score)
print(accuracy)
```

Listing 5.81 Random Forest for 14-Day Target

Unfortunately, the results don't look great. The accuracy is 76%, with the ROC AUC coming in at 0.60. We won't dig into the 14-day target or perform a grid search on it at this time. The goal was to check in to see how that target data performance compares to that of our 7-day target.

We'll pick back up with this use case during our GBM discussion in Section 5.5.3. That's where we'll determine whether we'll need additional feature engineering to generate a model that can be used in practice.

Use Case 3: Random Forest

We'll start where we left off with this use case in Section 5.5.1. We were a bit skeptical, but the decision tree got down to a max_depth of 17 before its performance stopped improving. As a first step, let's start by using the max_depth of 17 in the random forest algorithm to see what our results look like, as shown in Listing 5.82.

```
#load in libraries necessary for the modeling process
from sklearn.ensemble import RandomForestRegressor
from sklearn.model_selection import train_test_split
from sklearn.metrics import mean_absolute_error

#split data into what will be used to predict and what is being predicted
X = df.drop(columns=['DATE', 'AREA NAME', 'crime_count'])
y = df['crime_count']

#perform standard train test split
X_train, X_test, y_train, y_test = train_test_split(
    X,
    y,
    test_size=0.2,
    random_state=42
    )

#create and fit the random forest model
model = RandomForestRegressor(max_depth=17)
model.fit(X_train, y_train)

#predict with the modeling using the test data
y_pred = model.predict(X_test)

#calculate the mean absolute error
mae = mean_absolute_error(y_test, y_pred)

#calculate the mean absolute error divided by average of the target
mae_by_avg_target = mae / y_test.mean()

#print both results
print(mae)
print(mae_by_avg_target)
```

Listing 5.82 Random Forest Using a Max Depth of 17

The result is an MAE of 2.3 and an MAE divided by the average target of 9%. This improves upon the results of the decision tree model. At this point, however, you should have alarm bells going off in your head.

Take the step back and think about the use case. Is crime really *this* predictable? While LA is a large city, we've broken it up into roughly 20 areas by day, so something seems off. How is it that we're seemingly able to feed more and more data into the model and it continues to perform well, especially in the context of our use case? How is it that we can build such a simple model that predicts future crime so accurately? Here's a hint: are we *actually* predicting future crime?

We're not! We've forgotten an important step in the context of building a *predictive* model. Our models so far are more descriptive rather than predictive. This is because the information about the given day is on the same line as what we're trying to predict (the number of crimes). We need to introduce lagging, as we did with our second use case, and the delivery data for our target variable. This will allow the model to predict the number of crimes on a given day using information from the last day, week, or other interval we decide based on the use case.

Trust Your Instincts

What we just went through with the decision tree and linear regression model wasn't throwaway work. There were many times in my own work where I started down a path and then got the feeling that I'm missing something. When I was new to building models, I would suppress these concerns and just keep moving along, which often resulted in even more rework. As you build models, learn to trust that gut instinct when you start to feel like you might be missing something. I use two main tactics to approach this:

1. The first is to take a step back from the project and work on something else. This helps me come back to the project with a clear mind that often helps me see what I was missing.

2. The second is to get a literal fresh perspective. I'll review what I've done with someone else. It can be challenging to find the right person—someone who understands the business case and has sufficient technical expertise—but it's one of the most efficient ways to reset your view of the project.

One more point before we move on: Consider how much more rework we might have created if we kept going down the modeling path. We could've finished the model and presented it to the stakeholder, only to realize the flaw in how it was built. It's almost always better to identify these situations early in the process and correct them as soon as possible.

We'll now take a quick detour to lag our data. During this process, we learn from our stakeholder that scheduling occurs a week ahead of time. This tells us that we need to move the target variable up by 7 days to replicate a scenario in which a prediction must be made a week in advance.

If your thought process is shaped by SQL, you might automatically think you need to use a join for this. This isn't purely a shifting of the data; we have to ensure the data is shifted within the same area name. Your mind might also go to a for loop and iterating through each area name. That also isn't the best approach.

In Python, you can use the `groupby` and `shift` functionality together. The `groupby` function will ensure the shift occurs independently for each value in the column you select.

As you'll see in Listing 5.83, the steps for this are actually quite simple, especially if you compare this to join and for loop implementations. We've strung together the groupby function on AREA NAME, and then used the shift function to move crime_count back 7 rows. Lastly, we filter out the values that can't be shifted (because they're at the end of the dataset) using the notnull function.

```
#sort the values to ensure it's in ascending order
df = df.sort_values(by=['DATE', 'AREA NAME'])

#shift 7 days for each area name independently
df['crime_count_shifted'] = df.groupby('AREA NAME')['crime_count'].shift(-7)

#filter to remove what are now null values for the latest dates of the dataset
df = df[df['crime_count_shifted'].notnull()]
```

Listing 5.83 Shift the Target Variable by 7 Days

Now let's see what happens to the model results when we try to predict the crime for a specific area 7 days into the future (see Listing 5.84). The code is essentially the same, but now we need to switch out our target variable (set y to be the crime_count_shifted column) *and* remove the new target variable from the set of predictors (add crime_count_shifted to the list of columns in drop). I'll admit that when I wrote the code for this book, I initially forgot to do this and almost fell out of my chair reviewing the initial model results!

```
#split data into what will be used to predict and what is being predicted
X = df.drop(
    columns=['DATE', 'AREA NAME', 'crime_count', 'crime_count_shifted'])
y = df['crime_count_shifted']

#perform standard train test split
X_train, X_test, y_train, y_test = train_test_split(
    X,
    y,
    test_size=0.2,
    random_state=42
    )

#create and fit the random forest model
model = RandomForestRegressor(max_depth=17)
model.fit(X_train, y_train)

#predict with the modeling using the test data
y_pred = model.predict(X_test)
```

```
#calculate the mean absolute error
mae = mean_absolute_error(y_test, y_pred)

#calculate the mean absolute error divided by average of the target
mae_by_avg_target = mae / y_test.mean()

#print both results
print(mae)
print(mae_by_avg_target)
```

Listing 5.84 Predict Adjusting the Target Variable to Be 7 Days in the Future

We see an expected decrease in the metrics with this. We're back up to an MAE of 5.8, which is 22% of the average target value. These results better fit the expected behavior of a model.

In my time building models, I've never had a tree go beyond a `max_depth` of 10. That many layers in the model usually leads to overfitting. I could get down to a `max_depth` of 7 with this new target variable while getting essentially the same results.

We'll pick back up with this use case when we discuss GBM models in the next section, where we'll use the shifted target variable.

5.5.3   Gradient Boosting Machine

The last tree-based algorithm we'll discuss is the GBM. Just like a decision tree is the basis of a random forest algorithm, the underlying principles of GBM are the same as those of a random forest. However, GBM has an added layer of complexity that makes its predictions more accurate.

Like a random forest, a GBM consist of multiple models. However, each model is built in the context of errors made by the last one. In other words, each new model learns from the mistakes of the previous weaker model.

GBM Analogy

This concept can be a bit abstract, so here's an example using software development. Let's say you're an entrepreneur building a new app that you think has the potential to change the world. It's a productivity app that consolidates all your phone's apps—calendars, email, AI, etc.—into one easy-to-use interface. The idea is fantastic, as many people complain about having too many separate apps that don't talk to each other.

You get your first round of venture capital funding and need to start building the app, which also means prioritizing which integrations to start with. You focus on Microsoft and Apple apps for your beta version. You build this and launch it, and then you receive feedback that users want integrations with Google's workspace apps next. So, you build those and launch them, but now your users want integrations with their home security systems, like Ring and SimpliSafe. You build these and launch them, and now users want integrations with smart thermostats like Nest, Ecobee, and Honeywell, which you then build and launch. This process continues as you develop your product.

This iterative cycle of learning from your last version is how GBMs work! As a GBM builds its models, it identifies the error from the last version of the model and builds a new model that reduces that same error.

In a random forest, each tree is built independently. With a GBM, the trees are built upon each other. This means more care is required to prevent overfitting in a GBM. Random forests create and bring together many different independent trees, which reduces the risk of overfitting. Conversely, a GBM's goal is to reduce the error from the last model.

On the other hand, there are instances where the data just doesn't support reduction of error. But don't worry—without hyperparameter tuning, the GBM will do its best to find *something*. Sometimes your data just isn't detailed enough to allow the model to get precise, and with GBMs, great power comes with great responsibility. Hyperparameter tuning is incredibly important to ensure your model is not overfit.

This is why data scientists emphasize finding the simplest algorithm possible for the use case. If a random forest can predict your data well, then go with it—you won't have to deal with as many complexities or nuances as you would with a GBM. More predictive power comes with more problems.

Now, let's walk through modeling with GBMs and see what this complex technique can do for our use cases.

Gradient Boosting Machine Modeling

Let's look at a GBM in practice. At the risk of sounding like a broken record, the code in Listing 5.85 likely looks very familiar to begin with (aside from the new GradientBoostingClassifier function).

```
from sklearn.ensemble import GradientBoostingClassifier
from sklearn.model_selection import train_test_split
from sklearn.metrics import accuracy_score
from sklearn.datasets import load_digits

digits = load_digits()

X_train, X_test, y_train, y_test = train_test_split(
    digits.data,
    digits.target,
    test_size=0.2,
```

```
  random_state=42
  )

model = GradientBoostingClassifier(n_estimators=100, random_state=42)

model.fit(X_train, y_train)

y_pred = model.predict(X_test)

y_train_pred = model.predict(X_train)
accuracy = accuracy_score(y_train, y_train_pred)
print(f"Training Accuracy: {accuracy}")

accuracy = accuracy_score(y_test, y_pred)
print(f"Test Accuracy: {accuracy}")
```

Listing 5.85 GBM Example

The resulting accuracy scores are 100% on the training data (which indicates overfitting) and 97% on the testing data. When tuning hyperparameters to balance performance and overfitting, we'll still want to consider n_estimators, max_depth, and min_samples_split. However, now we'll also introduce two new hyperparameters: learning_rate and subsample.

> **The Problem of Predictive Power**
>
> As mentioned earlier, more predictive power comes with more problems. In this specific dataset example, you likely wouldn't use a GBM, as it's too powerful a model. For the sake of the example and continuity with the decision tree and random forest sections, we'll continue to use the same data. You'll see the performance impacts of the GBM later on in our use cases, so focus on learning the hyperparameters and understanding their impact as you go through this simpler example.

The learning_rate hyperparameter adjusts how much each tree contributes to the final model. The higher the number, the faster the model runs because learning is sped up—but then you risk overfitting. The risk of overfitting is reduced with a lower value. The lowest value is technically 0, and there is no upper limit on what you can input.

Think about raising or lowering the learning rate as a trade-off in how quickly the model makes an assumption based on the data it sees. A high learning rate means the model is more willing to trust the data it sees right away. A lower learning rate makes the model more skeptical of the data you give it. This runs the risk that your model never learns new information that may be valuable.

The `subsample` hyperparameter introduces randomness into each iteration of your model, which is another way to reduce overfitting. This number represents a percentage of the rows that each iteration of the model can use, so it's less abstract than the `learning_rate` hyperparameter.

Let's start with some simple adjustments to these new hyperparameters to see how the results are impacted, as shown in Listing 5.86.

```
from sklearn.ensemble import GradientBoostingClassifier
from sklearn.model_selection import train_test_split
from sklearn.metrics import accuracy_score
from sklearn.datasets import load_digits

digits = load_digits()

X_train, X_test, y_train, y_test = train_test_split(
  digits.data,
  digits.target,
  test_size=0.2,
  random_state=42
  )

model = GradientBoostingClassifier(
  n_estimators=100,
  learning_rate = .001,
  subsample =.75,
  random_state=42
  )

model.fit(X_train, y_train)

y_pred = model.predict(X_test)

y_train_pred = model.predict(X_train)
accuracy = accuracy_score(y_train, y_train_pred)
print(f"Training Accuracy: {accuracy}")

accuracy = accuracy_score(y_test, y_pred)
print(f"Test Accuracy: {accuracy}")
```

Listing 5.86 GBM with New Hyperparameters Adjusted

The training accuracy is down to 91%, which isn't surprising since we're restricting the model. However, the test accuracy is down to 86%. The gap between the training and test data suggests more overfitting. To dig into this, let's adjust only one hyperparameter at a time. For clarity when showing these iterations, we'll only show the model itself, but we're still running the full list of code to yield the results. In Listing 5.87, we've significantly increased the learning_rate and reset the subsample to its default.

```
model = GradientBoostingClassifier(
  n_estimators=100,
  learning_rate = .5,
  subsample = 1,
  random_state=42
  )
```

Listing 5.87 GBM with High Learning Rate

Wow, the results from this model are great! The training data accuracy is back to 100%, but now the test accuracy is up to 98%. Why is this? Let's go the other way and significantly reduce the learning_rate in Listing 5.88.

```
model = GradientBoostingClassifier(
  n_estimators=100,
  learning_rate = .001,
  subsample = 1,
  random_state=42
  )
```

Listing 5.88 GBM with Low Learning Rate

The results are worse again (88% on the training data and 83% on the test data). We can isolate this down to our learning_rate. As a reminder, the learning_rate determines how fast each iteration can learn. Each incremental tree (n_estimators) with a lower learning_rate can't make as much progress. This means we need to increase our n_estimators if we lower our learning_rate! As shown in Listing 5.89, we'll increase our n_estimators to 300.

```
model = GradientBoostingClassifier(
  n_estimators=300,
  learning_rate = .001,
  subsample = 1,
  random_state=42
  )
```

Listing 5.89 GBM with Low Learning Rate and Higher Number of Trees

Even by multiplying our number of trees by 3, we're still only seeing a 91% accuracy on the training data and 86% on the test data. If you haven't already started to make this connection, take a quick pause to let it sink in. Had we started with a GBM with hyper-parameter tuning, we'd see lower accuracy and more overfitting compared to a random forest with no hyperparameter tuning. Starting simple is important!

Another important point is the art of building models. While some tasks can be auto-mated away, it's still essential to understand how a model works and how to adjust it.

Let's get back to our example model. What if we continue ramping up how many trees the model has and slightly increasing the learning_rate, as shown in Listing 5.90?

```
model = GradientBoostingClassifier(
  n_estimators=600,
  learning_rate = .008,
  subsample = 1,
  random_state=42
  )
```

Listing 5.90 GBM with a Moderated Learning Rate and Even More Trees

The results are improved, but still not our best, with 100% accuracy on the training data and 96% accuracy on the test data. So, what can we learn from testing these hyperpa-rameters? At a general level, the data you use in your model can sometime behave dif-ferently. This specific data is not very complex, so a higher learning rate actually bene-fits the model and doesn't create overfitting issues.

We won't apply grid search to this dataset. As I tested the code while writing this chap-ter, I found that the hyperparameters in Listing 5.91 produced the best model, achieving 99.9% accuracy on the training data and slightly more than 98% on the test data. My interpretation is the overfitting was reduced by lowering subsample and increasing learning_rate.

```
model = GradientBoostingClassifier(
  n_estimators=300,
  learning_rate = .5,
  subsample = .75,
  random_state=42
  )
```

Listing 5.91 Best GBM Hyperparameters for Digits Dataset

Use Case 1: Gradient Boosting Machine

As we've done before for this use case, we'll start by running the model using the default hyperparameter settings. We already did some feature engineering on this data, so we're working with the same data we created in Section 5.4.3. Listing 5.92 shows this code while Figure 5.21 compares the actual test data against the prediction.

```
#load the necessary modeling package
from sklearn.ensemble import GradientBoostingRegressor

#create the model
model = GradientBoostingRegressor(n_estimators=100, random_state=42)

#fit the model to the data
model.fit(X_train, y_train)

#use the model to predict on the test data
y_pred = model.predict(X_test)

#use the model to predict on the training data
y_train_pred = model.predict(X_train)

#calculate and print MAE for the predictions on the training data
accuracy = mean_absolute_error(y_train, y_train_pred)
print(f" Training Accuracy: {accuracy}")

#calculate and print MAE for the predictions on the test data
accuracy = mean_absolute_error(y_test, y_pred)
print(f"Accuracy: {accuracy}")
```

Listing 5.92 GBM Default Hyperparameters

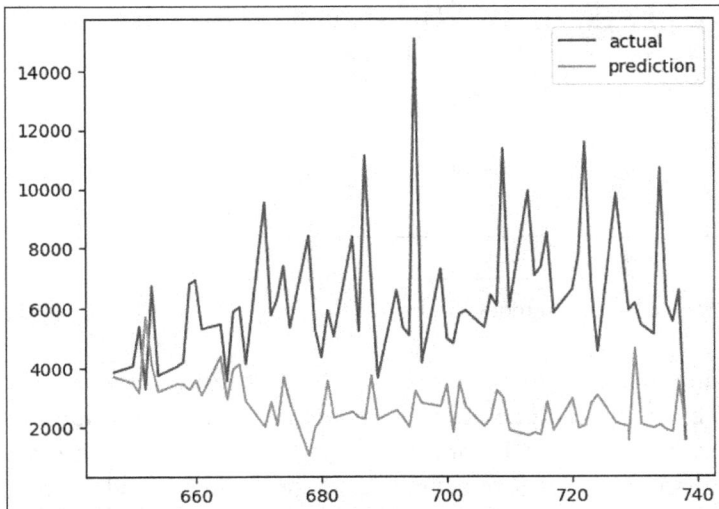

Figure 5.21 GBM Default Hyperparameters

Both our MAE and our graph metrics look promising! The training MAE is 546, with the testing MAE showing signs of overfitting at 3,661. While the model is overfitting, this is a better starting point than the random forest and decision tree algorithms. The graph shows the most promise. It's better at picking up where the variation in the days appears, which is something the previous algorithms struggled with. Let's try some manual hyperparameter tuning instead of using the default values for the model, as shown in Listing 5.93 and Figure 5.22. (The values selected have minimal rationale for being chosen; they're a starting point to see how the model behaves with hyperparameters beyond the default.)

```
#update to give the GBM new hyperparameters
model = GradientBoostingRegressor(
  n_estimators=300,
  learning_rate = .01,
  subsample = .75,
  random_state=42
  )
```

Listing 5.93 GBM First Manual Hyperparameter Tuning

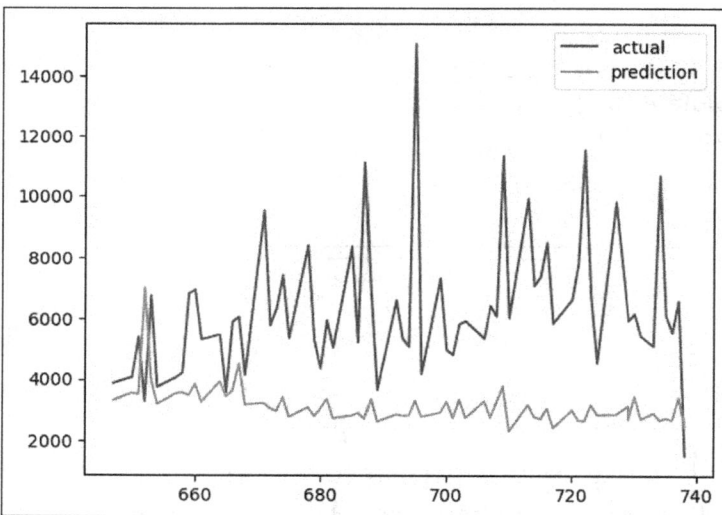

Figure 5.22 GBM First Manual Hyperparameter Graph

Unfortunately, the results didn't improve much. The MAE on the test set went down to 3,300 from 3,661, which isn't much of an improvement. Additionally, the graph isn't picking up the increasing trend. As a final approach, let's apply a grid search with a wide range of values for three primary hyperparameters to see if the model can find the trend of sales increasing over this time period (see Listing 5.94).

```
#load the necessary libraries
from sklearn.model_selection import GridSearchCV
from sklearn.ensemble import GradientBoostingRegressor
```

```
#create a parameter grid to search through
param_grid = {
  'n_estimators': [100, 300, 500],
  'learning_rate': [.01, .05, .1],
  'subsample': [.7, .8, .9]
  }

#create the model
model = GradientBoostingRegressor(random_state=42)

#set up the grid search
grid_search = GridSearchCV(
  estimator=model,
  param_grid=param_grid,
  cv=3,
  scoring='neg_mean_absolute_error'
  )

#execute the grid search
grid_search.fit(X_train, y_train)

#print the best parameters from the grid search
print("Best parameters:", grid_search.best_params_)

#print the best score from the grid search
print("Best score:", grid_search.best_score_)
```

Listing 5.94 GBM Grid Search

The results from this grid search indicate that these hyperparameters will provide the best results: learning_rate = 0.01, n_estimators = 100, and subsample = 0.9. If we run this through the same model workflow we've been using, as shown in Listing 5.95, we'll get the results in Figure 5.23.

```
#create model using best hyperparameter options from the grid search
model = GradientBoostingRegressor(
  n_estimators=100,
  learning_rate = .01,
  subsample = .9,
  random_state=42
  )

#fit the model to the data
model.fit(X_train, y_train)
```

```
#predict the test values
y_pred = model.predict(X_test)

#use the model to pick the training values
y_train_pred = model.predict(X_train)

#calculate and print the MAE for the training predictions
accuracy = mean_absolute_error(y_train, y_train_pred)
print(f" Training Accuracy: {accuracy}")

#calculate and print the MAE for the test predictions
accuracy = mean_absolute_error(y_test, y_pred)
print(f"Accuracy: {accuracy}")

#import the plotting library
import matplotlib.pyplot as plt

#plot the actual and prediction data
plt.plot(test_data['since_first_sale'], y_test, label = "actual")
plt.plot(test_data['since_first_sale'], y_pred, label = "prediction")
plt.legend()
```

Listing 5.95 GBM Use Best Grid Search Parameters

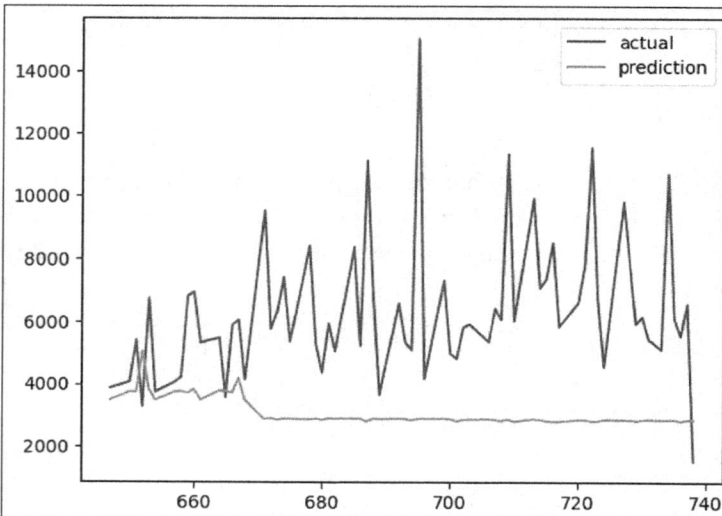

Figure 5.23 GBM Grid Search Best Hyperparameters Results

The graph helps tell the story here. In doing a grid search, we lost the variability we saw previously. There's an important lesson to take away from testing these various models.

It's useful to iterate on and test different models and tune their hyperparameters, but if the underlying data isn't available to support the prediction, no model can pick it up. In Chapter 6, we'll continue this use case and perform additional feature engineering on the model to see if it can pick up on the increasing sales trend.

Use Case 2: Gradient Boosting Machine

We haven't yet had a model for this use case that we feel good about using in practice. Will the GBM be able to find trends in the data that the other models haven't? We're about to find out!

As we do for each model, we'll split our data into the proper training and test datasets, as shown in Listing 5.96.

```
#load in train test split
from sklearn.model_selection import train_test_split

#create the columns to predict the data
X = df.drop(
  columns=[
    'target_7',
    'target_14',
    'Customer ID',
    'timestamp',
    'order_date'
    ]
  )

#using the 1 week lag as the target
y = df['target_7']

#execute the train test split
X_train, X_test, y_train, y_test = train_test_split(
  X,
  y,
  test_size=0.3,
  random_state=42
  )
```

Listing 5.96 GBM Train-Test Split

Now, we're able to build our GBM model. We'll start with the default hyperparameters to get a baseline understanding of how we can expect the model to perform (see Listing 5.97).

```
#load in packages for modeling
from sklearn.ensemble import GradientBoostingClassifier
from sklearn.metrics import accuracy_score
from sklearn.metrics import roc_auc_score

#create the model
model = GradientBoostingClassifier(random_state=42)

#fit the model to the data
model.fit(X_train, y_train)

#create both probability and binary predictions
y_pred_proba = model.predict_proba(X_test)[:, 1]
y_pred = model.predict(X_test)

#calculate the ROC AUC and accuracy scores
auc_score = roc_auc_score(y_test, y_pred_proba)
accuracy = accuracy_score(y_test, y_pred)

#print the ROC AUC and accuracy scores
print(auc_score)
print(accuracy)
```

Listing 5.97 GBM with Default Hyperparameters

With default parameters, we see the ROC AUC score at 0.62, while accuracy is 85%. It's the best starting point we've had, but not by much. We'll try some manual hyperparameter tuning next, but ultimately we'll have to apply grid search. Since the grid search for the random forest model gave us unexpected results, we should perform a grid search for GBM to avoid making assumptions with this dataset.

As a starting point, we'll apply a few hyperparameter options that allow the model to explore the data (max_depth above 10 and min_samples_split at 5 or less). We'll also get some variety in the model by ensuring it can't pick from all samples (subsample), as shown in Listing 5.98.

```
#create model with specific hyperparameters
model = GradientBoostingClassifier(
  learning_rate = .05,
  n_estimators = 500,
  subsample = 0.8,
  max_depth = 12,
  min_samples_split = 5,
  random_state=42
  )
```

```
#fit the model to the data
model.fit(X_train, y_train)

#create both the probability and binary predictions
y_pred_proba = model.predict_proba(X_test)[:, 1]
y_pred = model.predict(X_test)

#calculate both the ROC AUC and accuracy scores
auc_score = roc_auc_score(y_test, y_pred_proba)
accuracy = accuracy_score(y_test, y_pred)

#print both the ROC AUC and accuracy scores
print(auc_score)
print(accuracy)
```

Listing 5.98 GBM with Standard Hyperparameters

The resulting metrics are worse than before, with ROC AUC down to 0.60 and accuracy down to 84%. To test a theory, let's only change the subsample input. We'll change it back to 1, indicating all samples can be considered for each tree. This could be the cause of the low performance, so let's test it out using the model in Listing 5.99.

```
#create the model with new hyperparameters
model = GradientBoostingClassifier(
    learning_rate = .05,
    n_estimators = 500,
    subsample = 1,
    max_depth = 12,
    min_samples_split = 5,
    random_state=42
    )

#fit the model to the data
model.fit(X_train, y_train)

#create both probability and binary predictions
y_pred_proba = model.predict_proba(X_test)[:, 1]
y_pred = model.predict(X_test)

#calculate both the ROC AUC and accuracy scores
auc_score = roc_auc_score(y_test, y_pred_proba)
accuracy = accuracy_score(y_test, y_pred)
```

```
#print both the ROC AUC and accuracy scores
print(auc_score)
print(accuracy)
```

Listing 5.99 GBM with Subsample Input of 1

This is the value of testing theories! The results actually got slightly worse. They round up to the same results, but this indicates that the subsample hyperparameter wasn't the reason for poor performance.

Let's try one last manual attempt before trying the grid search. We'll increase the learning_rate next. The default is 0.1 and we decreased it to 0.05, so it's possible this is an important factor affecting the model's performance. Listing 5.100 shows the code for this.

```
#create model with updated hyperparameters
model = GradientBoostingClassifier(
   learning_rate = .25,
   n_estimators = 500,
   subsample = 1,
   max_depth = 12,
   min_samples_split = 5,
   random_state=42
   )

#fit model to the data
model.fit(X_train, y_train)

#create both the probability and binary predictions
y_pred_proba = model.predict_proba(X_test)[:, 1]
y_pred = model.predict(X_test)

#calculate both the ROC AUC and accuracy scores
auc_score = roc_auc_score(y_test, y_pred_proba)
accuracy = accuracy_score(y_test, y_pred)

#print both the ROC AUC and accuracy scores
print(auc_score)
print(accuracy)
```

Listing 5.100 GBM with Higher learning_rate

Unfortunately, the results got slightly worse again. Although it's only a rounding difference, the learning_rate also doesn't appear to be the main cause of the lower performance. This shows the value of grid search, where you can run all combinations of inputs to see what yields the best results.

You can usually get a feel for the dataset and discover which hyperparameters are important through manual exploration, as we just did. However, this dataset continues to present interesting behaviors. Good thing we have grid search to help us out!

The code for this iteration of the grid search is shown in Listing 5.101. We reduced the number of cv to 4. This keeps the number of combinations under 200, which will make our grid search run faster. It will take some time for this grid search to run, so it's okay to skip running the code yourself so you can continue on with the content instead of waiting for the code to execute.

```
#load the grid search library
from sklearn.model_selection import GridSearchCV

#set up the hyperparameter grid
param_grid = {
  'n_estimators': [300, 500],
  'max_depth': [5, 10],
  'min_samples_split': [5, 10],
  'learning_rate': [0.07, 0.1, 0.2],
  'subsample': [0.8, 1]
  }

#create the model
model = GradientBoostingClassifier(random_state=42)

#set the grid search
grid_search = GridSearchCV(estimator=model,
              param_grid=param_grid,
              cv=4,
              scoring='roc_auc',
              n_jobs=-1,
              verbose=1)

#execute the grid search
grid_search.fit(X_train, y_train)

#identify the best parameters for the model
best_model = grid_search.best_estimator_
print("Best Parameters:", grid_search.best_params_)
```

Listing 5.101 GBM First Grid Search

Unfortunately, the results still aren't good enough. This likely means we'll need to do more feature engineering to see what additional data we can provide to the model.

As we have with the other models, we'll quickly check on the 14-day target variable to see how it performs. Listing 5.102 shows the code to do so.

```python
#set up predictors for the model
X = df.drop(
  columns=[
    'target_7',
    'target_14',
    'Customer ID',
    'timestamp',
    'order_date'
    ]
  )

#set the target as the 2 week column
y = df['target_14']

#execute the train test split
X_train, X_test, y_train, y_test = train_test_split(
  X,
  y,
  test_size=0.3,
  random_state=42
  )

#create the model
model = GradientBoostingClassifier(
  learning_rate = .25,
  n_estimators = 500,
  subsample = 1,
  max_depth = 12,
  min_samples_split = 5,
  random_state=42
  )

#fit the model to the data
model.fit(X_train, y_train)

#create both the probability and binary predictions
y_pred_proba = model.predict_proba(X_test)[:, 1]
y_pred = model.predict(X_test)

#calculate both the ROC AUC and accuracy scores
auc_score = roc_auc_score(y_test, y_pred_proba)
```

```
accuracy = accuracy_score(y_test, y_pred)

#print both the ROC AUC and accuracy scores
print(auc_score)
print(accuracy)
```

Listing 5.102 GBM 14-Day Target

The metrics for the 14-day target return quite poorly again, which is further evidence that we need to reevaluate the data and add additional features. The ROC AUC was 0.57 and the accuracy was 74%. We'll pick this business case back up in Chapter 6, Section 6.6.1, where we'll spend additional time on feature engineering to see if we can get to a viable model.

Use Case 3: Gradient Boosting Machine

This use case taught us the importance of ensuring our target variable is appropriately lagged. As a starting point for the GBM model, we'll use the same max_depth of 7 from the random forest model in Section 5.5.2, as shown in Listing 5.103. In this section, we'll test more hyperparameters than we did for the random forest model.

```
#import necessary packages for the modeling process
from sklearn.ensemble import GradientBoostingRegressor
from sklearn.model_selection import train_test_split
from sklearn.metrics import mean_absolute_error

#split data into what will be used to predict and what is being predicted
X = df.drop(
  columns=['DATE', 'AREA NAME', 'crime_count', 'crime_count_shifted'])
y = df['crime_count_shifted']

#perform standard train test split
X_train, X_test, y_train, y_test = train_test_split(
  X,
  y,
  test_size=0.2,
  random_state=42
  )

#create and fit the GBM model
model = GradientBoostingRegressor(max_depth=7)
model.fit(X_train, y_train)

#predict with the modeling using the test data
y_pred = model.predict(X_test)
```

```
#calculate the mean absolute error
mae = mean_absolute_error(y_test, y_pred)

#calculate the mean absolute error divided by average of the target
mae_by_avg_target = mae / y_test.mean()

#print both results
print(mae)
print(mae_by_avg_target)
```

Listing 5.103 Testing GBM with Same max_depth of 7 from Random Forest Model

The results are only slightly better from what we saw from the last random forest model. The MAE is 5.7 and the MAE as a percentage of the average target variable is 22% (the random forest model resulted in 5.8 and 22%, respectively).

As a first step, let's see if increasing the number of estimators (trees) helps, as shown in Listing 5.104. Remember, the value of a GBM is that its trees build on each other, so theoretically we should see some incremental value by increasing the number of trees in a GBM model. (We'll also see an incremental increase in the time it takes to train the model.)

```
#split data into what will be used to predict and what is being predicted
X = df.drop(
  columns=['DATE', 'AREA NAME', 'crime_count', 'crime_count_shifted'])
y = df['crime_count_shifted']

#perform standard train test split
X_train, X_test, y_train, y_test = train_test_split(
  X,
  y,
  test_size=0.2,
  random_state=42
  )

#create and fit the GBM model
model = GradientBoostingRegressor(n_estimators = 300, max_depth=7)
model.fit(X_train, y_train)

#predict with the modeling using the test data
y_pred = model.predict(X_test)

#calculate the mean absolute error
mae = mean_absolute_error(y_test, y_pred)

#calculate the mean absolute error divided by average of the target
mae_by_avg_target = mae / y_test.mean()
```

```
#print both results
print(mae)
print(mae_by_avg_target)
```

Listing 5.104 Increase the Number of Estimators to 300

Well, the results are actually slightly worse. The MAE is 5.8 and the MAE as a percentage of the average target variable is 22%. In cases like this, you can default to grid search more quickly. As you saw in other use cases, it's worth manually testing the hyperparameters first to get a feel for how they work with the model. However, seeing no improvement from switching to a GBM from a random forest or tripling the number of trees indicates the model isn't going to behave differently. Rather than guessing, we can use the grid search process to get us there faster.

As you'll see in Listing 5.105, we'll pick two more extreme options for each hyperparameter in param_grid (n_estimator, max_depth, learning_rate, and subsample). There are two goals to this:

- Fewer options mean less time to get an answer. Since we're currently unsure of what to expect from the model, it's better to get an idea of the results faster. We can then test subsequent options using what we learned from the initial results.

- By testing the extremes, we can see which end of the spectrum to further explore. For example, if the model favors a max_depth of 7, we can interpret this as a preference for deeper trees and test additional values like 6 and 8.

The code will take some time to run, even with the limited hyperparameters given.

```
from sklearn.ensemble import GradientBoostingRegressor
from sklearn.model_selection import train_test_split, GridSearchCV
from sklearn.metrics import mean_absolute_error

#split data into what will be used to predict and what is being predicted
X = df.drop(
    columns=['DATE', 'AREA NAME', 'crime_count', 'crime_count_shifted'])
y = df['crime_count_shifted']

#perform standard train test split
X_train, X_test, y_train, y_test = train_test_split(
    X,
    y,
    test_size=0.2,
    random_state=42
    )

#create the model with random state
model = GradientBoostingRegressor(random_state=42)
```

```
#define the hyperparameter grid
param_grid = {
  'n_estimators': [100, 300],
  'max_depth': [3, 7],
  'learning_rate': [0.05, 0.2],
  'subsample': [0.8, 1.0]
  }

#set up GridSearchCV
grid_search = GridSearchCV(
  estimator= model,
  param_grid=param_grid,
  scoring='neg_mean_absolute_error',
  cv=3,
  n_jobs=-1,
  verbose=1
  )

#fit the grid search
grid_search.fit(X_train, y_train)

#best model from grid search
best_model = grid_search.best_estimator_

#predict and evaluate
y_pred = best_model.predict(X_test)
mae = mean_absolute_error(y_test, y_pred)
mae_by_avg_target = mae / y_test.mean()

#output results
print("Best Parameters:", grid_search.best_params_)
print("MAE:", mae)
print("MAE / Avg Target:", mae_by_avg_target)
```

Listing 5.105 First Grid Search on the GBM

The grid search results are quite disappointing. The best model shows no meaningful improvement with an MAE of 5.7 and an MAE divided by average target value of 22%. This is a good time to take a step back and consider the business case. The MAE divided by the average target variable of 22% is way too high a margin of error. Given that we've not seen any significant movement in the results by adjusting the hyperparameters, we don't want to spin our wheels trying to get small gains that won't result in a sufficient model for the use case.

Value of Distributed Computing

While you're waiting for the grid search to finish, let's discuss *distributed computing*. You may have heard of it before. The concept is simple, but the nuances are incredibly complex. Distributed computing means the job or task is divided into pieces, executed in parallel, and then recombined at the end. Grid search is a great example of this. Each version of the model can be trained independently before the final answers are brought back together.

If you're using your local computer or a free option online to run this code, it's unlikely you're using a distributed computing framework. Seeing how long something like grid search takes clearly illustrates the value of distributed frameworks.

At the time of writing, my day job is leveraging Microsoft Fabric to write Python code for building models and data pipelines. Within Microsoft Fabric, there are notebooks that are based on Spark, which is a common distributed framework. The difference in speed between running the same code in Microsoft Fabric in Anaconda is very noticeable. While working on this book, I've been tempted to abandon Anaconda for a different technology to speed up my writing process. However, I stayed committed to it to ensure that I experience what you would experience when running the code (how noble of me).

Assuming your grid search is still running, use the time to do some research on distributed frameworks and their history. I think you'll find it an interesting topic!

At this stage, we need to consider what else we should be doing to the data to improve the performance of the model rather than trying to adjust the hyperparameters. We'll explore this in Chapter 6!

5.6   Clustering

There's a running joke in model learning content that clustering always gets tacked on at the end. Having structured content myself, I now see the reason for this. Clustering is a bit of a one-off topic to discuss. It's certainly important, but it also doesn't fit neatly within the material. It's not predicting an outcome of the future, like sales; it's creating a new grouping mechanism for your rows based on your columns. In practice, it's more closely related to dimensionality reduction than a GBM algorithm.

The output of clustering is often used as part of a broader predictive analytics workflow. The common use case you'll see for clustering is segmenting (or clustering) customers into specific groups based on their characteristics, such as age, gender, or income. Taking information from many columns and condensing it into fewer should sound familiar—it's similar to what happens in dimensionality reduction.

We'll explore clustering in more detail in the following sections.

5.6.1   What Is Clustering?

People generally find clustering to be intuitive because it's closely aligned to how we think as humans. We're trying to simplify the world by identifying patterns and creating frameworks to group what we're observing. Let's start our clustering example by running the code in Listing 5.106.

```python
#import required libraries
import numpy as np
from sklearn.model_selection import train_test_split
from sklearn.cluster import KMeans
from sklearn.datasets import make_blobs
import matplotlib.pyplot as plt
from sklearn.metrics import silhouette_score

#create sample data
n_samples = 300
random_state = 42
X, y = make_blobs(
  n_samples=n_samples,
  random_state=random_state
  )

#execute train test split
X_train, X_test = train_test_split(
  X,
  test_size=0.3,
  random_state=random_state
  )

#set number of clusters to 3
n_clusters = 3

#create kmeans clustering model
kmeans = KMeans(
  n_clusters=n_clusters,
  random_state=random_state,
  n_init='auto'
  )

#fit the model to the data
kmeans.fit(X)

#create the predictions on the test data
test_labels = kmeans.predict(X_test)
```

```
#create the predctions on the training data
train_labels = kmeans.predict(X_train)

#calculate and print the silhouette score for the training data
silhouette_avg = silhouette_score(X_train, train_labels)
print(silhouette_avg)

#calculate the silhouette score for the test data
silhouette_avg = silhouette_score(X_test, test_labels)
print(silhouette_avg)
```

Listing 5.106 Clustering Example

The code is importing new packages for clustering. Kmeans() is the most common approach to clustering. While there are others, our focus will be on the kmeans() algorithm. Let's discuss this code in more detail:

- **Creating dataset**
  We're generating our own data for this example, and sklearn provides us with the make_blobs() function to do so. While we don't discuss generating our own data much in this book, it can be very useful when building out algorithms.

- **Slightly different splitting of data**
  Clustering is *unsupervised learning*, meaning it doesn't contain a *y* portion to split. This is why, we're only inputting and outputting the *x* components in our train_test_split() function.

- **Clustering algorithm**
  The clustering algorithm has two new hyperparameters we need to learn. The first is n_clusters, which says how many groups need to be created from the data. The next is n_init, which identifies how many different starting points the algorithm will try. You'll learn more about this in Section 5.6.3 when we discuss how clustering algorithms work behind the scenes.

- **Silhouette metric**
  We have a new metric from the silhouette_score() function. At a high level, this measures the quality of the clusters by comparing the distance of the data points within a cluster to the data points of the next closest clusters. The score can be anywhere from -1 to 1, with 1 representing very good clusters and -1 representing clusters that may be incorrect. A 0 usually represents clusters that overlap. The output from this code is 0.84 on the training data and 0.86 on the test data, which is pretty good!

Clustering tends to be visual, so we can also visualize our results as shown in Listing 5.107 with the output shown in Figure 5.24. Using the plt.scatter function, we take our centers dataset and provide it with the x and y coordinates (centers[:,0] and centers[:, 1]) to plot the data.

```
#create predictions for full dataset
all_labels = kmeans.predict(X)

#plot the data in a scatterplot
plt.scatter(X[:, 0], X[:, 1], c=all_labels, s=50, cmap='viridis')

centers = kmeans.cluster_centers_
plt.scatter(centers[:, 0], centers[:, 1], c='red', s=200, alpha=0.75)
plt.title('K-means Clustering')
plt.xlabel('Feature 1')
plt.ylabel('Feature 2')
plt.show()
```

Listing 5.107 Clustering Visualization Code

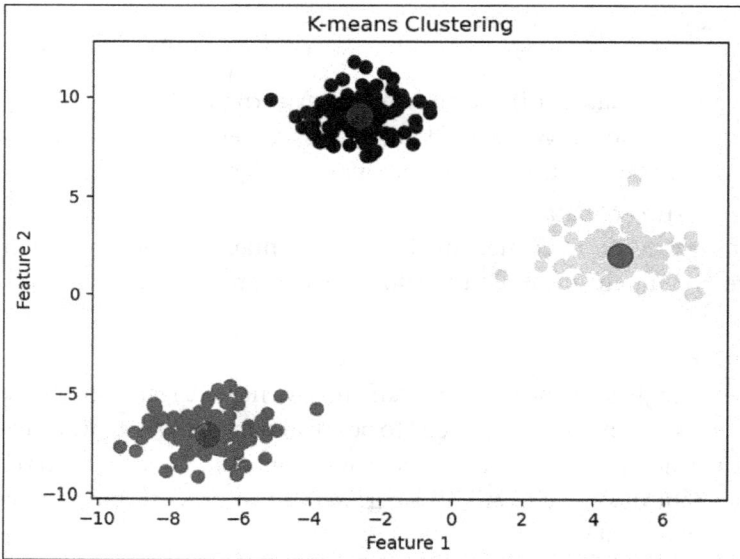

Figure 5.24 Clustering Visualization

You likely have a number of other questions after seeing this code and visualization, so let's go into more detail on clustering and how it works in practice.

5.6.2 Picking the Number of Clusters

Think of picking the number of clusters as a hyperparameter optimization exercise. You're trying to identify the proper number of clusters for your data, which the model won't automatically optimize. If you've done clustering before, you may have heard of the *elbow method*. This entails plotting the data and identifying how many clusters should be used. It's called the elbow method because you're looking for the "elbow" in the plot. This can be quite subjective. Instead, try thinking from the perspective of optimizing the silhouette score.

Using the same dataset as before, you can set up a loop to do this, as shown in Listing 5.108. After creating the dataset with the make_blobs function, we use a workflow similar to that of the machine learning models. We run through the train_test_split function, but we don't have a y or target to select and test on. This is because clustering is an unsupervised learning approach, so there isn't a target variable. Once the data is split into X_train and X_test, we create a set of options for the number of clusters and then loop through each one. Within the loop, we're using the KMeans() function to create clusters. Finally, we generate the silhouette score using the silhouette_score function and compare each cluster option against each other. Table 5.3 shows the results.

```python
#load required libraries
import numpy as np
from sklearn.model_selection import train_test_split
from sklearn.cluster import KMeans
from sklearn.datasets import make_blobs
import matplotlib.pyplot as plt
from sklearn.metrics import silhouette_score

#create sample data
n_samples = 300
random_state = 42
X, y = make_blobs(n_samples=n_samples, random_state=random_state)

#execute train test split
X_train, X_test = train_test_split(
  X,
  test_size=0.3,
  random_state=42
  )

#create list of cluster options
cluster_options = [2, 3, 4, 5, 6]

#loop through the full clustering process
for n_clusters in cluster_options:
  kmeans = KMeans(
    n_clusters=n_clusters,
    random_state=random_state,
    n_init='auto'
    )
  kmeans.fit(X)
  test_labels = kmeans.predict(X_test)
```

```
train_labels = kmeans.predict(X_train)
silhouette_avg = silhouette_score(X_train, train_labels)
print(n_clusters)
print("Training: ", silhouette_avg)

silhouette_avg = silhouette_score(X_test, test_labels)
print("Test:", silhouette_avg)
print("")
```

Listing 5.108 Looping Through Cluster Options

Clusters	Training Silhouette Average	Test Silhouette Average
2	0.69	0.71
3	0.84	0.86
4	0.65	0.73
5	0.50	0.49
6	0.54	0.54

Table 5.3 Cluster Options Silhouette Average Results

These results shouldn't be surprising, given that the data was synthetically generated and there are very clearly 3 clusters. However, the results show you how you can use the silhouette average to find the optimal number of clusters. This metric confirms 3 clusters is the best option.

Another consideration is your use case. If you're working in the marketing world and using clustering to segment your customers, even if the optimal number of clusters is 12, it's unlikely the marketing team wants to create profiles and materials for 12 segments. As with any modeling use case, you need to accommodate the realities of your stakeholders and their use case. If you don't, it's likely they won't end up using what you've built.

5.6.3   Behind the Scenes of Clustering

So, how does clustering actually work? Understanding this helps you interpret your results and recognize its limitations. Here's the high-level process:

1. Based on the number of clusters you defined, the algorithm randomly choose that many data points to serve as a starting point. These are called *centroids*.
2. Once the centroids are identified, all other data points are assigned to the nearest centroid based on their distance from each other.
3. New centroids are then calculated to minimize the distance between data points within the clusters.

4. The second and third steps will repeat until the centroids are stabilized and don't change significantly.

There's an element of randomness with this approach because the starting point of each centroid could impact the final result. This is why establishing the `random_state` helps distinguish true changes from randomness caused by different starting points.

> **Where Are Our Use Cases?**
> We won't leverage clustering for our use cases, as none of them naturally have a need for it. Think of this section on clustering as bonus content!

5.7 Summary

This chapter focused on how aggregation and feature engineering transform raw data into meaningful signals for machine learning. You learned that models don't "think"—they detect patterns in numbers. The job of the person building these models is to shape those numbers into something the model can learn from. Here's a recap of what you've learned:

- You learned about how noisy transaction level data needs to be aggregated. Not only does this enable you to predict future values like a day, it also stabilizes patterns in the data so the model can pick up trends.

- You learned about the concept of lagging variables, both for the target variable and as input variables. Executing this can be tricky—it's difficult to remember which direction you should lag the data. (When writing the code for the book, I did it incorrectly more than once.) This is a key reason that data leakage occurs. Data leakage is the kiss of death for your model and should be avoided at all costs.

- You learned the importance of understanding your use case when it comes to effectively executing feature engineering. Knowing the context around the use case helps you add more value through feature engineering at a faster pace.

- You learned about the most commonly used models. We discussed each of their pros and cons and showed them in practice for our use cases. We found that ultimately, GBM is usually the best performing model, which is why it's often the default approach in practical settings.

- Lastly, we discussed clustering. Although we didn't apply it to any of the use cases, it's an algorithm with many practical uses, especially when working with customer data.

You now have foundational knowledge of the steps required to prepare data for your model as well as how the models themselves work. The next chapter will cover the different approaches to validating your model, while carrying each of our use cases forward in a manner that reflects real-world practice.

Chapter 6
Evaluating the Model and Iterating

We've gone through a variety of algorithms and already performed some light iterations on them. In this chapter, we'll take a brief step back and discuss foundational concepts to evaluate your model before diving into the iterative process of model development.

A core concept within the machine learning process is learning how to evaluate the quality of your model, which translates to learning how to iterate on your model. The good news is we've already started to cover these topics in the previous chapters. In Chapter 5, we u^sed metrics like accuracy and mean absolute error (MAE) to build our models, and we iterated on the models to some extent as well—but there's much more to learn.

This chapter is split into three categories:

1. An overview of validation metrics
2. An introduction to a new topic, machine learning interpretability
3. Further iteration on our modeling work from Chapter 5

Validation metrics are foundational to your understanding of machine learning. A quick warning: They're more math-heavy and are well established (that is to say, the content is a bit dry, though I've done my best to make it engaging). Machine learning interpretability is a topic I've found fascinating. As we've discussed earlier in this book, the biggest knock on machine learning algorithms is that they're black boxes. This content will show you tools you can leverage to gain insight into what your model finds predictive. In addition to making explanations to stakeholders and business users easier, this also helps with iterating on your models.

The third part of the chapter is about iterating on our model for each use case, ultimately showing how it can be saved. By the end, you'll have your models saved in a file, ready to make predictions on truly new data!

6.1   Importance of Picking Validation Metrics

The magic of the modeling process is how objective it can make discussions about the appropriate methodology or approach to a problem. Without a validation metric, the modeling process wouldn't work! You'll come across many examples of subjective approaches being leveraged to make decisions, especially in the business world. Not having a metric to ground the conversation in the right approach can lead to some challenging situations. Let's review an example from personal experience.

Measuring Recruiting Demand

We'll start with a quick talent acquisition lesson for those who aren't familiar with this space. Even if a business isn't trying to grow the number of employees they have, and they want to stay at the same headcount, the recruiting teams still need to recruit for and fill roles from individuals leaving the company or moving within the company.

A highly urgent request came from one of the recruiting leaders to calculate the estimated demand for their business. The goal was to justify an ask for more headcount. Even without looking at the data, it was clear they needed more recruiters to meet the demand. However, the decisionmakers were stuck on the business strategy of "headcount is flat." To the decisionmakers, this translated into "we don't need to do any hiring."

Due to the tight timeline, we used an arithmetic approach to calculate the demand instead of building a predictive model. This is where the challenges started. With the arithmetic approach, we couldn't train a model to see how accurate the methodology would have been last year. We were only using logical arguments to determine what went into the methodology and what didn't. This meant instead of grounding ourselves in an objective metric like last year's data, the loudest voice in the room (the decisionmakers) determined what did and didn't go into the methodology. This resulted in poorly informed decisions being made and the entire group being frustrated.

The takeaway from this story is that the quality of a decision relies heavily on the metric or form of validation in use. In this case, the metric was the decisionmakers' opinion, which isn't objective (especially in this specific example). We can demonstrate with two more examples:

- **When accuracy isn't sophisticated enough**

  There is a concept in machine learning called *imbalanced data*. This phenomenon occurs with classification problems when one outcome is much more likely than the other outcome. The imbalance in the data can create challenges for models since one outcome is much more likely than the other.

  So, why does this matter in the context of picking your validation metric? Let's use the example of a fraud detection model. This is an example where non-fraudulent activity is significantly more likely than fraudulent activity (hopefully), making this a highly imbalanced dataset. If you used something as simple as accuracy, you could create a simple decision tree model with 99.9% accuracy that predicted all activity is non-fraudulent. The model is about as accurate as it can be, but it has no value.

- -10,000 + 10,000 = 0

  When evaluating your model, be aware that aggregating your predictions can hide evidence of poor performance. You'll see many regressor-based metrics leverage either an absolute value (making all numbers positive) or a square (a negative times a negative is positive) to ensure the error metrics are all positive before aggregating them into a number, usually by taking an average.

  Here's a way to think about this: When you're making predictions with the model, you're making a prediction for one row of data at a time and often making a decision from that single instance. For example, in Table 6.1, if you're predicting sales for each day and making decisions based on those predictions, keep in mind that even if the model has 100% accuracy, each individual day has an error of at least 150.

Day of Week	Actual Sales	Predicted Sales	Error
Monday	500	800	300
Tuesday	400	150	-250
Wednesday	350	650	300
Thursday	600	400	-200
Friday	800	650	-150
Total	2650	2650	0

Table 6.1 Sales Prediction Error Example

6.2   Validation Metrics

Now that we've discussed the importance of the metrics, let's review the most commonly used ones. For each metric, we'll start with a foundational explanation and then consider its practical uses.

The mathematical explanation of each metric is intentionally excluded as a standalone topic and embedded within the sections where necessary. A robust understanding of the underlying math isn't necessary, given the functions in scikit-learn (sklearn) do all this for you.

> **Further Resources**
>
> If you're interested in learning more, the sklearn documentation (*https://scikit-learn.org/stable/api/sklearn.metrics.html*) has more specifics on how each of the metrics are calculated.

6.2.1   Accuracy

The best way to think about accuracy is the number of correct predictions as a percentage of all predictions. This makes it a classification-specific metric.

Accuracy is usually applied to classification algorithms, meaning we're flagging whether the prediction was right or wrong. We're not predicting a number, as we would when forecasting sales, where the focus is on the distance between the actual outcome and the predicted outcome. Table 6.2 is an example of predicting employee turnover. Four out of the five employees' turnover behavior was accurately predicted, giving the model 80% accuracy.

Employee ID	Actual	Prediction	Correct?
1	0	0	Yes
2	0	0	Yes
3	0	1	No
4	1	1	Yes
5	0	0	Yes

Table 6.2 Employee Turnover Accuracy Example

In practice, accuracy is rarely used to train your model. It lacks a degree of sophistication that real-world datasets often require. However, it can be an effective metric to communicate to stakeholders. The simplicity of this metric presents a risk when training your model. However, it can be a major benefit when presenting your results, especially with nontechnical audiences.

Many of the mathematical operations done by the calculations transform the original data to improve the quality of the metric. This means the metric's outcome can differ from the original dataset, which can add a layer of complexity and potential confusion when explaining the model and its results.

Consider using a different metric to build your model and then reverting to accuracy when presenting to stakeholders (especially nontechnical ones). This creates a win-win situation where you're able to build a model you feel more confident in and can avoid a complicated conversation with your stakeholders.

6.2.2   Confusion Matrix

A confusion matrix consists of four metrics used for classification tasks. As you can see in Table 6.3, each box represents a distinct metric.

Predictive Outcome	Actually Occurred	Did Not Actually Occur
Occurring	True positive	False positive
Not occurring	False negative	True negative

Table 6.3 Confusion Matrix

The values in a confusion matrix sum to 100%. As you're building your model, you can use the confusion matrix to quickly see where your model may be performing well or missing the mark. A number of metrics are then derived from the confusion matrix numbers. In this section, we'll talk through each of the four components to a confusion matrix. The following sections will then discuss the metrics they can create: precision (Section 6.2.3), recall (Section 6.2.4), and F1 score (Section 6.2.5).

The components of confusion matrix are as follows:

- **True positive**
  The four confusion matrix components are easy to understand, but a true positive is probably the easiest. Of all the cases where the outcome actually occurs, it shows how many the model correctly predicts. In isolation, this metric tells part of the story. The value in looking at this in isolation will vary by use case. For example, if your use case involves relatively rare events, like fraud detection, this metric can help you understand the mode's performance and communicate it with stakeholders. Say you're looking for fraud across 100,000,000 transactions, but only 1,000 are fraudulent. The true positive result will give you an idea of how well the model picks up the small amount of fraud amongst the vast majority of legitimate transactions.

- **True negative**
  True negative is the inverse of true positive. Rather than indexing on the number of outcomes that did occur, it indexes on the number that didn't occur. Again, leveraging this depends on your use case. In practice, this metric is rarely used in isolation due to the typical setup of the target in your model.

  In classification, when predicting a binary outcome, you set up your target outcome to be a 1 or 0 (1 for the outcome occurring and 0 for it not). Anything important to the use case is usually coded as 1 and the rest is coded as 0. In our examples of fraud detection and employee turnover, you would code fraud or an employee leaving as 1. If almost the entire the dataset is negative, like in a fraud detection use case, the true negative percentage is less useful on its own.

- **False positive**
  The term *false positive* has worked its way into common language, especially in the corporate world. In a model context, it states the number of times the model predicted an outcome that didn't actually occur. The implications of this vary by use case.

If you're a business predicting fraud, you have to balance the cost of incorrectly flagging a fraud charge. If a transaction is flagged as fraudulent, your customer's card is frozen and someone from your company must call them. All those results have a cost to your company. A similar dynamic occurs in the medical world when screening for illness. Is it better to err on the side of incorrectly flagging something as serious as cancer? This can add additional stress to the patient's life and increase costs for the healthcare system.

False positive is almost always thought about in the context of balancing with false negative.

- **False negative**

False negative is the opposite of false positive. How many times did the outcome occur when the model predicted that it wouldn't? Let's use the same examples of fraud detection and medical outcomes to show how false negative and false positive are inversely related.

In fraud detection, there's a cost associated with incorrectly flagging a transaction as fraud. What about a scenario where the model doesn't flag a transaction as fraud, even though it is? This also has an associated cost. You have an unhappy customer, need to spend time rectifying the situation, and may develop a reputation for being reactive to fraud, which is more costly than being proactive. The balance depends on your use case. Using the medical screening example, it's a similar dynamic. If you're screening for something as serious as cancer and incorrectly say the patient doesn't have cancer, you're putting the patient at risk.

In both of these use cases, it's often best to accept more false positive and less false negative situations because the vast majority of outcomes are negative. Both domains have implemented additional follow-ups and subsequent reviews to assess the initial prediction.

Balancing False Positive and False Negative in the Real World

In high-stakes environments, people are hesitant to hand over all the decision-making power to predictive models and artificial intelligence (AI). Many companies are moving towards a cyborg approach that leverages the best qualities of man and machine. Models are great at looking at large amounts of data, but they often lack the ability to understand nuance and common sense (we'll see how well this ages). Humans are limited in how much information they can comb through and digest at one time, but they're skilled at making nuanced decisions using context that isn't always directly related to the decision. This is rooted in the balance between false positive and false negative and avoiding the risks associated with each.

6.2.3   Precision

Precision is the first of three metrics that are derived from the confusion matrix boxes. Here's the math for precision:

True positive ÷ (True positive + False positive)

Said in another way, it's the true positive divided by all positive predictions. By normalizing the true positive to all the positive predictions, you're able to understand how good the model is at avoiding false alarms. In practice, you're more likely to use precision than the true positive to identify false positive errors in your model, since the true positive does not account for false negatives by itself.

You'll want to use precision in cases where the risk of a false positive prediction is high, either in frequency or in impact to your use case. An email spam filter is a good example of precision being the preferred approach. You don't want to miss important messages (a false positive error), so a spam filter model heavily avoids incorrectly flagging an email as spam.

6.2.4   Recall

Recall is the true positive divided by all actual positive outcomes. Here's the formula:

True positive ÷ (True positive + False negative)

This identifies how well the model identifies the actual outcomes occurring. In practice, recall is the measure for capturing false negative errors in your model.

An example of where recall is prioritized over precision is early warning systems for natural disasters. The goal is to ensure you don't miss a potentially catastrophic event, which can have significant and devastating consequences.

Recall and precision measure opposite sides of the error spectrum. Recall accounts for false negatives while precision accounts for false positives. Depending on your approach, it may make more sense to lean on one or the other. But what about situations where you want to balance both? That's where the F1 score comes in.

6.2.5   F1 Score

The F1 score is the answer to balancing both precision and recall. The math behind it doesn't directly translate back to the underlying data, which can complicate discussions of the F1 score with your stakeholders. However, as a measure to build your model, it provides a single metric to optimize. Here's the formula:

2 × (Precision × Recall) ÷ (Precision + Recall)

This equation balances the impact of recall and precision. For example, when using grid search, the goal is to find the combination of hyperparameters that balance the highest recall and highest precision.

If you had to pick one metric to represent the confusion matrix, it would be the F1 score. Its goal is to balance each of the four confusion matrix outcomes in a single metric.

6.2.6 Area Under the Curve

If this metric sounds familiar, it's because we've been using it for the second use case throughout the book. The key to understanding the area under the curve (AUC) is that all classification problems are actually probability problems. Whether it's logistic regression or a tree-based classification model, the model isn't actually calculating a binary outcome of a 0 or 1. Rather, it's generating a probability of that outcome occurring, and then identifying the optimal probability cutoff that sets the 1 or 0 prediction.

Probabilistic Thinking

The concept of thinking probabilistically is actually quite important as an analyst or data scientist. Some of the best business leaders think in this way, but certainly not all of them. When building models, it's critical you think probabilistically because all models are wrong. However, if a model tends to be more right than wrong, then it can be useful.

Think about this from the perspective of the end user when building your models. If you're building a model that predicts whether an employee will leave in the next 12 months, representing that as a 1 or 0 (yes or no) is very rigid. Using the underlying probabilities ends up being more valuable to the use case. Within all the 1s, there will be varying degrees of flight risk. Breaking up the results into multiple groups (high risk, medium risk, and low risk, for example) allows users to focus on how they use the data to make decisions.

Beyond thinking with this mindset in the context of predictive modeling, adopting this mindset more broadly will increase your decision-making skills and your value to an organization. At the very least, think about being asked by your boss or manager to make a decision with a lot of uncertainty. Thinking about the decision in terms of probabilities, and then sharing them that way, will at the very least ensure you're communicating your expectations of the situation more effectively. Sharing your view on the likelihood of success or degree of uncertainty provides your stakeholders with another layer of information. Rather than just saying "Let's make X decision," framing it as "Let's make X decision, which I believe has a moderate chance of success" encourages more dialogue and understanding from the individuals you're working with.

AUC is the metric I use the most. It isn't magic, but it does handle many of the complexities you'll come across in the predictive modeling world quite well. When researching AUC, you'll also see it associated with something called the receiver operating characteristic (ROC) curve. AUC is the *metric* that summarizes the performance of the model, and the ROC curve consists of the *plotted lines* of the true positive rate against the false positive rate.

One of the biggest benefits of AUC is its ability to handle imbalanced data well. All the nuance and complexity we discussed with the confusion matrix becomes less of a concern because of the underlying math behind AUC. However, what does the AUC metric actually mean in practice? This becomes a challenge because it doesn't directly translate back to the data.

The output of AUC is as follows:

- The results range from 0 to 1.
- A result of 0.5 indicates the model is no better than random chance.
- A result of 1 indicates the model predicts the results perfectly.
- A result between 0 and 0.5 usually indicates you've done something wrong with your model.

This leads to the loosely agreed-upon interpretation of the results:

- 0.6 means the model has some predictive power, but it definitely needs more work. It's better than a random guess, but not by much.
- 0.7 is generally the minimum threshold for a model to be considered usable. While it's not great, it can start adding value at this stage (especially for complex predictions).
- 0.8 is a good model, regardless of your use case.
- 0.9 is an excellent model! This can be quite difficult to achieve in some real-world applications.
- Anything greater than 0.95 should cause concern. Is it possible you're feeding in data into the model that it otherwise wouldn't have when you're trying to make a prediction?

With these interpretations, you can educate stakeholders on AUC. Since AUC is expressed as a percentage on a number line, it can be a useful, stakeholder-friendly metric, especially for those you work with frequently.

6.2.7   R-Squared

The R-squared metric is commonly used with regression techniques, where the goal is to predict a number rather than classify data. The concept itself is quite simple: R-squared tells you what percentage of the variation in what you're trying to predict is captured by the model. A higher R-squared means more of the variation is captured. In practice, this can be helpful when comparing one model to the next, as a higher R-squared value generally indicates a better model. However, this isn't always the case. You should also consider other metrics like mean squared error (MSE) or MAE, which we'll discuss next.

The R-squared metric is a number between 0 and 1, where 0 means the model isn't capturing any of the variation and 1 means the model captures all the variation. The threshold for a good model depends on the use case:

- **Business and social science domains**

  Any time you involve humans, the model's ability to capture most of the variation becomes much harder. This is reflected in what's usually considered acceptable when evaluating the R-squared metric for business and social science domains. Anywhere between 0.3 to 0.6 is acceptable, with higher numbers being preferable. (Until it's closer to 0.9—then you need to consider whether you've made an error in the modeling process).

- **Engineering and science domains**

  If your domain is rooted in the laws and theories of science, the expectation becomes much higher, and models are expected to capture most of the variation. This translates to a generally acceptable minimum R-squared of 0.8 and above.

Assuming you're a business user, you'll want to be in the range of 0.5 to 0.7 to feel good about the model.

Sometimes chasing an R-squared and other error metrics can be a bad idea, especially with human-centric use cases. The following story illustrates how efforts to improve R-squared can raise ethical and fairness concerns.

Do You Always Want the Best Model?

In one of my roles, I was working on a recruiter capacity model. The approach was simple in practice: Use historical data to determine what the workload of a recruiter should be. Various historical data attributes were considered about the types of jobs being worked (skillsets, job level, etc.). However, we struggled to get R-squared above 0.6 when only looking at the data for the jobs that were being recruited. We decided to look at *who* was filling the jobs (the recruiters). We then saw significant improvements in the R-squared, which reached closer to 0.75.

Intuitively, this makes sense. The performance and abilities of recruiters will vary, so without this context, there's unexplained variance in the model. The model was better, but is this the right approach? Think about the use case of a recruiter capacity model. The use is for managers to identify who to assign work to. If you're at the same level as another recruiter, should you be expected to fill more roles because you've historically performed better? No!

We made the decision to ensure performance was not included and instead add in the level of the recruiter. Higher-level recruiters will have higher expectations because of their title and compensation. This is a fair and justifiable inclusion. While the R-squared only ended up being around 0.65, the model usage was ethical and fair.

6.2.8   Mean Squared Error

MSE is one of the most common regression error metrics. The name of the metric tells you how it works:

- Each of the predictions are compared to their actual outcome, giving you a list of positive or negative errors for each prediction.
- The error numbers are then squared. This accomplishes two things:
  - All numbers are positive, ensuring a negative and positive number don't cancel each other out to make the model appear like it's performing better than it is.
  - More weight is applied to bigger errors.

The challenge with MSE usually comes down to interpretability. Squaring the number makes it hard to translate back to your original data. This means the metric is better for comparing one version of the model to another than for sharing with stakeholders.

6.2.9   Mean Absolute Error

MAE is very similar to MSE, but it doesn't square the errors. Instead, it takes the absolute value of the error results, so you get the benefit of avoiding positive and negative errors cancelling each other out without the lack of interpretability.

This is personally my favorite metric and the one I default to the most with regressor predictions. It's also the primary metric we used in our first and second use cases. It's often easier to work with a model when you can relate its output back to the original data. This is also a metric that stakeholders can understand. Once they understand the concept of absolute values, they actually view this as a very strict metric (in a good way). This viewpoint can help with the stakeholder expectations, as they understand it's challenging to get the metric close to zero.

6.2.10   Metric Summary

There are many other metrics available within `sklearn.metrics` that you can explore. We focused on some of the most common ones to get you started. From here, you can continue to explore new metrics for your use cases!

Table 6.4 summarizes the metrics we discussed as a quick reference.

Metric	Type of Prediction	Pros	Cons
Accuracy	Classification	Easy to understand	Simplicity can lead to misleading results
True positive	Classification	Easy to understand; great when predicting rare outcomes	Tells part of the story in isolation
True negative	Classification	Easy to understand; great when predicting common outcomes	Tells part of the story in isolation
False positive	Classification	Easy to understand; ideal when use case needs to avoid incorrectly predicting an outcome occurring	Tells part of the story in isolation
False negative	Classification	Easy to understand; ideal when use case needs to avoid incorrectly missing a prediction	Tells part of the story in isolation
Precision	Classification	More robust at measuring how the model avoids false alarms	Adds complexity and doesn't address false negative error
Recall	Classification	More robust at measuring how the model identifies the outcome occurring	Adds complexity and doesn't address false positive error
F1 score	Classification	Balances precision and recall into one metric to optimize	Adds complexity and likely can't be used with nontechnical stakeholders
AUC	Classification	Handles imbalanced data and can be understood by stakeholders	Doesn't translate directly back to the dataset
R-squared	Regression	Relatively simple to understand and helpful when comparing models	Often not ideal to use as a standalone metric
MSE	Regression	Handles large errors well and addresses negative numbers	Hard to interpret compared to the original data
MAE	Regression	Interpretable and can be used with stakeholders	Doesn't penalize large errors as much as MSE

Table 6.4 Metric Summary Table

6.3   K-Fold Cross-Validation

We snuck this approach into our discussion of grid search in Chapter 5, Section 5.5, without explaining much about it. Now, we'll go deeper. First, we'll explain exactly what it is, and then we'll go through why you should use it. It's a tricky concept, so give yourself some grace if you have to review this content a few times before you feel like you understand it.

Cross-fold validation splits your data up into k datasets. K is the hyperparameter that you pass to the function. This is a similar concept to the train-test split (see Chapter 5, Section 5.3); however, instead of just doing one split, you're splitting the data into k equally sized datasets. You can do a traditional train-test split alongside a k-fold cross validation, which creates multiple variations of the training set for your model to build upon. This improves the model's generalizability by giving it multiple versions of the training data to identify the most important trends.

It can be helpful to think about cross-fold validation as an additional train-test split within the model training process. For each split of the data, the remaining splits are used to train the model, while the last split serves as the testing dataset. This process is repeated for all splits.

So, why do this? Think about how much your model relies on one split of the data. If you select a k of 5, you're getting 5 models brought together. This means the training data has less risk of overfitting.

In practice, this process is obscured behind the functions you use. Think back to when we were built the initial models for our use cases and introduced grid search. Cross-fold validation was as simple as putting in a number for the CV hyperparameter. The primary downside of cross-fold validation is the processing and load times. The introduction of both grid search and cross-folds creates an exponentially increasing number of models that need to be built. It's best to limit your number of cross-folds earlier in the process when you have a larger number of grid search options. (Generally, it's best to stay below 5.) As you refine the model and have a smaller grid to search, it can be beneficial to increase your number of cross-folds (aim between 5 and 10).

6.4   Business Validations

Up to this point, we've been discussing metrics that are used universally across different applications of machine learning and predictive modeling. Now, we'll discuss the softer (but arguably more important) legal and ethical validations you should consider when building and testing models.

This is likely something you'll be required to do as part of any model governance process, especially if you work for a large company. Once the legal team gets involved, you'll likely won't have to come up with the validations; you'll execute them instead. If you work for a small company, it's possible your company's policy and governance around models isn't as mature. This means you'll likely be taking a more active role in what considerations should be reviewed.

Throughout my roles, I've worked with a number of lawyers on these questions. Some have been great to work with, while others are more challenging because they want to assume no risk. To be fair, their job as a corporate attorney is to minimize and mitigate risk for the company. We can separate their decision-making into two categories: legal and ethical. In more prescriptive language, they are asking two questions:

1. Will this get the company sued?
2. If this goes poorly, will it be on the front page of the New York Times?

The first question directly translates into legal considerations, which we'll discuss first. The second question is a blend of ethical and reputational risk. Doing something unethical can definitely land you on the front page of the New York Times, even if it's not illegal. However, there are also situations in which the company does nothing illegal or unethical, and it still hurts their reputation. For example, what if a retailer decided to use your purchasing behavior from their e-commerce data to help make decisions when hiring? If they submitted all the risk data usage and privacy disclosures for their accounts, they're not doing anything illegal or unethical. However, the risk of poor public perception remains high.

6.4.1   Legal Considerations

While both legal and ethical considerations are very important, legal considerations should be the priority. If your model is breaking the law, you'll be in a bad spot. There are many new laws about letting AI and machine learning be the sole decisionmaker in a business process, so it's essential that your model doesn't contain legally challengeable concerns. Let's see an example from the recruiting space.

Resume Scoring Models

I've worked on and with many resume scoring models and projects. One of the common threads among them has been the focus of avoiding racial or gender bias. The companies I worked for never wanted ethnicity and gender to be decisionmakers in the hiring process. So, how does this relate to modeling?

One of the challenges with machine learning is that we can't explicitly understand what data the models rely on. In the context of ethnicity and gender, there's a risk that the model can find a proxy for these factors in other features. This means it's not enough to prove you're not including ethnicity and gender into your model explicitly; you have to prove through a deeply complex model that it's not being brought into the model's knowledge base from other variables.

This is an instance where you want to focus on the outcomes. To prove the model isn't perpetuating any bias it may have been trained on, you can cut your model's results by ethnicity and gender. Compare them to the results of the actual hiring process to identify whether your model is discriminating based on ethnicity or gender.

This will get a bit deeper into the recruiting analytics space, but the best comparison for actual results is that the proportion of applicants to your jobs should match the proportion of individuals who are hired. There are a number of exceptions to this (sponsorship requirements being one), but in general the distribution of candidates should be consistent throughout the process. If your model is aimed at identifying qualified candidates, you should see a similar phenomenon in its results.

6

Given that pretty much all companies have lawyers, it's best to find one familiar with your model's domain as soon as you start the project. Contrary to some people's opinions, lawyers are still human beings with feelings. Involving them in the project from the beginning can help avoid roadblocks when you get to the model deployment stage. They'll also be more invested into the project and are more likely to provide you with a workable solution rather than a flat "No, you can't do this."

6.4.2 Ethical Considerations

This isn't a book on ethics, but unfortunately, it's not a topic you can escape in the landscape of predictive models and AI. Just because you can technically do something doesn't mean you should.

Personally, I prioritize maintaining the ethical and moral high ground on my projects. This often gets me into trouble with stakeholders who just want something delivered, but it also allows me to sleep at night.

What is considered unethical but not illegal can be hard to generalize. It often goes back to the use case and a question of fairness. In the story about the recruiting capacity model, is it fair to take into account someone's past performance on their capacity expectation? Probably not. Is it fair to show different prices to different people based on what you believe their price sensitivity is, or is that just good business? Also probably not, but this is a very common practice.

This advice is meant to be practical, but admittedly it could come across as a bit jaded. *In general*, in the analytics space, you'll be asked to cut corners on best practices to expedite timelines regardless of whether you're working on a predictive model, statistical analysis, or experiment. Part of being good at navigating your company and their culture while also being able to sleep at night is finding the balance of where and when you'll put your foot down. If you state you're uncomfortable with the approach for a project too often, you'll develop a reputation as someone who is inflexible. If you don't ever practice this, you'll be sacrificing your morals, which will begin to weigh on you over time. This is one of the unfortunate realities you'll most likely encounter and need to address as you advance in your analytics career.

6.5 Machine Learning Interpretability

We discussed machine learning interpretability briefly throughout this book, but let's dive further into the topic so you have a robust understanding of why machine learning models are commonly referred to as black box models. There are two main use cases for generating interpretability:

1. **Stakeholder explanation**

   It's very common for stakeholders to want to understand which variables drive the prediction. Sometimes, this stems from a desire to ensure the model is being developed properly, so they're checking that the model's best predictors make sense. Other times, the inner workings of the model are helpful to stakeholders, depending on their use cases. Simply knowing that a variable is important can be helpful to stakeholders.

2. **Model development**

   It can be incredibly helpful to check what your model finds predictive. This can guide your feature engineering efforts as well as identify potential errors in how your model works. For example, if one column seems way more important than every other column, you may want to verify there isn't an issue with the data you're feeding into your model.

In the following sections, we'll explain how interpretability works in simple regression model scenarios and then explore options for interpreting tree-based models.

6.5.1 Regression Models

To understand the topic of machine learning interpretability, we first need to discuss regression models. Regression produces something called *coefficients*, as we saw in Chapter 5. Coefficients represent the relationship between the variable and what is being predicted. Table 6.5 is an artificial example from a model used to predict the risk of someone having a stroke or heart attack. Emphasis on *artificial*: This isn't medical advice, as the data is not real!

Feature Name	Coefficient	P-Value
Age	.04	<0.001
Family history	1.2	<0.001

Table 6.5 Coefficients for Risk of Stroke or Heart Attack

Feature Name	Coefficient	P-Value
Average daily exercise	-0.8	0.002
Other conditions	1.5	<0.001

Table 6.5 Coefficients for Risk of Stroke or Heart Attack (Cont.)

There are a few important items to note with interpreting these results:

- **Does the p-value show statistically significant results?**
  The p-value is the probability that the results are due to random noise in the data. The standard for evaluating p-values is that anything less than 0.05 is considered statistically significant. There are many schools of thought on p-values that we won't explore here, but consider this a starting point. If a value is above the 0.05 threshold, it's an indication you may need to remove the feature from the model.

- **Coefficients capture direction and magnitude**
  When interpreting coefficients, their value and whether they're positive or negative provide additional information. For age, you can interpret this to mean the risk of a stroke or heart attack increases by 0.04 (4%) each year. This is a small positive increase in risk. In contrast, your risk goes down by 80% with each additional hour of daily exercise.

  You can see the value in this level of interpretability. From the model, you're able to identify how important each of the variables is to a high degree of specificity.

- **Assuming all other variables remain constant**
  This can be a bit difficult to wrap your mind around, but each of these coefficients assume all other variables remain constant. This means you can see notable changes in the coefficients as you add or remove features. For example, it may seem counterintuitive that age has such a small coefficient. However, the other variables soak up the predictive influence on the risk. This means that someone who is 70 years old, has no family history, regularly exercises, and doesn't have any other underlying conditions can have a relatively low risk. The non-age factors end up being more powerful predictors than age alone.

- **Dummy coded variables versus numeric variables**
  Numeric variables and dummy coded variables are technically interpreted in the same way, even if they don't appear to be. Age, for example, is a number that increases in increments of 1 year, whereas the family history variable would be coded as a 0 or 1. There is only 1 increase possible for the predictor, which translates into the full value of the coefficient.

  This means the interpretation of the variable's coefficient is important. While family history and the other conditions variable are both quite high, they're also binary—they're either present or not, rather than increasing on an infinite number line.

6.5.2   Tree-Based Models

Coefficients are not available in the tree-based models that we learned about in Chapter 5. If you think about a tree-based model in its simplest form, this makes sense. For example, a single decision tree can split on the same variable in multiple branches and levels. Therefore, it's impossible to isolate the variables' impact on the outcome in the simple sense of "with each increase in x, y increases by z."

So, what options *are* available to you? We'll cover them next.

Feature Importance

The first and simplest option is feature importance within sklearn. Using a synthetic dataset to generate a model, we can build a graph that shows the importance of each feature. As shown in Listing 6.1, we're creating an artificial dataset using the `make_classification` function, and then building and fitting a model. A new function we haven't discussed yet is `feature_importances`, which shows how important each column in our data is for the model's predictions. This is then plotted in a bar graph using `plt.bar`. Figure 6.1 shows the results.

```python
from sklearn.ensemble import RandomForestClassifier
from sklearn.datasets import make_classification
import matplotlib.pyplot as plt
import numpy as np

X, y = make_classification(
    n_samples=1000,
    n_features=10,
    n_informative=5,
    n_redundant=0,
    random_state=42
    )

model = RandomForestClassifier(random_state=42)
model.fit(X, y)

importances = model.feature_importances_

indices = np.argsort(importances)[::-1]

plt.figure(figsize=(10, 6))
plt.title("Feature Importances")
plt.bar(range(X.shape[1]), importances[indices], align="center")
plt.xticks(range(X.shape[1]), indices)
```

```
plt.xlabel("Feature Index")
plt.ylabel("Importance Score")
plt.show()
```

Listing 6.1 Synthetic Data, Model, and Feature Importance Graph Code

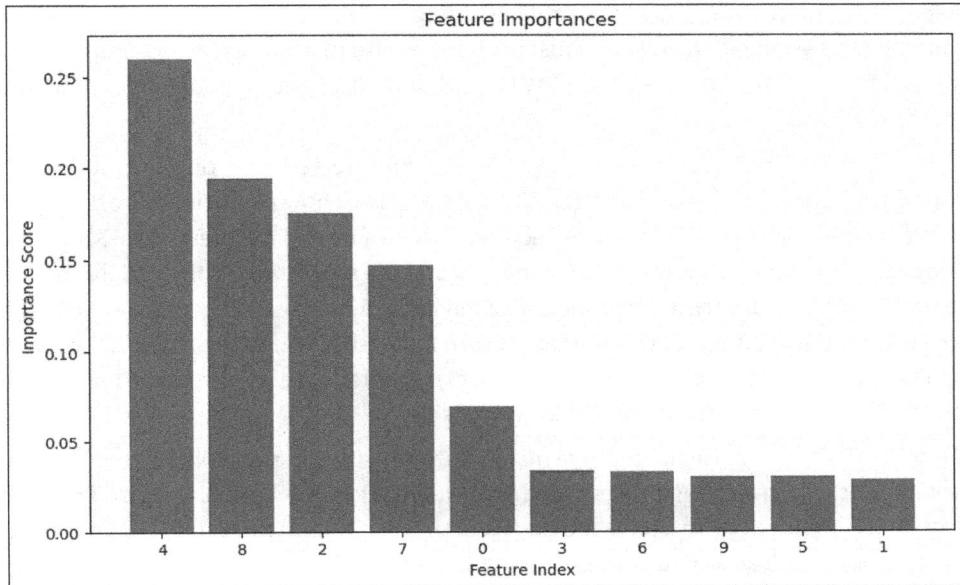

Figure 6.1 Feature Importance Graph

The code to generate the graph is relatively simple, but what do the numbers mean? All features are normalized to add up to 1, so this can be interpreted as a percentage of importance. There are two factors considered for this number:

1. **Reduction of uncertainty**
   This is an aggregation of how much uncertainty the variable removes from the data. As you recall, this is how a decision tree makes its decisions about which variable to split on. This means part of the feature importance calculation uses this same mechanism to identify the importance of the variable.

2. **Split frequency**
   This is more straightforward, as it's simply the number of times the feature is used in the model creation process.

The largest benefit to feature importance is its simplicity. The next option we'll discuss, Shapley Additive exPlanations (SHAP) values, provides more detail—but that detail comes at the cost of time. When you're developing models, it's valuable to be able to see what features are the most important at a quick glance.

In my personal workflow, I include the feature importance graph as part of each model iteration. This becomes another guide for me to understand what variables the model considers most important, which informs my next steps.

Shapley Additive exPlanations Values

When machine learning and AI started becoming popular, the SHAP values framework was adopted as one of the more robust ways to explain these complicated models. One of its differentiating factors is its ability to work with different types of models, making it a flexible tool.

Rooted in cooperative game theory, values are interpreted based on how much each variable contributes to the end result. This differs from how coefficients work, as you can't create a direct relationship between the SHAP value and the prediction. SHAP creates a relative score to show both the direction and the magnitude of the variable on the prediction. For example, a coefficient of 1.2 indicates that for each 1-unit increase in the column's value, you'd expect the prediction to increase by 1.2. With SHAP values, there isn't a specific number being generated. You can generally see the direction and magnitude of the column's impact to the target variable.

How SHAP values are calculated give insight into why they're so helpful. The SHAP process has its own modeling process on the backend. It looks at each feature and runs all possible combinations to isolate the impact of that variable. In a non-regression setting, this is about as close as you can get to a coefficient.

This benefit comes with a significant downside. Whenever you run through all possible combinations, the processing time and computing power required are large. The run-time to generate the SHAP values can be extensive. This is why it's best to reserve the SHAP value generation for the later stages of the model development process. You can generate a feature importance chart almost instantly, but generating the SHAP values will take around the same amount of time as it did to train your original model, if not a bit longer.

SHAP values should be used to get insights into the model. The generated graph can be helpful; however, most stakeholders find it confusing. Instead of presenting the SHAP visuals, it's usually easier to provide written insights about the model.

Listing 6.2 and Figure 6.2 show the code and output to generate a SHAP value summary plot. Take some time to read through the plot specifically to see if you can figure out how to interpret its values.

```
import shap
from sklearn.ensemble import RandomForestRegressor
from sklearn.model_selection import train_test_split
from sklearn.datasets import make_regression
```

```
X, y = make_regression(n_samples=2000, n_features=5, noise=2, random_state=42)

X_train, X_test, y_train, y_test = train_test_split
  (X, y, test_size=0.2, random_state=42)

rf_model = RandomForestRegressor(n_estimators=100, random_state=42)
rf_model.fit(X_train, y_train)

explainer = shap.TreeExplainer(rf_model)

shap_values = explainer.shap_values(X_test)

shap.summary_plot(shap_values, X_test)
```

Listing 6.2 SHAP Example

The steps to build a SHAP plot are similar to a predictive model—until the end, where the output is a graph instead of a set of predictions.

We then use the `TreeExplainer` function, which allows SHAP to understand the model that's been built:

```
explainer = shap.TreeExplainer(rf_model)
```

Now it understands the model, you can calculate the SHAP values using the `shap_values` function:

```
shap_values = explainer.shap_values(X_test)
```

In this case, it's being done on the `X_test` data, which gives us insight into how the model interprets the results of our test data. The same could be done for the `X_train` data as well.

Both of these lines of code will be relatively consistent. Regardless of how you want to use the SHAP values, you'll run these two lines of code.

The last piece of code is the summary plot, as follows, which you see generated in Figure 6.2:

```
shap.summary_plot(shap_values, X_test)
```

This is my personal favorite SHAP plot as it provides a macro-level perspective on what the model is thinking. SHAP is considered one of the most popular, if not the most popular, approaches to gaining insight into the black box. Another approach is called Local Interpretable Model-Agnostic Explanations (LIME). We won't go through LIME here, but you can check out another resource that discusses it at the following URL: *https://christophm.github.io/interpretable-ml-book/lime.html*.

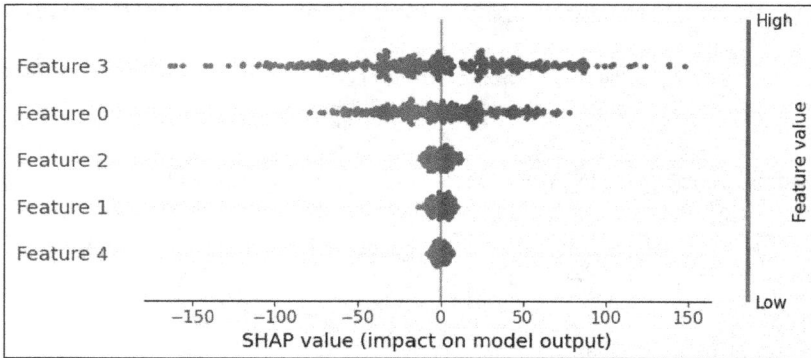

Figure 6.2 SHAP Example Summary Plot

If you haven't seen this type of graph before, it can be intimidating. However, once you've used it a few times, you'll start to recognize the gold mine of information this plot contains:

- **Dots**

  Each dot in the plot represents a single instance in your data (1 row).

- **Y axis**

  Each column in your data is represented on the Y axis. For each row in your data, there will be 1 dot within each feature. The position of the dots within each feature is called *jittering*, and it provides insight into how many instances occur in the same area. You don't want all these dots to be right on top of each other, as 100 instances would look the same as 1 instance.

- **Color**

  This is where people get most confused. The *feature value* legend is the feature's actual value relative to its possible range, not its importance or value to the model. For example, if feature 3 represents home value, the darker reds indicate multimillion-dollar houses, while the blues represent older, run-down houses.

- **X axis**

  The last component is the actual SHAP value, which is represented on the X axis. This shows the impact that feature has on the model for that specific instance. Points further to the right indicate a positive increase in the output, while points further to the left signify a decrease. If this model was predicting the probability of selling a house, then dots further to the right would indicate the owners are more likely to sell the house, while the dots further to the left indicate they're less likely to sell the house.

You'll see this more clearly when we get to our use cases, but its value really shows in the combination of the color and X-axis. If feature 3 represents home value, the combination of color, X-axis position, and dot trends show that higher home values lead to a high probability of selling the house.

We'll return to the topic of SHAP values and apply them to our use cases in Section 6.7.

6.6   Iterating on the Model

There are many important parts of the modeling process. In 99 out of 100 projects, you'll spend most of your time preparing your data (per our discussion in Chapter 4). The next most time-intensive stage of the modeling process is iteration. Although it takes time, it's an important step in building effective predictive models. Your first model is rarely your best. This is especially true for complex models like Gradient Boosting Machines (GBMs), where there are more hyperparameters to optimize.

Developing the first version of a model can provide a big dopamine hit. You spent all that time preparing the data, and now you get to see the results of the model!

Then, as you proceed with your initial iterations, you may feel a bit of a letdown. The excitement of your initial achievement fades, which can be demotivating. How can you stay engaged throughout the iteration process?

It's difficult to answer this question for everyone, but here are some thoughts based on my personal experience:

- Set high expectations for your own work—it can help you stay engaged when you know your results can be better.
- Remember that reaching the success threshold gives your stakeholders a valuable model.
- Focus on the continuous improvement mindset and apply it to the modeling process.

The process of iterating on your model can vary, but it fits into these four main categories:

- **Grid search**
  We discussed grid search when we introduced the various algorithms and their hyperparameters in Chapter 5, so we won't go into more detail here. This does not mean it's any less critical! When we iterate on the models in Section 6.7, we'll use grid search concepts heavily.

- **Feature engineering**
  We briefly touched on feature engineering in Chapter 5; however, we'll go into a bit more detail in this section to illustrate how important it is for both regression and tree-based techniques. There's a common misconception that tree-based models will do all of this for you, which won't yield the best results.

- **Removing variables**
  Sometimes you need to remove data from your model because it's simply adding noise. The more columns you feed into your model, the more processing power and time are required to train it. Sometimes less is more!

- **Adding new data**

  Sometimes a model just doesn't have the right data to understand what's going on. In these instances, you need to consider what new data you could add to give the model additional context.

Keep in mind that these categories often aren't linear; sometimes you need all of them, sometimes you only need a few. In the following sections, after discussing these topics with some light examples, we'll dive into the use cases and apply these concepts to the models we've already started.

6.6.1   Feature Engineering

People with a statistics background who understand predictive modeling tend to have a more refined perspective on feature engineering than those who started with tree models. While tree models do some feature engineering for you, it can be a bit of a trap to think you don't need to do any feature engineering yourself. In contrast, individuals trained in statistics are accustomed to curating variables to set up their model for success.

As we discussed before, feature engineering is a data transformation step where we adjust our existing data to change the perspective of the model. There are many categories feature engineering can fit into, as it's a wide-ranging topic. We'll focus on three of the most common ones in the following sections.

Revisiting the Business Context

Regardless of what features you're engineering, it's never a bad idea to make sure you're grounded in the business context. This is why revisiting your business case and its associated context is the first recommendation. What does this look like in practice? Consider the following anecdote.

Take-Home Assignment

I once interviewed for a job where one of the steps involved a take-home assignment. The goal was to generate a forecast for insurance claims data. I went through my usual process of exploring the data before going into the model development. Displayed in a line graph, the data showed a number of very clear trends.

I was using a feature-based model approach instead of a traditional time series model approach, so I needed to find ways for the model to pick up on the time-based context (month, year, etc.). The tree-based models would eventually pick up on this by the month and year information; however, this expends layers to the model and reduces the remaining sample size for other splits. My solution was to create a set of new variables that identified when these significant trends occurred, coding before as 0 and after as 1. This significantly increased the model's accuracy.

What's the takeaway from this example? By doing additional feature engineering, you can unlock additional value and accuracy from your model. Let's go a bit deeper into why this is.

For this explanation, let's assume you have two columns that represent the month and year of your data (remember you can't explicitly feed in a date, since it's not a numeric variable). For the model to identify a potential trend that occurs in a given month, it requires two splits: the first on year and the second on month. This already uses half the layers in your model, especially if your dataset is smaller and you have to set a smaller max_depth of 4. By performing feature engineering, you've allowed the model to explore other parts of the data and make an additional split that it otherwise may not have.

Creating Composite Variables

The combination of two variables into one is called a *composite variable*. The previous example combined two variables together (month and year) to form a composite variable. Another common example is the creation of an index or score that combines multiple variables together. Again, the goal is to do the exploration for the model and provide it as a feature, so the model only takes one split of the data instead of two or more.

An example of an index or score is a health risk score. This can be generated from a combination of body mass index (BMI), blood pressure, cholesterol, and smoking status to create a numeric variable the model can use to split. By providing what you already know about each metric and combining them, you can input existing knowledge into the model while reducing the effort required from the model to do so.

When your dataset is smaller (hundreds or thousands of rows), creating more composite variables becomes increasingly important. The more splits required for the model, the more likely it is to overfit. This becomes a problem for a smaller dataset because you combat overfitting by limiting how deep the model can go and how many samples are required to execute a split. When your dataset is larger (at least tens of thousands of rows), you're more likely to get by with allowing the model to go deeper because you have more data for the model to build upon.

Switching Variable Data Types

This technique is a sneaky yet effective approach you can test out when building your models. To explain it, we'll need to get a bit meta when it comes to data types.

To start, what is a numeric variable?

- It's a number.
- The numbers are relative and measurable to each other.
- The basic foundations of math can be applied to them to extract meaning from their relationship.

Next, what is a categorical variable?

- In a model, it's represented as a 1 or 0.
- Each unique value requires its own column.

While some variables are relatively straightforward (e.g., a fruit is categorical, not numerical), the grey area between what is a numeric or categorical variable is larger than you may think. Let's take a look at two examples: job levels and department names.

It may seem more logical that a job level could fit into either category. Job levels are usually represented in a non-numeric fashion; however, they do have a natural order to them. Say the company you work for uses alphabetical notation, with A as the lowest-level job and H as the chief executive officer (CEO). Each letter has a combination of steps within it, indicating incremental increases within that letter (A1, A2, A3, B1, B2, etc.).

By itself, this is not numeric. However, it can be converted into numbers by making A1 equal to 1, A2 equal to 2, and so on. You should consider how the job levels relate to each other, especially in the context of one level to another. Is the jump from one level to the next always equal? If you've worked in the corporate world, you'll know the answer is likely no. Another consideration is how this translates into the dimensionality (number of columns) of your data. If there are more than 20 unique job levels, including this as a categorical variable adds more than 20 columns for your model to sift through (before any dimensionality reduction techniques). The last consideration is the amount of value in keeping the relative nature of each job level. When using job level as a numeric variable, even if the gaps between levels aren't consistent, is it better to keep some relative nature? For example, as a numeric variable, the gap between entry-level, managers, and executives is preserved as context the model can use within the column. When used as a categorical variable, this relative relationship is completely lost.

On the more controversial side, is it possible to make the names of departments a numeric variable? You could identify a metric or metrics that show how similar the departments are and quantify from there. You could also look into building a model that creates this type of metric, where the objective is identifying the similarity of the departments.

The goal of covering this isn't necessarily to encourage you to make everything a numeric variable. Instead, it's to make you think about your data in new ways.

6.6.2   Remove Variables

Removing columns can be an effective way to improve the speed at which your model can be trained. By removing the columns, you're removing how many decisions a model may need to make. Columns are most commonly removed as part of the iteration process, as well as when you're looking at feature importance and/or SHAP values.

Identify the low value features, remove them, and then retrain the model to see if there's any impact on the accuracy.

The approach to remove a single column or multiple columns is simple. Listing 6.3 shows an example with an artificial dataset.

```
import pandas as pd

data = {'col_1': [1, 2, 3], 'col_2': [4, 5, 6], 'col_3': [7, 8, 9]}
df = pd.DataFrame(data)


df_dropped_column = df.drop('col_2', axis=1)
df_dropped_column

df_dropped_multiple = df.drop(['col_1', 'col_3'], axis=1)
df_dropped_multiple
```

Listing 6.3 Removing Columns

After creating the dataset, the following line will drop a single column by specifying its name:

```
df_dropped_column = df.drop('col_2', axis=1)
```

The axis parameter must be set to 1 to drop a column. If it's set to 0, it will drop a row.

If you have multiple columns to drop, you can pass via a list of columns, as follows:

```
df_dropped_multiple = df.drop(['col_1', 'col_3'], axis=1)
```

This is quite simple to execute once you've identified which columns you need to drop.

6.6.3   Add New Data

This section is intentionally placed last in our discussion of the iterative process of building your model. When I started building predictive models, I was always very quick to add new data to my models. This shouldn't be your first step when you're iterating—the feature engineering of your existing data should be your first priority. However, you'll eventually start to see diminishing returns from your feature engineering, and at that point you should consider adding new data that isn't already directly or indirectly included in your data. The goal is to find a balance between extracting what you can from your existing data spinning your wheels for too before you consider other data sources.

When you do start adding new data, consider the following:

- **How will you prioritize which data you add?**
  You won't know exactly which data has the most value to add; however, you should make an educated guess based on your use case. Prioritize the data that you think will have the most impact and isn't already accounted for in the model. The tricky part to consider is whether the data you already have is a proxy for, or is correlated with, the data you're thinking about adding.

  For example, say you're predicting the sales of a seasonal item like clothing. You've already included the date in your data, and you consider adding temperature next. However, temperature data is already correlated with the date information included in your model. While temperature and weather data may provide more detailed information, including a completely new, unrelated dataset—such as recent social media trends—will likely provide more new information for predicting clothing sales. It's a stronger candidate to prioritize including in the model.

- **How will you join in the data?**
  Considering how the data will be integrated is a critical component when adding new data into your model. This is highly dependent on your use case, but your starting point will be the level of data currently used in your model. Compare it to the new data you're considering adding. The easiest scenario is when they're both already at the same level of detail (daily sales and daily weather, for example). This is a simple join based on the join key between both datasets. The next easiest scenario is when your existing dataset is less granular than the new dataset. For example, if one dataset is at the day level, while the other is at the month level, you can aggregate the new dataset to the same level as your existing dataset before joining them together. The most challenging scenario is when your existing dataset needs to become more granular to match the new dataset. This likely means you need to refactor all your data inputs to accommodate a lower level of detail. This can be a very time-consuming process, so it's best to ensure this new data is highly valuable to your predictions before considering its inclusion.

- **Should you bring in the data?**
  This is the ethical perspective on what data to use. Ask yourself: *Should* you be including this new data? This question is distinct from the value it may add to the accuracy of your model. As we discussed in Section 6.4, just because data may make your model more accurate doesn't mean it should be included.

Here is a real-world example that highlights some of the factors to consider when adding new data to a model.

Retirement Risk Model

I was working on a development project where I was tasked with predicting the risk of retirement for all the employees. This was a complement to the overall employee

---

turnover model, knowing the underlying reasons an employee would retire are different from the reasons they would voluntarily leave the company for another job.

I was able to get relatively positive results, but that was partly due to the nature of the decision making for an employee to retire. Any employee under the age of 50 essentially has no risk of retirement (you don't need a model to tell you that). Once I began looking at employees in their mid-to-high 50s, the risk started showing up.

An alternative to a predictive model is a rules-based model. With retirement, this could equate to a rule-based formula, which is easier to create than a predictive model. The formula would include a combination of time with the company and age, which together determine when the legacy pension plan could kick in and when the employee would become eligible for Social Security payouts. The model I built didn't perform any better than this rule-based approach. I built a tree-based model, and the lack of transparency into how risk was assigned really tipped the scales against it. So, I needed to start exploring other datasets.

The most logical one was payroll. Could you identify how much an employee contributes to their 401(k) from their paycheck? In theory, this is where a machine learning approach thrives, because there are likely many factors and outcomes that could mean different things. For example, if someone is already set for retirement, it's possible they're not maxing out to their full catchup amount as they near retirement age. But in another situation, this could indicate a lack of readiness for retirement. For this use case, it makes a lot of sense to include payroll information in a model that predicts retirement risk. However, is it appropriate to bring in this data? Ultimately, the answer is no. Generally, payroll information is viewed as non-analytics data, only to be used for its intended purpose. The sensitivity of this data goes beyond the decisions individuals are making with their 401(k). It can reveal other personal, non-work-related information, such as wage garnishments. Overall, the payroll data is messy and complicated, and I determined that it shouldn't be included in the model.

I went through the same process with employee experience data. There is likely predictive value in seeing who is less engaged, as they might be more likely to retire on time than someone who is still highly engaged and enjoys their work. Ultimately, I decided not to include the dataset for reasons similar to those with the payroll data. When taking the employee experience survey, employees expect anonymity. While using the data does not directly violate this, including it in a model that may be used to make decisions could indirectly impact an individual.

After this exploration, I decided not to move forward with the model. While I was disappointed in the moment after spending quite a bit of time on it, it ultimately was the right decision. We found that associates discussed retirement openly with their managers, meaning the value of a predictive model was relatively small. It would never be able to reach the level of context that a manager has.

6.7   Application to Use Cases

Now that we've discussed the iteration process, let's get back to our use cases! We'll focus on the code, with commentary and points of emphasis throughout.

To ensure clarity of the examples, some of the code from the previous chapter when the models were introduced is revisited. This creates an all-inclusive and continuous example of the modeling process.

6.7.1   Use Case 1

For the first step, let's load in our sales data (see Listing 6.4).

```
#import pandas and numpy
import pandas as pd
import numpy as np

#load in the data
df_cleaned = pd.read_csv("df_cleaned.csv")
```

Listing 6.4 Load in Data

Now, let's do the same feature engineering of creating a 3-month lagged variable, identifying the day of the week, and creating a variable that identifies the timeline of events (refer to Chapter 5, Section 5.4.3; see Listing 6.5).

```
#aggregate the data
df_data_grouped = df_cleaned.groupby('invoicedate').agg(
    {
      'Description': 'nunique',
      'Customer ID': 'nunique',
      'Country': 'nunique',
      'Quantity': 'sum',
      'Price':'sum'
    }
    ).reset_index()

#import date specific libraries
from datetime import datetime
from dateutil.relativedelta import relativedelta

#create date columns
df_data_grouped['invoicedate'] = pd.to_datetime(
  df_data_grouped['invoicedate']
  ).dt.date
```

```python
#create lagged date column
df_data_grouped['invoicedate_minus_3'] =
  df_data_grouped['invoicedate'] + relativedelta(months=3)

#create data frame with just the date and price
df_for_model = df_data_grouped[['invoicedate', 'Price']]

#create dataframe for lagging
df_minus_3 = df_data_grouped[[
  'invoicedate_minus_3',
  'Description',
  'Customer ID',
  'Country',
  'Quantity',
  'Price'
  ]]

#create month feature
df_minus_3['month'] = pd.to_datetime(
  df_minus_3['invoicedate_minus_3']
  ).dt.month

#create first sale date feature
df_minus_3['since_first_sale'] = pd.to_datetime(
  df_minus_3['invoicedate_minus_3']) -
  min(pd.to_datetime(df_for_model['invoicedate']))

#update datatype of the first sale date to be a number
df_minus_3['since_first_sale'] = df_minus_3['since_first_sale'].dt.days

#join the data frames together
df_for_model = pd.merge(
  df_for_model,
  df_minus_3,
  how = 'left',
  left_on = 'invoicedate',
  right_on = 'invoicedate_minus_3',
  suffixes = ('','_3months')
  )

#remove any rows with a blank for the lagged data
df_for_model = df_for_model[pd.notna(df_for_model['invoicedate_minus_3'])]
```

```
#create the day of the week feature
df_for_model['day_of_week'] = pd.to_datetime
  (df_for_model['invoicedate']).dt.weekday
```

Listing 6.5 Repeating Feature Engineering

Finally, to get our data caught up to where it was previously, we need to split it into training and testing datasets (see Listing 6.6, and refer to Chapter 5 for more details).

```
#identify the max date for splitting into the train and test data
max_data_date = max(df_data_grouped['invoicedate']) - relativedelta(months=3)

#create the training and test data
training_data = df_for_model[df_for_model['invoicedate'] <= max_data_date]
test_data = df_for_model[df_for_model['invoicedate'] >= max_data_date]

#print the number of rows and columns for the training and test data
print(training_data.shape)
print(test_data.shape)

#specify the columns for the training data
training_data = training_data[[
  'Description',
  'Customer ID',
  'Country',
  'month',
  'since_first_sale',
  'Price_3months',
  'day_of_week',
  'Price'
  ]]

#specify the columns for the test data
test_data = test_data[[
  'Description',
  'Customer ID',
  'Country',
  'month',
  'since_first_sale',
  'Price_3months',
  'day_of_week',
  'Price'
  ]]
```

```
#select only the columns used as predictors
X_train = training_data.iloc[:, :-1]
X_test = test_data.iloc[:, :-1]

#select only the target column
y_train = training_data.iloc[:, -1]
y_test = test_data.iloc[:, -1]

#create the holiday season feature
X_train['holiday_season'] = np.where(X_train['month'] >= 9, 1, 0)
X_test['holiday_season'] = np.where(X_test['month'] >= 9, 1, 0)

#dummy code the day of the week feature for the training data
train_weekday_dummies = pd.get_dummies(
  X_train['day_of_week'],
  prefix='weekday'
  )

#add the dummy coded values onto X_train
X_train = pd.concat(
  [X_train, train_weekday_dummies],
  axis = 1
  )

#dummy code the day of the week feature for the test data
test_weekday_dummies = pd.get_dummies(
  X_test['day_of_week'],
  prefix='weekday'
  )

#add the dummy coded values onto X_test
X_test = pd.concat(
  [X_test, test_weekday_dummies],
  axis = 1
  )
```

Listing 6.6 Split into Training and Testing Data

Now our dataset is set up for modeling. Let's rebuild the last model we created in Chapter 5, Section 5.5.3 as a starting point (see Listing 6.7 and Figure 6.3). As a reminder, we've created a starting point for a GBM model, but as we can see from Figure 6.3, the model isn't performing well, which is also reflected in the MAE value.

```
#load in necessary libraries
from sklearn.ensemble import GradientBoostingRegressor
from sklearn.metrics import mean_absolute_error

#create the model with the last hyperparameters used in the last chapter
model = GradientBoostingRegressor(
  n_estimators=100,
  learning_rate = .01,
  subsample = .9,
  random_state=42
  )

#fit the model to the data
model.fit(X_train, y_train)

#create predictions on the test data
y_pred = model.predict(X_test)

#create predictions on the training data
y_train_pred = model.predict(X_train)

#calculate and print the MAE for the training data
accuracy = mean_absolute_error(y_train, y_train_pred)
print(f" Training Accuracy: {accuracy}")

#calculate and print the MAE for the test data
accuracy = mean_absolute_error(y_test, y_pred)
print(f"Accuracy: {accuracy}")

#import the plotting library
import matplotlib.pyplot as plt

#plot both the actual and prediction data
plt.plot(test_data['since_first_sale'], y_test, label = "actual")
plt.plot(test_data['since_first_sale'], y_pred, label = "prediction")
plt.legend()
```

Listing 6.7 Initial GBM Model

This reminds us that our model isn't appropriately predicting the pattern. Using our new tools, where do you think is the best place to start?

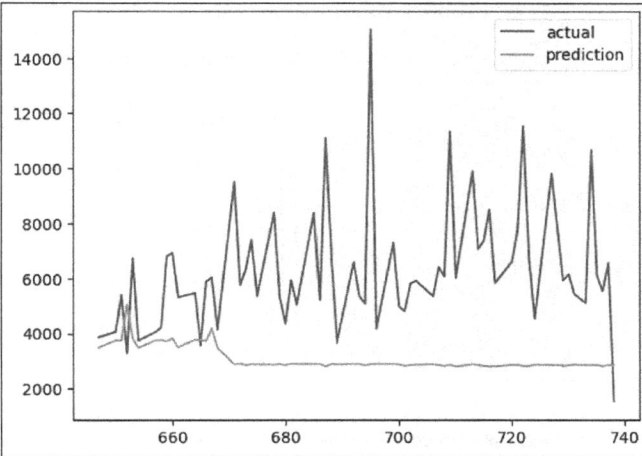

Figure 6.3 Initial GBM Model Result

Let's review the feature importance chart first to see what the model currently indexes on, with the hope that there will be insights we can pull from. Listing 6.8 shows the code, where we extracted the feature importance values using the feature_importance_ function. Then, we added it to a DataFrame and plotted it in a bar graph. Figure 6.4 shows the output.

```
#identify the importance of the features
importances = model.feature_importances_

#put feature importance and names of the features into a data frame
feature_importances = pd.DataFrame(
  {'Feature': X_test.columns,
  'Importance': importances}
  )

#sort the data frame to show the important features first
feature_importances = feature_importances.sort_values(
  'Importance',
  ascending=False
  ).reset_index(drop=True)

#plot the feature importance
feature_importances.plot(
  x='Feature',
  y='Importance',
  kind='bar'
  )
```

Listing 6.8 Initial GBM Feature Importance

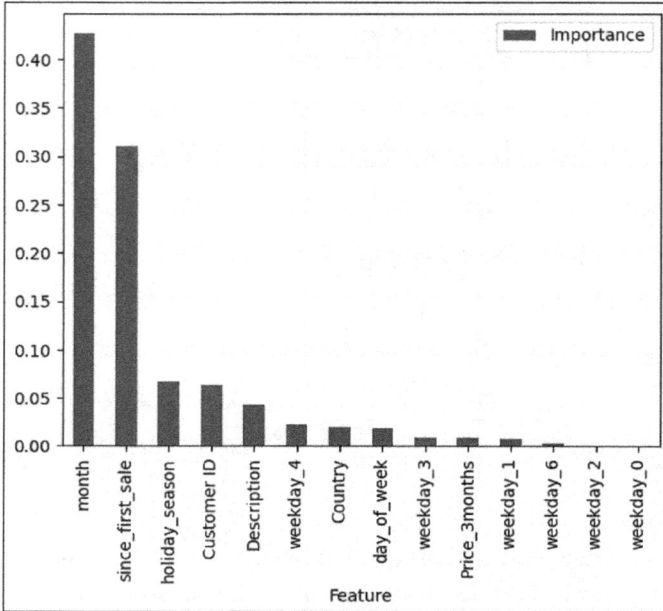

Figure 6.4 Output of Initial GBM Feature Importance

We can see that the month and since_first_sale (the amount of time since the first sale) are the two most important features in the model. This may help explain why the model isn't performing well on the test data. The month granularity is too fine, and the trend over the entire dataset isn't steadily increasing; the increase occurs mostly towards the end of each year. As a reminder, Figure 6.5 shows the trend graph.

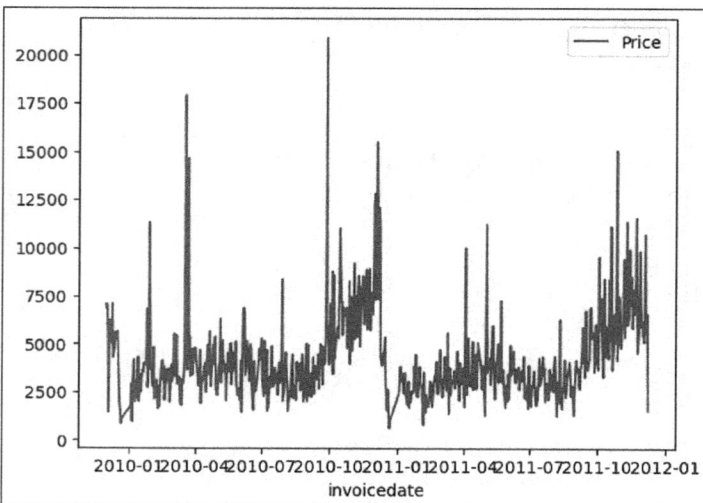

Figure 6.5 Data Trend

If you think about the underlying data, this also means the model only has one instance to learn the trend occurring at the end of the year. This is a use case for additional feature engineering. Let's identify the day of the year for each date (1–365) with the hope that the model picks up on this trend and provides a more accurate prediction at the end of the year. The code to do this is relatively simple:

```
#create day of the year feature
df_for_model['day_of_year'] = pd.to_datetime
  (df_for_model['invoicedate']).dt.dayofyear
```

6

Now we need to recreate our training and test data to incorporate this. The primary adjustment required is to select the new day_of_year column (see Listing 6.9).

```
#identify the max date for the train test split
max_data_date = max(df_data_grouped['invoicedate']) - relativedelta(months=3)

#create the training and test data
training_data = df_for_model[df_for_model['invoicedate'] <= max_data_date]
test_data = df_for_model[df_for_model['invoicedate'] >= max_data_date]

#print the number of rows and columns for the training and test data
print(training_data.shape)
print(test_data.shape)

#select the columns for the training data
training_data = training_data[[
  'Description',
  'Customer ID',
  'Country',
  'month',
  'since_first_sale',
  'Price_3months',
  'day_of_week',
  'day_of_year',
  'Price'
  ]]

#select the columns for the test data
test_data = test_data[[
  'Description',
  'Customer ID',
  'Country',
  'month',
  'since_first_sale',
  'Price_3months',
```

```
   'day_of_week',
   'day_of_year',
   'Price'
   ]]

#select the predictors to become the X_train and X_test data frames
X_train = training_data.iloc[:, :-1]
X_test = test_data.iloc[:, :-1]

#select the target column for both train and test data
y_train = training_data.iloc[:, -1]
y_test = test_data.iloc[:, -1]

#create the holiday season feature
X_train['holiday_season'] = np.where(X_train['month'] >= 9, 1, 0)
X_test['holiday_season'] = np.where(X_test['month'] >= 9, 1, 0)

#dummy code the weekday feature
train_weekday_dummies = pd.get_dummies(
  X_train['day_of_week'],
  prefix='weekday'
  )

#add the dummy coded weekday feature to X_train
X_train = pd.concat(
  [X_train, train_weekday_dummies],
  axis = 1
  )

#dummy code the weekday feature
test_weekday_dummies = pd.get_dummies(
  X_test['day_of_week'],
  prefix='weekday'
  )

#add the dummy coded weekday feature to X_test
X_test = pd.concat(
  [X_test, test_weekday_dummies],
  axis = 1
  )
```

Listing 6.9 Adding in New Day of Year Column

Next, we'll retrain our model using the same hyperparameters we used in our last model (see Listing 6.10 and Figure 6.6).

```
#load necessary libraries
from sklearn.ensemble import GradientBoostingRegressor
from sklearn.metrics import mean_absolute_error

#create the model
model = GradientBoostingRegressor(
  n_estimators=100,
  learning_rate = .01,
  subsample = .9,
  random_state=42
  )

#fit the model to the data
model.fit(X_train, y_train)

#create predictions for the test data
y_pred = model.predict(X_test)

#create predictions for the training data
y_train_pred = model.predict(X_train)

#calculate and print the MAE for the training data
accuracy = mean_absolute_error(y_train, y_train_pred)
print(f" Training Accuracy: {accuracy}")

#calculate and print the MAE for the test data
accuracy = mean_absolute_error(y_test, y_pred)
print(f"Accuracy: {accuracy}")

#plot the actual and prediction results
import matplotlib.pyplot as plt
plt.plot(test_data['since_first_sale'], y_test, label = "actual")
plt.plot(test_data['since_first_sale'], y_pred, label = "prediction")
plt.legend()
```

Listing 6.10 GBM with Day of Year Column

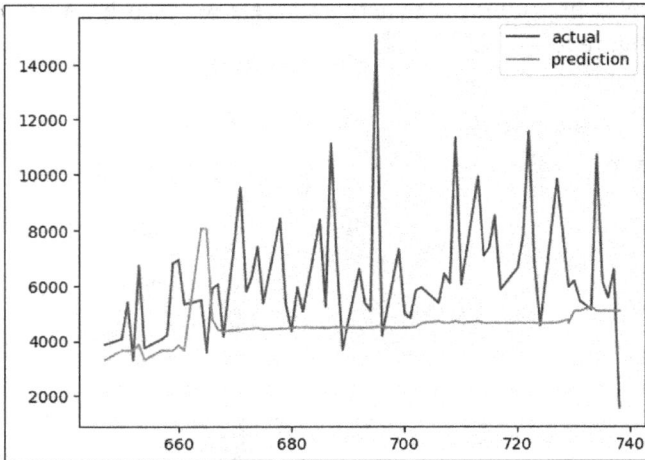

Figure 6.6 GBM with Day of Year Column

Before even looking at the MAE result, you can probably guess the model is much improved. The MAE is down to 2,200 from 3,300 in Chapter 5, Section 5.5.3. It's still underpredicting, but it's much closer than our previous iteration of the model. Now let's check in on our feature importance results (see Listing 6.11 and Figure 6.7).

```
#identify the importance of the features
importances = model.feature_importances_

#put the feature importance values and column names into a data frame
feature_importances = pd.DataFrame(
  {'Feature': X_test.columns,
  'Importance': importances}
  )

#sort the data frame so the most important features are at the top
feature_importances = feature_importances.sort_values(
  'Importance',
  ascending=False
  ).reset_index(drop=True)

#plot the feature importance data
feature_importances.plot(
  x='Feature',
  y='Importance',
  kind='bar'
  )
```

Listing 6.11 Feature Importance After Day of Year

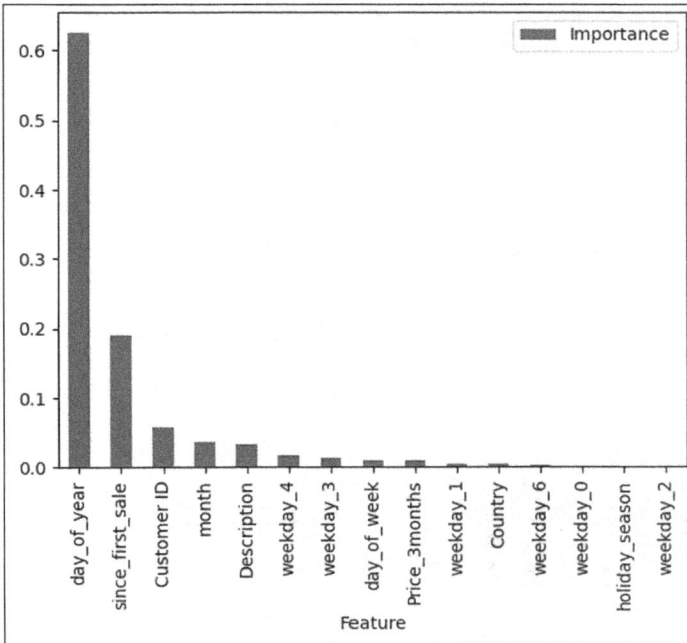

Figure 6.7 Output of Feature Importance After Day of Year

Wow, there's a lot to unpack here! The new day_of_year variable is now by far the most important feature in the model. Why are we seeing such a change in a feature like month? The same information available in month is also available in the day_of_year variable. However, the day_of_year variable appears to contain more valuable information for the model than month, so the model uses day_of_year instead. It still appears to find value in the number of days since the first sale (since_first_sale), indicating it's using this to pick up on the overall trend in the dataset.

Look back at Figure 6.6. What do you think the model is missing that's causing it to underpredict? It might still be missing the week-over-week variance in the results. Let's try a similar approach as we just did for the day_of_year variable, but this time, we'll focus on the week of the year (yes, that means rerunning our data pipeline). The following line is similar to day_of_week and day_of_year. We'll add it to the broader data pipeline we've been building upon:

```
#create week of the year feature
df_for_model['week_of_year'] = pd.to_datetime(
  df_for_model['invoicedate']
  ).dt.isocalendar().week
```

Then, if we rerun out model using the same hyperparameter set, we get essentially the same MAE. If we run the feature importance plot, we find something interesting, as shown in Figure 6.8.

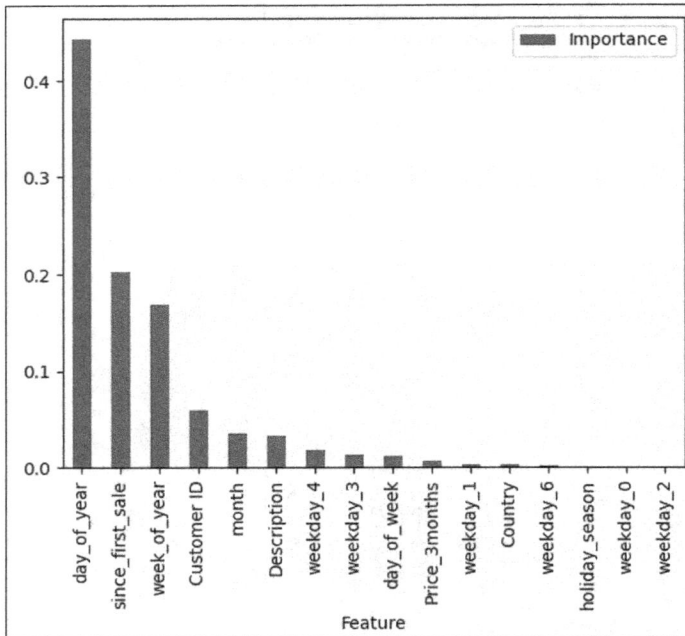

Figure 6.8 Adding in Week of Year

Now that we've added in week_of_year as a feature, the day_of_year dependency is reduced. This is interesting, because this indicates there should have been some performance improvements to the model, but we're not seeing that. In that case, it's time to go back to the grid search to see if there are additional hyperparameter options to consider now that we've added a few new columns. In Listing 6.12, we've added the grid search approach back into our model with GridSearchCV. We're starting with a wide range of hyperparameter values for our param_grid to see which combination produces the best results.

```
#load necessary libraries
from sklearn.model_selection import GridSearchCV
from sklearn.ensemble import GradientBoostingRegressor

#create grid of hyperparameters to test
param_grid = {
  'n_estimators': [100, 300, 500],
  'learning_rate': [.01, .05, .1],
  'subsample': [.7, .8, .9]
  }

#create the model
model = GradientBoostingRegressor(random_state=42)
```

```
#set up the grid search
grid_search = GridSearchCV(
  estimator=model,
  param_grid=param_grid,
  cv=3,
  scoring='neg_mean_absolute_error'
  )

#execute the grid search
grid_search.fit(X_train, y_train)

#print the best parameters from the grid search
print("Best parameters:", grid_search.best_params_)

#print the best score from the grid search
print("Best score:", grid_search.best_score_)
```

Listing 6.12 Grid Search After Adding Some New Features

The recommended hyperparameters returned from this are essentially the same, except the n_estimators parameter recommendation is now 300 instead of 100. When running the model with this adjustment, we get our best MAE at 2,100. The graph also appears to show our model picking up some variation (see Figure 6.9).

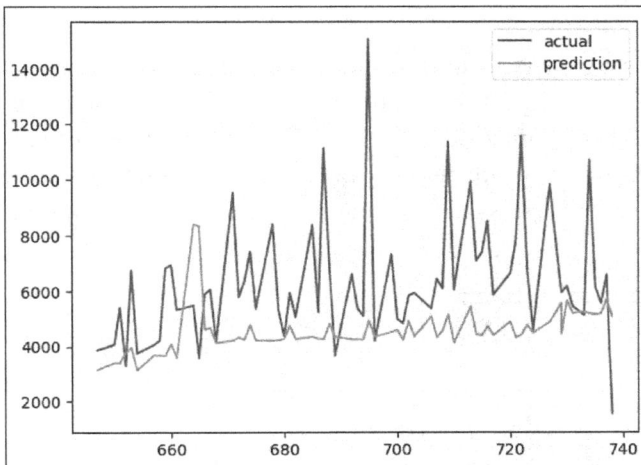

Figure 6.9 After Using Grid Search Recommendations

These results are promising. Let's do another round of grid search with more refined criteria centered around the values from the grid search process we picked previously (see Listing 6.13).

341

```
#updated grid search parameters
param_grid = {
  'n_estimators': [250, 300, 350],
  'learning_rate': [.005, .01, .15],
  'subsample': [.85, .87, .9, .92]
  }
```

Listing 6.13 Second Iteration of Grid Search

The resulting hyperparameters don't change, with the exception of the number of n_estimators. This likely means we're reaching the end of our iterations. The last adjustment we're going to make is to adjust the number of cross-folds. Given how small our dataset is, it's possible that increasing the number of cross-folds will result in positive improvements to the model. Within our GridSearchCV function, let's adjust the cv to 5, as shown in Listing 6.14.

```
#set up grid search
grid_search = GridSearchCV(
  estimator=model,
  param_grid=param_grid,
  cv=5,
  scoring='neg_mean_absolute_error'
  )
```

Listing 6.14 Adjusting Cross-Fold to 5

Our grid search has returned some new results! It's adjusted down the learning_rate to 0.005 and bumped the subsample up to 0.92. If we put these results into the model and train it, we unfortunately don't get a much better result. As you can see in Figure 6.10, there's less variation in the day-to-day predictions of the model. I'd feel more comfortable using the previous hyperparameters than I would these, given the difference in variation identification.

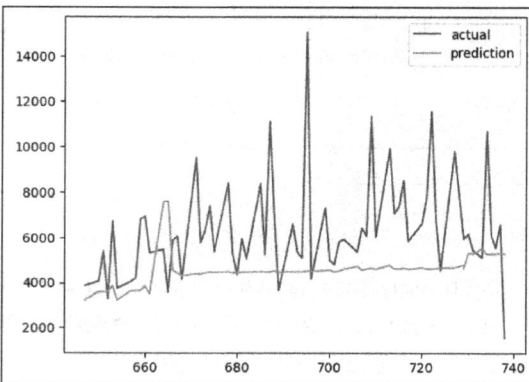

Figure 6.10 After Increasing cv for New Hyperparameters

Before landing on this model, let's revisit one more assumption. The train-test split was based on a time, not a randomized split. This approach is generally recommended if you have multiple years' worth of data. However, we only have about a year of training data for this use case. So, what if we revisit the assumption of how we split the data and see what happens when we randomize the split?

To do this, we need to refactor our data pipeline. As shown in Listing 6.15, we're recreating our dataset from our df_for_model object. This allows us to use the train_test_split function to randomize the split of our training and test datasets.

```
#load in train test split
from sklearn.model_selection import train_test_split

#create the predictors for the model
X = df_for_model[[
  'Description',
  'Customer ID',
  'Country',
  'month',
  'since_first_sale',
  'Price_3months',
  'day_of_week',
  'day_of_year',
  'week_of_year'
  ]]

#select the target variable
y = df_for_model['Price']

#create the holiday season feature
X['holiday_season'] = np.where(X['month'] >= 9, 1, 0)

#dummy code the weekday feature
x_weekday_dummies = pd.get_dummies(
  X['day_of_week'],
  prefix='weekday'
  )

#add the dummy coded weekday feature on X
X = pd.concat(
  [X, x_weekday_dummies],
  axis = 1
  )
```

```
#execute the randomized train test split
X_train, X_test, y_train, y_test = train_test_split(
  X,
  y,
  test_size=0.2,
  random_state=42
  )
```

Listing 6.15 Randomized Split

Now that we have the new training and testing datasets set up, we can rerun the modeling process. We made a few adjustments to the hyperparameters of our GBM model in Listing 6.16, and the reasoning will become clear when you review the graph in Figure 6.11.

```
from sklearn.ensemble import GradientBoostingRegressor
from sklearn.metrics import mean_absolute_error

#create model
model = GradientBoostingRegressor(
  n_estimators=250,
  learning_rate = .01,
  subsample = .9,
  random_state=42
  )

#fit the model to the data
model.fit(X_train, y_train)

#predict the test values
y_pred = model.predict(X_test)

#predict all values
y_pred_all = model.predict(X)

#predict the training values
y_train_pred = model.predict(X_train)

#calculate and print the MAE for the training data
accuracy = mean_absolute_error(y_train, y_train_pred)
print(f" Training Accuracy: {accuracy}")

#calculate and print the MAE for the test data
accuracy = mean_absolute_error(y_test, y_pred)
```

```
print(f"Accuracy: {accuracy}")

#plot the actuals and predictions
import matplotlib.pyplot as plt
plt.plot(X['since_first_sale'], y, label = "actual")
plt.plot(X['since_first_sale'], y_pred_all, label = "prediction")
plt.legend()
```

Listing 6.16 Randomized Test Split

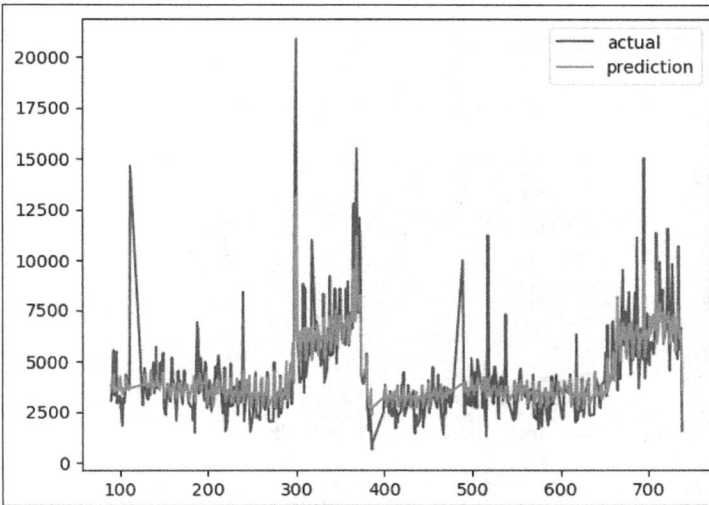

Figure 6.11 After Randomized Train-Test Split

This graph certainly passes the eye test! First, the MAE is now down to just over 1,000. The rationale for a randomized split is to ensure the model sees the full scope of the data, rather than being trained on only one year's cycle.

To show the graph this way, we created a prediction for the entire dataset. This really allows you to see how the model fits but doesn't overfit. We didn't focus on this metric comparison when we reviewed the time-based train-test split, but the MAE for the training and test set are much closer together on this model, so the overfitting is less of a concern.

This may feel a bit like magic. It's healthy to have some skepticism—and you should have some after seeing these results flip so quickly. First, let's review our feature importance graph in Figure 6.12.

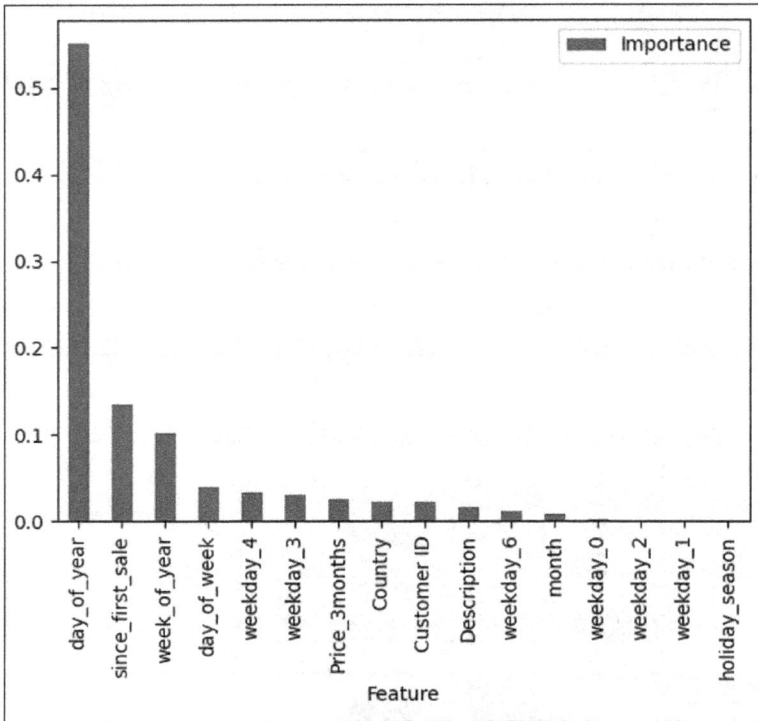

Figure 6.12 Feature Importance After Randomized Train-Test Split

The underlying feature importance hasn't changed much here. What we're concerned about is *leakage*, which in this case means the model is being told the answer by the data we're giving it. The risk of this occurring is especially heightened in cases where we're lagging the data because we're moving data elements to new rows. We're looking for cases where only one variable appears important, because leakage typically causes the model to focus solely on the column where the leakage occurs. However, this hasn't changed much from our previous model built on the time-based train-test split.

Now we can pull out the SHAP visual to verify this as well. In Listing 6.17, we've applied the SHAP value steps to this data, resulting in the output shown in Figure 6.13.

```
#import shap library
import shap

#create the SHAP explainer using the last model we created
explainer = shap.TreeExplainer(model)

#create the shap values for the test data
shap_values = explainer.shap_values(X_test)
```

```
#show the summary plot
shap.summary_plot(shap_values, X_test)
```

Listing 6.17 Building the SHAP Summary Plot

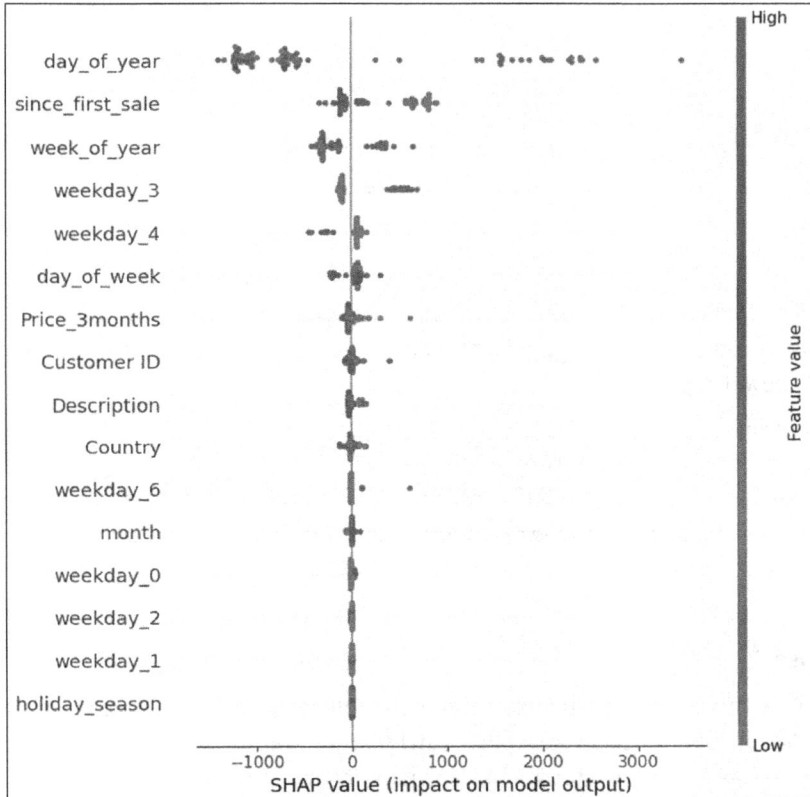

Figure 6.13 SHAP Summary Plot

Nothing in the SHAP summary plot indicates model leakage (the mixture of the dot columns in the day_of_year row confirms this). Instead, it appears that the purchase volumes from one year to the next are generally consistent. This explains how the variation from day to day and the trend over time is captured. Additionally, if you back at Figure 6.11, you won't see a perfect fit of the data on certain days; it follows the trends, but it doesn't match them perfectly. It also isn't perfectly fit to the extreme highs and lows. This is another indicator that model leakage isn't occurring.

Taking a step back, it's quite an accomplishment that we were able to get a model that can predict this well on one year of data. While Chris may say otherwise, the amount of data available doesn't really support a best-in-class predictive model. This is where the business case and realities of the real-world clash with model building. This foreshadows topics we'll cover Chapter 7, but we'll want to work with the predictions and then add some level of interpretation on them to ensure we're appropriately predicting the future.

The last component to this use case in this chapter is saving the model. This is called *pickling*, and it can be done easily with the pickle library, as shown in Listing 6.18.

```
#load pickle library
import pickle

#create model.pkl file and save the model to it
with open('model.pkl', 'wb') as file:
  pickle.dump(model, file)
```

Listing 6.18 Pickling Model

The mechanics of this are simple. You're creating a file called `model.pkl` and saving the `model` object to it. You can load it again later on (which we'll do in Chapter 7) to make predictions on new data.

If you've read other resources in the machine learning or predictive modeling space, pickling your model to save it for future predictions is not always a step discussed. Personally, I only learned about the modeling process through retraining my models each time, which isn't always practical in the real world when you're working with big models that take hours to train.

6.7.2   Use Case 2

When we left off with this use case, the underlying model metrics were poor, so we came to the conclusion that we need to take additional feature engineering steps. To start, we'll replicate the same steps we took when initially testing the models to create our target variables (refer to Chapter 5, Section 5.4.4). We'll first do this for the 7-day target, as shown in Listing 6.19.

```
import pandas as pd

#read in data
df = pd.read_csv("use_case_2_cleaned.csv")

#select columns for lagging
df_lagging = df[['Customer ID', 'order_date']]

#import specific date libraries
from datetime import datetime, timedelta

#create new date column for the lagging
df_lagging['order_minus_8'] =
```

```
    pd.to_datetime(df_lagging['order_date']) - timedelta(days=8)

#rename the date column
df_lagging['max_date'] = df_lagging['order_date']

#drop the old date column name
df_lagging = df_lagging.drop(columns=['order_date'])

#join the data
df_lag_7 = pd.merge(
  df[['Customer ID', 'order_date']],
  df_lagging,
  how = 'inner',
  on = 'Customer ID'
  )

#filter to only the rows to keep
df_lag_7 = df_lag_7[
  (df_lag_7['order_date'] > df_lag_7['order_minus_8']) &
  (df_lag_7['order_date'] < df_lag_7['max_date'])
  ]

#select only the customer and order date
df_lag_7 = df_lag_7[['Customer ID', 'order_date']]

#remove duplicates
df_lag_7 = df_lag_7.drop_duplicates()

#create a column of all 1's for the target
df_lag_7['target_7'] = 1

#join the lagged data onto the original dataset
df = pd.merge(
  df,
  df_lag_7,
  how = 'left',
  on = ['Customer ID', 'order_date']
  )

#fill any blank values with 0
df['target_7'] = df['target_7'].fillna(0)
```

Listing 6.19 Create 7-Day Target Variable

Next, we'll do the same for the 14-day target variable, as shown in Listing 6.20.

```python
#select only the customer and order date
df_lagging = df[['Customer ID', 'order_date']]

#create the new date for the 2 week lag
df_lagging['order_minus_15'] =
  pd.to_datetime(df_lagging['order_date']) - timedelta(days=15)

#rename the order date column
df_lagging['max_date'] = df_lagging['order_date']

#drop the old column
df_lagging = df_lagging.drop(columns=['order_date'])

#join the data
df_lag_14 = pd.merge(
  df[['Customer ID', 'order_date']],
  df_lagging,
  how = 'inner',
  on = 'Customer ID'
  )

#keep only the necessary columns
df_lag_14 = df_lag_14[
  (df_lag_14['order_date'] > df_lag_14['order_minus_15']) &
  (df_lag_14['order_date'] < df_lag_14['max_date'])
  ]

#select only the customer and order date
df_lag_14 = df_lag_14[['Customer ID', 'order_date']]

#drop the duplicates
df_lag_14 = df_lag_14.drop_duplicates()

#create the target variable column with 1's
df_lag_14['target_14'] = 1

#join the data to the original dataset
df = pd.merge(
  df,
  df_lag_14,
  how = 'left',
  on = ['Customer ID', 'order_date']
  )
```

```
#fill the blank target values with 0
df['target_14'] = df['target_14'].fillna(0)
```

Listing 6.20 Create 14-Day Target Variable

Now that we have our target variables set up, let's check in on the feature importance to see what the model currently focuses on before we do any additional feature engineering. To get the feature importance information, we need to train a model. If you recall from Chapter 5, the model performed best with the default hyperparameters, so we'll use those, as shown in Listing 6.21. As a reminder, we've created and trained a GBM model to predict the probability of a customer ordering again in the next week, using the ROC AUC as our accuracy metric to evaluate the model.

```
#load necessary libraries for the modeling process
from sklearn.model_selection import train_test_split
from sklearn.ensemble import GradientBoostingClassifier
from sklearn.metrics import accuracy_score
from sklearn.metrics import roc_auc_score

#select the predictors for the model
X = df.drop(
  columns=[
    'target_7',
    'target_14',
    'Customer ID',
    'timestamp',
    'order_date'
    ]
  )

#select the 1 week target column
y = df['target_7']

#execute the train test split
X_train, X_test, y_train, y_test = train_test_split(
  X,
  y,
  test_size=0.3,
  random_state=42
  )

#create the model
model = GradientBoostingClassifier(random_state=42)
```

```
#fit the model to the data
model.fit(X_train, y_train)

#calculate both the probability and binary predictions
y_pred_proba = model.predict_proba(X_test)[:, 1]
y_pred = model.predict(X_test)

#calculate the ROC AUC and accuracy scores
auc_score = roc_auc_score(y_test, y_pred_proba)
accuracy = accuracy_score(y_test, y_pred)

#print both the ROC AUC and accuracy scores
print(auc_score)
print(accuracy)
```

Listing 6.21 Model with Default GBM Hyperparameters

Now we can look at the features and their importance. Listing 6.22 shows the code to do this, and Figure 6.14 shows the resulting importance plot. Spend some time reviewing the feature importance plot and think about what the results could mean.

```
#extract feature importance from the model
importances = model.feature_importances_

#create a data frame of the feature importance and column names
feature_importances = pd.DataFrame(
  {
    'Feature': X_test.columns,
    'Importance': importances
  }
)

#sort the data frame so most important features are at the top
feature_importances = feature_importances.sort_values(
  'Importance',
  ascending=False
  ).reset_index(drop=True)

#select only the top 10 important features
feature_importances = feature_importances.head(10)

#create the feature importance plot
feature_importances.plot(
  x='Feature',
```

```
y='Importance',
kind='bar'
)
```

Listing 6.22 Feature Importance Plot with Default Hyperparameters

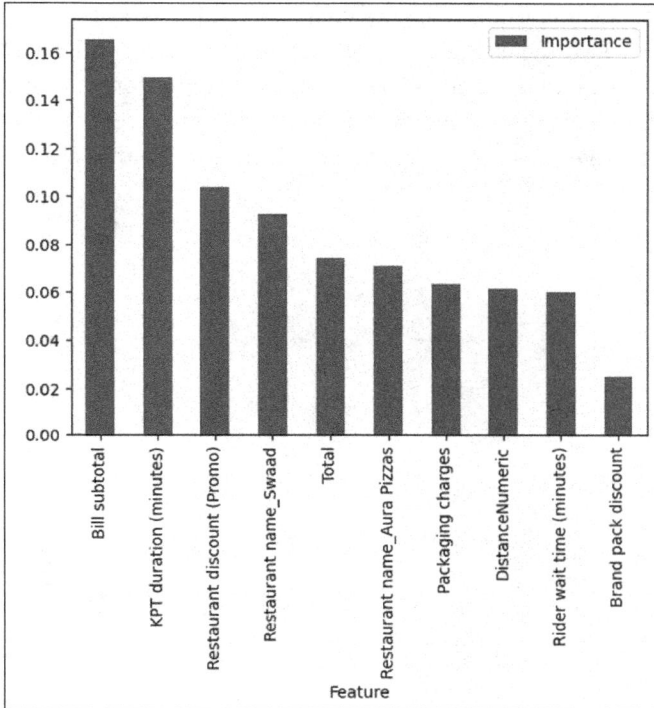

Figure 6.14 Output of Feature Importance Plot with Default Hyperparameters

The best point of reference is the plot from our first use case. The most important features had higher importance scores compared to the rest of the features. Here, the difference between the most important feature and the 9th-most important feature is barely about 2x. For the last iteration of the feature importance in our first use case, this difference was around 5x.

What does this mean? One possibility is the predictive power of our data simply isn't there, meaning the model is more or less grasping at straws to see what it can find. Given our results with the previous modeling attempts, this further justifies the need for more feature engineering to bring more predictive features into the dataset. Before moving to the next paragraph, spend some time thinking about what type of feature engineering makes sense to you.

One idea is to leverage a type of feature engineering we've already done for this dataset: lagging variables. To create the target variable, we had to employ lagging from the future into the past. For a variable to be a predictor, we need to do the opposite—lag information from the past into the present. This is equivalent to checking whether someone has ordered previously to predict whether they'll order again. We'll create this lag in a few different ways, testing the model and the feature importance each time to see how the model reacts to the changes.

This will be similar to the target variable lagging. We'll create two different DataFrames from the original DataFrame so we can execute our calculations. The first will be used to create the actual lags, and the second joined as an intermediate step before adding it to the dataset for the model. The purpose of the second DataFrame is to give us a list of customers by day, which we'll then add the lagged data to. These DataFrames are created in Listing 6.23.

```
#select only the customer, date, and bill subtotal
df_last_7_lag = df[[
  'Customer ID',
  'order_date',
  'Bill subtotal'
]]

#add 7 days to the data
df_last_7_lag['max_date'] =
  pd.to_datetime(df_last_7_lag['order_date']) + timedelta(days=8)

#select only the customer id and order date
df_last_7_lag_join = df[['Customer ID', 'order_date']]

#rename the columns
df_last_7_lag_join.columns = ['Customer ID', 'base_order_date']
```
Listing 6.23 Create New DataFrames for Executing the Lag

Now we can merge these DataFrames together. We're joining only on customer ID but then filtering down based on the dates. This is because Python doesn't have a good mechanism to account for greater-than or less-than logic when joining the datasets together. Listing 6.24 shows the merge and filtering.

```
#join the data
df_last_7_lag = pd.merge(
  df_last_7_lag_join,
  df_last_7_lag,
  how = 'left',
  on = ['Customer ID']
  )
```

```
#keep only the rows that fit the date range
df_last_7_lag = df_last_7_lag[
  (df_last_7_lag['base_order_date'] > df_last_7_lag['order_date']) &
  (df_last_7_lag['base_order_date'] < df_last_7_lag['max_date'])
  ]
```

Listing 6.24 Merge DataFrames and Then Filter Based on Dates

From a logical perspective, someone can have multiple orders in the last 7 days. We saw this when lagging our target variable. The way we account for this will slightly differ when creating variables for prediction. We'll use the groupby functionality to aggregate the dataset up to customer ID and order date level. This will create a single record that we can join onto our main dataset. Listing 6.25 shows the grouping and merging onto the final dataset.

```
#aggegrate the data
df_last_7_lag = df_last_7_lag.groupby(
  ['Customer ID', 'base_order_date']
  ).agg(
    mean=('Bill subtotal', 'mean'),
    count=('Bill subtotal', 'count')
    ).reset_index()

#rename the columns
df_last_7_lag.columns = [
  'Customer ID',
  'order_date',
  'mean_order_subtotal_last_7',
  'count_order_last_7'
  ]

#join to original dataset
  df = pd.merge(
  df,
  df_last_7_lag,
  how = 'left',
  on = ['Customer ID', 'order_date']
  )

#fill any blank values with 0
df = df.fillna(0)
```

Listing 6.25 Final Steps to Create Lagged Predictor Variables for Previous 7 Days

Now that we've created these variables, let's test them by adding them one at a time and checking the feature importance plots.

First, we'll only keep the count field (count_order_last_7) when running the model. This is done by removing the mean field before it's added in the modeling step, as shown in Listing 6.26. These are the same steps we took to create the model for this use case; the key difference is we now need to remove the mean_order_subtotal_last_7 column from the list of columns to train the model on.

```
#select the columns to be predictors in the model
X = df.drop(
  columns=[
    'target_7',
    'target_14',
    'Customer ID',
    'timestamp',
    'order_date',
    'mean_order_subtotal_last_7'
    ]
  )

#select the one week target variable
y = df['target_7']

#execute the train test split
X_train, X_test, y_train, y_test = train_test_split(
  X,
  y,
  test_size=0.3,
  random_state=42
  )

#create the model
model = GradientBoostingClassifier(random_state=42)

#fit the model to the data
model.fit(X_train, y_train)

#create both the probability and binary predictions
y_pred_proba = model.predict_proba(X_test)[:, 1]
y_pred = model.predict(X_test)
```

```
#calculate both the ROC AUC and accuracy scores
auc_score = roc_auc_score(y_test, y_pred_proba)
accuracy = accuracy_score(y_test, y_pred)

#print both the ROC AUC and accuracy scores
print(auc_score)
print(accuracy)
```

Listing 6.26 Run Model with Count of Orders in Previous 7 Days

We see a notable increase in the ROC AUC—it's up to 0.68!

It seems apparent that the order history is important, so let's see if the feature importance plot tells us the same story. We'll generate the feature importance plot using the code in Listing 6.27. Figure 6.15 shows the resulting plot.

```
#extract the feature importance from the model
importances = model.feature_importances_

#put the feature importance into a dataframe with the column names
feature_importances = pd.DataFrame(
  {'Feature': X_test.columns, 'Importance': importances}
  )

#sort the values of the dataframe
feature_importances = feature_importances.sort_values(
  'Importance',
  ascending=False
  ).reset_index(drop=True)

#select only the top 10 features
feature_importances = feature_importances.head(10)

#plot the features in a bar graph
feature_importances.plot(x='Feature', y='Importance', kind='bar')
```

Listing 6.27 Feature Importance Plot with Last 7 Days of Orders

It's safe to say the model finds this new feature quite helpful! When I ran it for the first time, I was skeptical about how I approached the lag, worrying about potential leakage. I reviewed the code to ensure the lag only pushed data into the past, not up to the current date, which would create leakage.

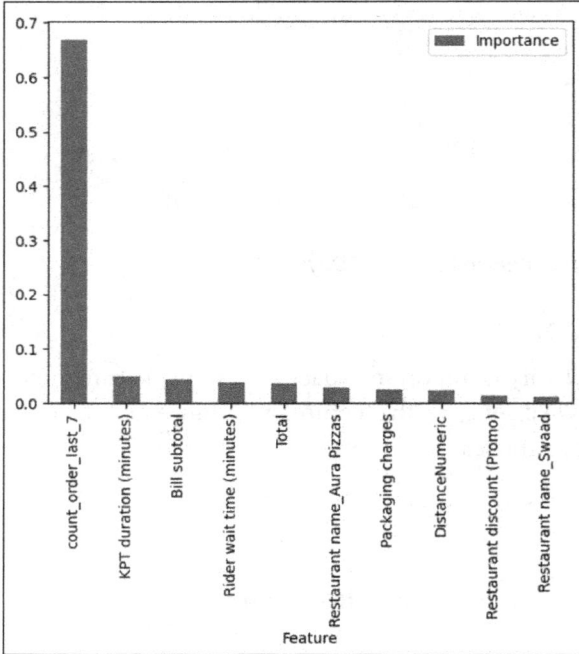

Figure 6.15 Feature Importance Plot with Count of Orders from Last 7 Days

Let's try running this again, but we'll keep the mean of the last 7 days subtotal as shown in Listing 6.28. The code to generate the feature importance plot doesn't change. Figure 6.16 shows the result.

```
#select the columns to be predictors in the model
X = df.drop(
  columns=[
    'target_7',
    'target_14',
    'Customer ID',
    'timestamp',
    'order_date'
    ]
  )

#select the one week target variable
y = df['target_7']

#execute the train test split
X_train, X_test, y_train, y_test = train_test_split(
  X,
  y,
```

```
  test_size=0.3,
  random_state=42
  )

#create the model
model = GradientBoostingClassifier(random_state=42)

#fit the model to the data
model.fit(X_train, y_train)

#create both the probability and binary predictions
y_pred_proba = model.predict_proba(X_test)[:, 1]
y_pred = model.predict(X_test)

#calculate both the ROC AUC and accuracy scores
auc_score = roc_auc_score(y_test, y_pred_proba)
accuracy = accuracy_score(y_test, y_pred)

#print both the ROC AUC and accuracy scores
print(auc_score)
print(accuracy)
```

Listing 6.28 Rerun Model with Mean of Order Subtotal in Last 7 Days

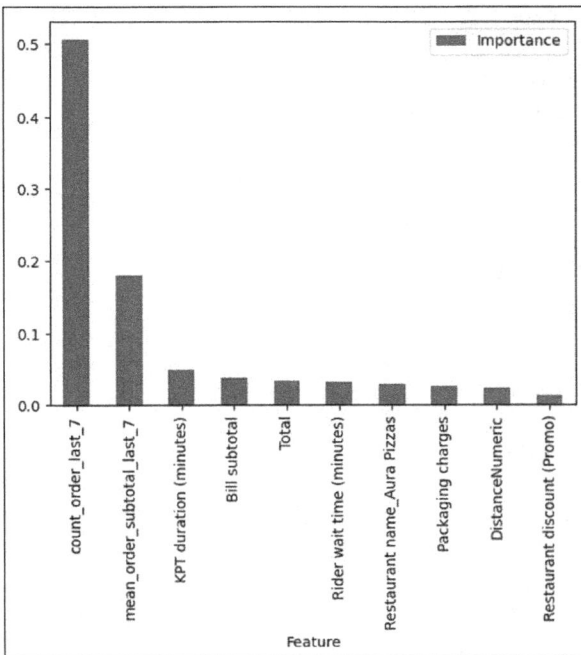

Figure 6.16 Updated Feature Importance with Both 7-Day Lagged Variables

The ROC AUC didn't change from 0.68, but the feature importance did. What does this tell us? Both feature lags help determine the predictions, as the count and average give the model different information.

While we've improved our model, it's still not viable. We need to be closer to 0.75 before we can confidently use this model for any type of decision making. For this use case, the last things we'll explore are the time-based attributes. If you recall way back to when we explored the data in Chapter 4, Section 4.2.2, we found a trend in the days of the week (you can revisit this table in Figure 6.17).

day_of_week	day_name	Order ID
0	Monday	99.818182
1	Tuesday	131.136364
2	Wednesday	139.863636
3	Thursday	130.863636
4	Friday	154.681818
5	Saturday	186.809524
6	Sunday	134.454545

Figure 6.17 Revisiting the Trend in the Days of the Week

Based on this trend, let's add this day of the week feature and run the model (see Listing 6.29), and then evaluate the results shown in Figure 6.18.

```
#create day of the week feature
df['day_of_week'] = pd.to_datetime(df['order_date']).dt.weekday

#select the columns to be predictors in the model
X = df.drop(
  columns=[
    'target_7',
    'target_14',
    'Customer ID',
    'timestamp',
    'order_date'
    ]
  )

#select the one week target variable
y = df['target_7']

#execute the train test split
X_train, X_test, y_train, y_test = train_test_split(
  X,
```

```
y,
test_size=0.3,
random_state=42
)

#create the model
model = GradientBoostingClassifier(random_state=42)

#fit the model to the data
model.fit(X_train, y_train)

#create both the probability and binary predictions
y_pred_proba = model.predict_proba(X_test)[:, 1]
y_pred = model.predict(X_test)

#calculate both the ROC AUC and accuracy scores
auc_score = roc_auc_score(y_test, y_pred_proba)
accuracy = accuracy_score(y_test, y_pred)

#print both the ROC AUC and accuracy scores
print(auc_score)
print(accuracy)
```

Listing 6.29 Rerun Model with the Weekday Feature Included

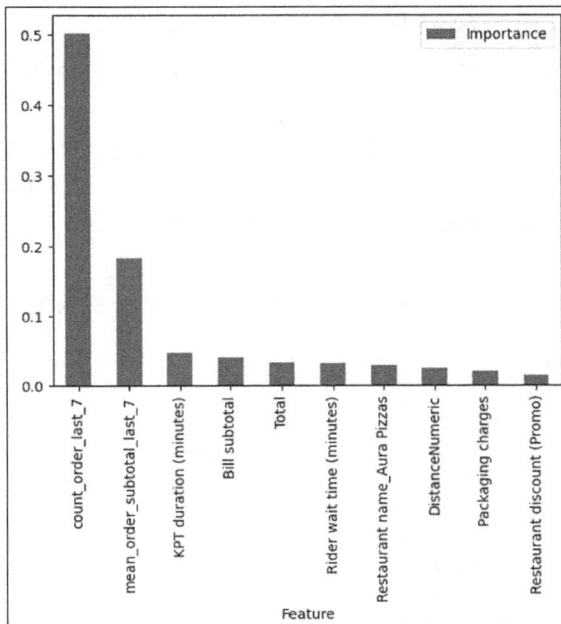

Figure 6.18 Rerun Model with Weekday Feature

The ROC AUC result is still 0.68, and the new day_of_week feature didn't crack the top 10. What does that mean for the model? Unfortunately, we'll need to go back to our stakeholder and explain we can't build a viable model at this time.

This can be a difficult conversation, but in this situation, we benefit from having a stakeholder like Mary. When you tell your stakeholders that a model isn't feasible, it's important to ensure you're communicating what the action items are for next steps. In this case, the model shows some promise on less than 6 months of data. After another few months, it's entirely possible we'd see different results.

In the real world, another option is to discuss joining additional data sources into your dataset. These could include:

- Marketing campaigns on specific dates that could be driving customer behavior
- Website or app usage data to provide more customer-level detail for the model to pick up on buying behavior trends
- Publicly available information, like local event calendars or weather data, that may influence consumer behaviors

This use case represents an important lesson: Sometimes, the data just can't support a model. How you communicate this and approach next steps is what will differentiate you with your stakeholders.

6.7.3   Use Case 3

When we left off with this use case, we didn't see any movement in our model's metrics by adjusting the hyperparameters. The next step is to consider additional feature engineering actions to improve the accuracy of the model. Luckily, we have some low-hanging fruit!

The first and most obvious step is to create features about the dates themselves, like we did with our first use case. If you think all the way back to Chapter 4, Section 4.2.2, when we were analyzing the data, the crime numbers dropped significantly in the latter half of the dataset.

First, let's load in our data and get it into the same state we left it in Chapter 5, Section 5.4.3, as shown in Listing 6.30.

```
import pandas as pd

#read in summarized data
df = pd.read_csv("df_summarized")

#sort the values to ensure it's in ascending order
df = df.sort_values(by=['DATE', 'AREA NAME'])
```

```
#shift 7 days for each area name independently
df['crime_count_shifted'] = df.groupby('AREA NAME')['crime_count'].shift(7)

#filter to remove what are now null values for the earliest dates of the dataset
df = df[df['crime_count_shifted'].notnull()]
```

Listing 6.30 Load In Data and Apply Target Variable Lagging

Now, we'll create various features about the dates, as shown in Listing 6.31. This includes adding the day of the week (1–7), day of the month (1–31), week of the year (1–52), and month of the year (1–12).

```
#we found some bad date values we need to remove
df = df[df['DATE'] != '0']

#convert the DATE column to pandas date for ease of use
df['DATE'] = pd.to_datetime(df['DATE'])

#create numeric day of week column
df['day_of_week'] = df['DATE'].dt.dayofweek

#create numeric day of month column
df['day_of_month'] = df['DATE'].dt.day

#create numeric week of year column
df['week_of_year'] = df['DATE'].dt.isocalendar().week

#create numeric month of year column
df['month_of_year'] = df['DATE'].dt.month
```

Listing 6.31 Create Date Features

Before we consider more features, let's test these in a GBM to see what impact they may have on performance. As a starting point, we'll use the best parameters from our grid search in Chapter 5, Section 5.5.3 (see Listing 6.32).

```
from sklearn.ensemble import GradientBoostingRegressor
from sklearn.model_selection import train_test_split
from sklearn.metrics import mean_absolute_error

#split data into what will be used to predict and what is being predicted
X = df.drop(
  columns=['DATE', 'AREA NAME', 'crime_count', 'crime_count_shifted'])
y = df['crime_count_shifted']
```

```
#perform standard train test split
X_train, X_test, y_train, y_test = train_test_split(
    X,
    y,
    test_size=0.2,
    random_state=42
    )

#create and fit the GBM model
model = GradientBoostingRegressor(
    learning_rate = 0.05,
    max_depth = 7,
    n_estimators = 100,
    subsample = 0.8
    )
model.fit(X_train, y_train)

#predict with the modeling using the test data
y_pred = model.predict(X_test)

#calculate the mean absolute error
mae = mean_absolute_error(y_test, y_pred)

#calculate the mean absolute error divided by average of the target
mae_by_avg_target = mae / y_test.mean()

#print both results
print(mae)
print(mae_by_avg_target)
```

Listing 6.32 Initial GBM with New Date Features

While they're an improvement, the results aren't quite as positive as we would have hoped. The MAE is down to 5.6, and the MAE divided by the average target is down to 21%.

Before proceeding, let's take a look at the feature importance plot to see what the model is focusing on (see Listing 6.33 and Figure 6.19).

```
import matplotlib.pyplot as plt
import pandas as pd
import numpy as np

#get feature importances from the trained model
importances = model.feature_importances_
```

```
feature_names = X.columns

#create a DataFrame for easy sorting and plotting
feat_imp_df = pd.DataFrame({
  'Feature': feature_names,
  'Importance': importances
  }).sort_values(by='Importance', ascending=False)

#filter to top 10
feat_imp_df = feat_imp_df.head(10)

#plot
plt.figure(figsize=(10, 6))
plt.barh(feat_imp_df['Feature'], feat_imp_df['Importance'], color='skyblue')
plt.xlabel('Feature Importance')
plt.title('Gradient Boosting Feature Importances')
plt.gca().invert_yaxis() # Highest importance at the top
plt.tight_layout()
plt.show()
```

Listing 6.33 Feature Importance Plot to See What the Model Is Focusing On

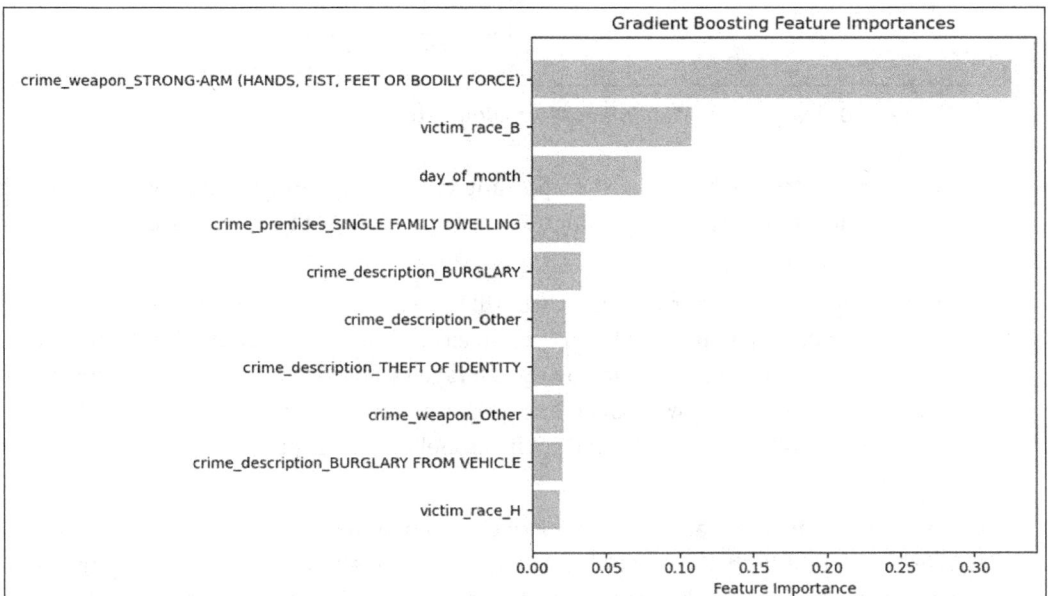

Figure 6.19 Feature Importance Plot After First Feature Engineering Additions

It looks like only one of our new features made it into the model (day_of_month). Given the overall trend we saw where the latter half of the data had lower volumes, let's see if

365

adding the year as a feature makes any impact. We'll rerun the same model, with the only change being this new column in the dataset.

When we rerun the same code, the MAE is down to 5.4 and the MAE divided by the average target is down to 21%. We're seeing some progress! Let's look at the updated feature importance plot to see what has changed (see Figure 6.20).

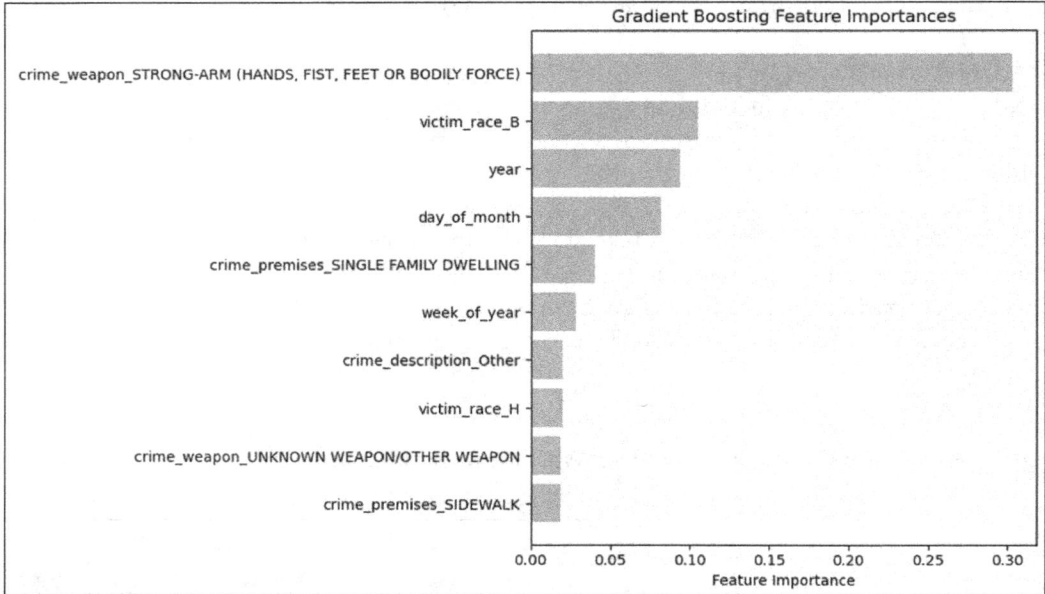

Figure 6.20 Feature Importance Plot After Adding the Year as a Feature

The year is now the third most important variable. Given the improvement in the model's metrics with the only change being the addition of the year variable, this is the expected result.

Even though we saw improvements, we still have quite a way to go. While there isn't an official threshold you need to hit, it'll be an easier sell to the stakeholder if you can get closer to 15% for the MAE divided by the average target metric. This isn't exactly scientific, but selling a model with an error of 20% is much harder than selling one that *feels* closer to 10%. What else can we add to the model now to improve it? Let's give it historical context!

It seems reasonable that the last week of crime could be predictive of the next week of crime, so let's add that as a variable. To do this, we'll take an average of the target variable for the last 7 days. This is another situation where if you think from the perspective of SQL, you may be jumping to calculate the average and perform a join. We can use the same groupby approach for the rolling function as we did previously with the shift function.

To do this for the last week, we'll use the original target variable column `crime_count`. We use this instead of our predictor variable because we don't want to take averages of future values. By using `crime_count`, we're ensuring we only use the historical values to avoid any leakage into the model. Listing 6.34 shows the code for executing this.

```
#create rolling 7 days feature
df['last_7_days'] = df.groupby('AREA NAME')['crime_count']
  .rolling(window=7, min_periods=1)
  .mean()
  .reset_index(level=0, drop=True)
```

Listing 6.34 Create Variable to Take Average of crime_count Over Last 7 Days

Let's run the model again to see what type of value we get from this new feature (see Listing 6.35).

```
#split data into what will be used to predict and what is being predicted
X = df.drop(columns=['DATE', 'AREA NAME', 'crime_count', 'crime_count_shifted'])
y = df['crime_count_shifted']

#perform standard train test split
X_train, X_test, y_train, y_test = train_test_split(
  X,
  y,
  test_size=0.2,
  random_state=42
  )

#create and fit the GBM model
model = GradientBoostingRegressor(
  learning_rate = 0.05,
  max_depth = 7,
  n_estimators = 100,
  subsample = 0.8
  )
model.fit(X_train, y_train)

#predict with the modeling using the test data
y_pred = model.predict(X_test)

#calculate the mean absolute error
mae = mean_absolute_error(y_test, y_pred)

#calculate the mean absolute error divided by average of the target
mae_by_avg_target = mae / y_test.mean()
```

```
#print both results
print(mae)
print(mae_by_avg_target)
```

Listing 6.35 Run GBM with New Average of Crime in the Last Week Feature

We see another notable shift! The MAE is down to 5.1, and the MAE divided by the average target is down to 19%. We'll rerun the same feature importance code to see how this variable stacks up against the others. Figure 6.21 shows the result of this run of our feature importance plot. What sticks out to you?

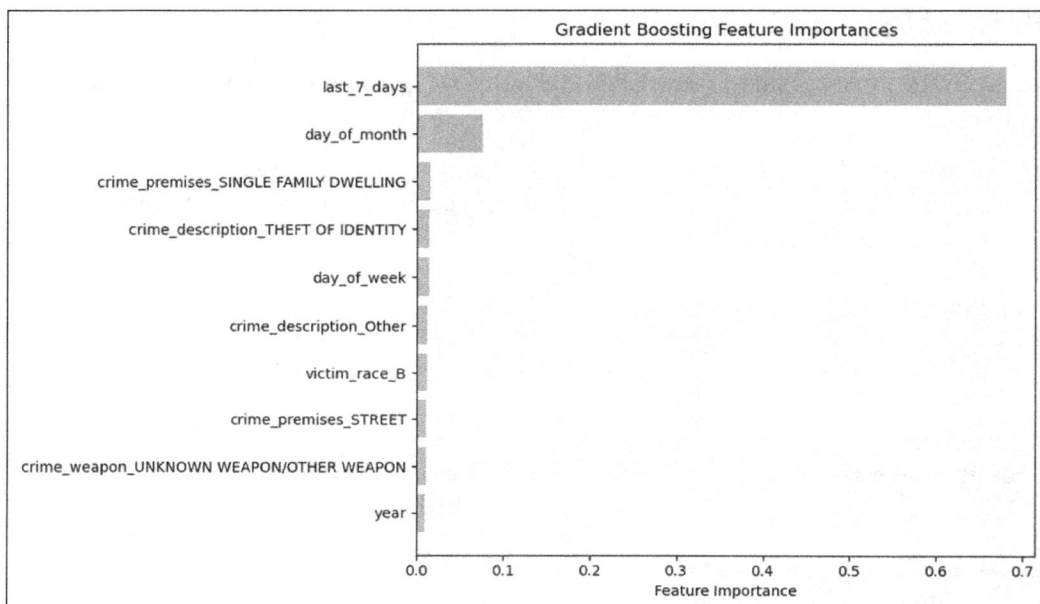

Figure 6.21 Feature Importance with the Last Week of Average Crime Added

Two thoughts likely come to your mind, and both can be true at the same time. The first is that it should be pretty intuitive to assume that crime in the last 7 days will likely predict crime in the next 7 days. The second thought is that we should be skeptical any time a single variable carries this much weight in a model. To address this, let's run the code shown in Listing 6.36 to validate how this new feature operates (see the results in Figure 6.22).

```
#check the results
df[df['AREA NAME'] == 'Central'][['DATE', 'crime_count', 'last_7_
days']].head(14)
```

Listing 6.36 Preview for a Single Area to Validate Last 7 Days Feature

DATE	crime_count	last_7_days
2020-01-08	36.0	36.000000
2020-01-09	29.0	32.500000
2020-01-10	47.0	37.333333
2020-01-11	49.0	40.250000
2020-01-12	40.0	40.200000
2020-01-13	28.0	38.166667
2020-01-14	27.0	36.571429
2020-01-15	35.0	36.428571
2020-01-16	44.0	38.571429
2020-01-17	49.0	38.857143
2020-01-18	46.0	38.428571
2020-01-19	33.0	37.428571
2020-01-20	37.0	38.714286
2020-01-21	33.0	39.571429

Figure 6.22 First 14 Days of the Central Area Name

Once you have these results, you can copy their values into Excel to check whether the operation is working. You can use the AVERAGE function in Excel to select the last 7 days to see if the average from Python aligns, which validates that the operation has been done properly. For this use case, it's okay that the first 6 records are not a full average of 7 values. The math checks out, so it does appear that this feature really is *that* impactful. That means we should probably add more!

To get an idea of what's valuable, we'll keep running these variable lags with different time intervals one at a time. We'll check the feature importance plots along the way to see what shifts may have occurred. First up, we'll add an average of the last 30 days as shown in Listing 6.37. While the code for it isn't shown (it would be redundant to keep showing you the same code), we'll also rerun the model and feature importance chart. Figure 6.23 shows the resulting feature importance chart.

```
#create 30 day lag feature
df['last_30_days'] = df.groupby('AREA NAME')['crime_count']
  .rolling(window=30, min_periods=1)
  .mean()
  .reset_index(level=0, drop=True)
```

Listing 6.37 Add Average of Last 30 Days Feature

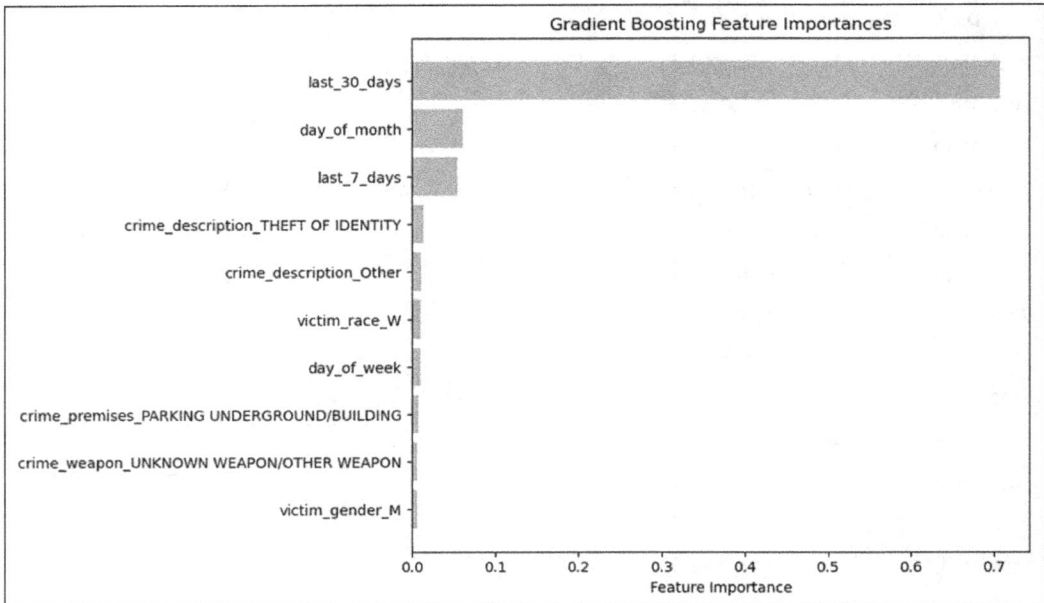

Figure 6.23 Feature Importance Chart After Adding Average of the Last 30 Days Feature

Now the last_30_days feature has overtaken the last_7_days feature. Could this be due to the variation from week to week? And if so, how can we account for that?

The answer in this case is standard deviation, because its purpose is to measure how spread out the data is. The approach we're using with the rolling function is flexible, so we can use the same approach with any aggregation function. While the standard deviation isn't a sum or average, it is a means to summarize our data. Let's try aggregating the last 30 days with the standard deviation to see what happens (see Listing 6.38).

```
#create 30 day standard deviation feature
df['last_30_days_std'] = df.groupby('AREA NAME')['crime_count']
  .rolling(window=30, min_periods=1)
  .std()
  .reset_index(level=0, drop=True)
```

Listing 6.38 Add Standard Deviation of the Last 30 Days

If you try running the model with the exact same code, you'll now receive an error regarding not a number (NaN) values. This is because the standard deviation function requires at least two numbers to calculate. We'll need to add a snippet of code to filter out these rows before running our model (see Listing 6.39).

```
#filter out blank values
df = df[df['last_30_days_std'].notna()]
```

Listing 6.39 Filter Out NaN Values

When we run this model, we continue to see improvement, with the MAE down to 4.9 and the MAE divided by the average target at 19%. As you can see in Figure 6.24, the feature importance plot still heavily focuses on last_30_days, but the standard deviation of the last 30 days (last_30_days_std) is the second most important.

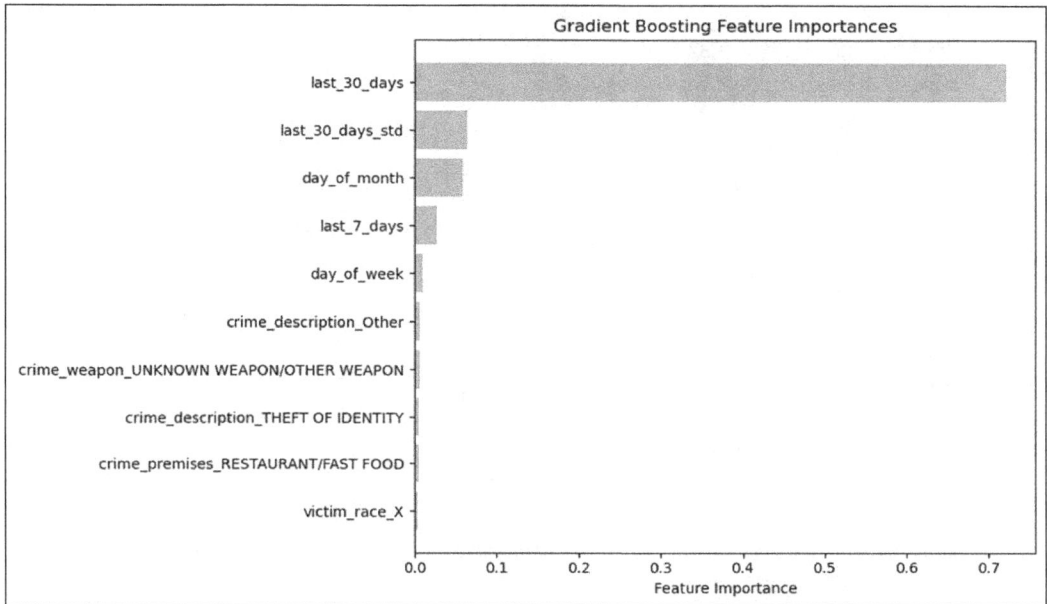

Figure 6.24 Feature Importance After Adding Standard Deviation of Last 30 Days

This brings us to another "take a step back" moment. We've introduced some new features into the model that didn't previously exist, so it's entirely possible our hyperparameters need updating. Let's make some brief manual adjustments to see if they have an impact on our results, as shown in Listing 6.40. We're increasing n_estimators, reducing the learning_rate, increasing the max_depth, and setting subsample back to 1. Now that we have the new, very important column last_30_days, the added trees may give us more value than they have in the past. By decreasing the learning rate, we're encouraging the model to take its time. The hope is that it won't assume it's found the right answer too quickly. Increasing how deep the trees can go could also encourage more learning from the features that aren't as important.

```
df = df[df['last_30_days_std'].notna()]

#split data into what will be used to predict and what is being predicted
X = df.drop(columns=['DATE', 'AREA NAME', 'crime_count', 'crime_count_shifted'])
y = df['crime_count_shifted']

#perform standard train test split
X_train, X_test, y_train, y_test = train_test_split(
```

```
X,
y,
test_size=0.2,
random_state=42
)

#create and fit the GBM model
model = GradientBoostingRegressor(
    learning_rate = 0.02,
    max_depth = 10,
    n_estimators = 300,
    subsample = 1
    )
model.fit(X_train, y_train)

#predict with the modeling using the test data
y_pred = model.predict(X_test)

#calculate the mean absolute error
mae = mean_absolute_error(y_test, y_pred)

#calculate the mean absolute error divided by average of the target
mae_by_avg_target = mae / y_test.mean()

#print both results
print(mae)
print(mae_by_avg_target)
```

Listing 6.40 Updating Hyperparameters to See If It Impacts the Model, Given the New Features

Unfortunately, our results came back with essentially the same performance metrics (although with a much longer time to train the model). So, what should our next steps be?

In a situation like this, it's best to go back to our stakeholder, Jen, to have a conversation. This use case represents a murky middle ground between the first two use cases. In the first use case, we were able to create a model we could feel good about, especially given the limited data. In the second use case, it became clear that the data we had wasn't going to support a reliable model, and we need more time to build up the dataset. In this use case, we're getting close to what seems like an acceptable model, but it's not quite there. This is where being a good communicator with your stakeholder becomes critical. Jen is a more technically inclined individual, so you can have an open dialogue with her. Think about how you'd have that conversation with a stakeholder before you move to the next paragraph.

You have the conversation with Jen, and she's actually surprised you've been able to create a model with this level of accuracy. She's an expert on this data and knows how much the crime volumes can fluctuate and seemingly pop up without warning. Given her technical understanding, she points out that if you were to just take the average of the error values instead of calculating using an absolute value, your last model's error would only be -0.12. Listing 6.41 shows the code to perform this calculation.

```
import numpy as np

#subtract the arrays
error = np.subtract(y_test, y_pred)

#calculate the mean of the differences (error)
error.mean()
```

Listing 6.41 Calculate the Average Error

While you wouldn't want to do this as you're building the model, it does give you some perspective and context on how the model performs in the context of the use case. If the model generates the right predictions on average, it can still be useful in a use case like this.

It's also helpful to remember that, as Jen pointed out, what we're trying to predict can have unexpected drivers that we can't always account for. Translation: It's not always possible to create a model with near-perfect accuracy. In a use case like this, having a model to help make decisions is better than purely deciding based on opinions and feelings.

A final consideration is the level at which we're predicting. The daily level is quite detailed, but if we summarized our results up to the weekly level, the average of these results would likely look much better.

With all this context, Jen is comfortable with us saving the model so it can start being used! To save the file, we'll use the `pickle` package like we did with the first use case (see Listing 6.42).

```
#package specific for saving in .pkl format
import pickle

#create file name and open the file
file_name = 'model.pkl'
with open(file_name, 'wb') as file:
  #save the model object to the file name
  pickle.dump(model, file)
```

Listing 6.42 Saving the Model

We've now saved this model and will use it to create predictions in Chapter 7.

6.8   Summary

In this chapter, we took a step back to consider what makes a good machine learning model. Then, we covered how you can improve your model and what that looks like in a practical setting. Here's a recap of what you've learned:

- You learned that picking the right validation metric turns the modeling process from pure guesswork into a more objective debate. We covered the various validation metrics that you can use. Which validation metric you choose depends on what you're trying to predict (classification or regression), as well as the type of data you have. Each metric has its pros and cons, so it's important to match the right metric to your use case.

- You learned about the legal and ethical considerations when building a machine learning model. Your takeaway should be: Just because you can doesn't mean you should.

- You learned about the various approaches to interpreting a model. For the traditional regression approaches of linear and logistic regression, the model will give you coefficients. For machine learning models, you have to use feature importance plots or SHAP values.

- You learned more about the hyperparameter tuning process. It's best to use grid search after you've done some reasonability checks, given how long it can take to execute.

- You learned about the iterative process and how you'll often need to revisit your model and execute additional feature engineering to improve it. However, playing with your hyperparameters will only get you so far; sometimes the model needs to be given new information.

- Finally, you learned about pickling (not the canning process that involves preserving vegetables). This is how you save your model so you can load it again without needing to retrain it each time you generate predictions.

The next chapter will show you how to load your model and use it to generate predictions.

Chapter 7

Implementing, Monitoring, and Measuring the Model

We've reached the final stages of applying machine learning. This chapter wraps the modeling content in a bow so you can take it out in the real world. We also cover some additional considerations, like proving the importance of your model.

This chapter represents the last mile of the modeling process. It covers a set of critical topics that are not often discussed. The goal of this chapter is to show how you can take your model from a pickle file to making predictions—and ultimately demonstrate its value. If you've just gone through a grueling process to get your model together, these final steps can feel a bit trivial. However, if you don't get this step right, your modeling process has failed.

We'll cover the logistics of loading your pickle file and creating predictions with your model, using our use cases as examples. Then, we'll discuss the model monitoring process, which will only become more important with the rise of artificial intelligence (AI) rules and regulations. Lastly, we'll discuss the ways in which you can measure the impact of your model. This is an underrated topic that can really help you differentiate yourself with stakeholders and in job interviews.

7.1 Implementing Your Model for Predictions

You've built a model and saved it, and now you want to use it to predict future data. What does that look like? Like most of the modeling process, it's relatively consistent from one use case to another. The primary difference is the method you use to give the model data for its predictions.

For example, let's take a sales prediction. You need to structure the data in a way that mimics the input the model expects and incorporate any required feature engineering. The tricky part about this situation is you don't have the dates to simply adjust your existing data; you need to create a starting point for the dataset with a new DataFrame that contains the dates you're predicting on.

Now let's discuss the best practices for implementing your predictions and then apply them to our use cases.

7.1.1   Don't Train the Model Each Time

It's important to make the mental shift from training your model each time you want to make a prediction to calling the same model and making a prediction on it without having to train it. We've already done the first step in Chapter 6, Section 6.7, to save our model through pickling.

There's a fun devil's advocate scenario to explore here, though. Why can't you train your model each time you want to make a prediction? Here are some of the reasons:

- **Time**
  Creating a model from scratch can be quick, but it's never as quick as simply loading your pickled model. Even if you have all the code ready to train the model, you still need to review the inputs to ensure that there isn't any overfitting and that the training process is working.

  For large datasets, the time to train your model can also be a significant factor. As a personal example, a model I helped with took more than 24 hours to train from beginning to end. The code to make predictions took a few minutes to run.

  When you're working with smaller datasets, the line between retraining your model each time and referencing your pickled model is blurred. Keep in mind that even if you can train your model in less than a minute, you still need to do the proper validations!

- **Consistency**
  Training a model comes with a degree of randomness, and with randomness comes variation. Imagine how difficult it would be to explain to a stakeholder why predictions for similar situations were different across three consecutive days. There's value to consistency and limiting your retraining of the model to fixed periods of time with longer intervals. This also ensures that the model's results don't change significantly from day to day for business users who review its output and make decisions from it.

- **Model governance**
  As you work within organizations, you'll encounter various policies and guidelines for models, which often include submitting documentation for your model. Retraining your model could trigger this process each time, which is not feasible. At many companies, the model governance process can take many hours to complete. Apart from this not being the most exciting work, your ability to build and maintain multiple models is limited if you're dedicating a significant amount of time to model governance each week.

7.1.2   Predictions for Our Use Cases

Now, let's put the models we've developed to work. We'll implement our successful use case models to make predictions in the following sections. (Since we weren't able to create a viable model for the second use case on customer retention, we'll skip it in this section.) Let's see how they do!

Use Case 1 Predictions

Thinking back to our extract, transform, load (ETL) process for this use case, we lagged our variables by 3 months. This was intentional because now we can use the same ETL process to predict 3 months into the future, which is what Chris is looking for. We'll use the same code with slight alterations to generate the lagged data we need for the next 3 months of prediction dates.

Coding Like a Novice

One of the most impactful pieces of feedback I ever received was when I interviewed for my first data science role. I bombed the technical interview, and the team told me I wrote code like a novice.

I was taking each step in the process and writing the code in a linear way, similar to how you'd do it in Excel. However, as a data scientist, you're expected to think about things from a programmatic perspective and use coding best practices. The team could spot this difference easily and it prompted them to give me that feedback.

As you learn about predictive modeling, I would encourage you to code like a novice. It will help you understand the predictive modeling space more quickly, so you won't have to split your focus on learning coding techniques. As you get comfortable with building models, you can move on to writing more efficient code.

I mention this because the proper way to approach your ETL process is to write it as a function you can use for model training and predictions. This is more efficient because you're able to ensure consistency between your ETL process for both training and predictions, which is critical for delivering proper predictions. However, learning the code this way can be a bit abstract, so instead we're focusing on the concept.

First, we need to load in our data, as shown in Listing 7.1. This is the same process we followed when training the model in Chapter 5, Section 5.3.

```
#import libraries
import pandas as pd
import numpy as np

#read in data
df_cleaned = pd.read_csv("df_cleaned.csv")

#group the data
df_data_grouped = df_cleaned.groupby(
  'invoicedate'
  ).agg({
```

```
    'Description': 'nunique',
    'Customer ID': 'nunique',
    'Country': 'nunique',
    'Quantity': 'sum',
    'Price':'sum'
    }).reset_index()

#import date specific packages
from datetime import datetime
from dateutil.relativedelta import relativedelta

#create date
df_data_grouped['invoicedate'] = pd.to_datetime(
    df_data_grouped['invoicedate']
    ).dt.date

#add 3 months to the date
df_data_grouped['invoicedate_minus_3'] =
    df_data_grouped['invoicedate'] + relativedelta(months=3)
```

Listing 7.1 Load in Data

Now we need to create dates that don't already exist in our dataset, as shown in Listing 7.2. We're using the date_range function to create a list of dates and then using the pd.DataFrame function to put the columns into a DataFrame. This gives us a DataFrame with a single column of dates.

```
#create date range for the future predictions
dates = pd.date_range(
    df_data_grouped['invoicedate'].max(),
    df_data_grouped['invoicedate'].max()+relativedelta(months=3),
    freq='d'
    )

#put data into a data frame
df_for_model = pd.DataFrame(
    dates,
    columns=['invoicedate']
    )

#convert date to a pandas date time column
df_for_model['invoicedate'] = pd.to_datetime(
    df_for_model['invoicedate']
    )
```

Listing 7.2 Create Next 3 Months of Dates

Next, we'll join with the original data, as shown in Listing 7.3. This allows us to get the lagged data that the model requires to generate predictions.

```
#select specific columns
df_minus_3 = df_data_grouped[[
  'invoicedate_minus_3',
  'Description',
  'Customer ID',
  'Country',
  'Quantity',
  'Price'
  ]]

#create month feature
df_minus_3['month'] = pd.to_datetime(
  df_minus_3['invoicedate_minus_3']
  ).dt.month

#create days from first sale feature
df_minus_3['since_first_sale'] = pd.to_datetime(
  df_minus_3['invoicedate_minus_3']) -
  min(pd.to_datetime(df_for_model['invoicedate'])
  )

#change data type of days from first sale feature
df_minus_3['since_first_sale'] = df_minus_3['since_first_sale'].dt.days

#convert date to pandas date time
df_minus_3['invoicedate_minus_3'] = pd.to_datetime(
  df_minus_3['invoicedate_minus_3']
  )

#join the data
df_for_model = pd.merge(
  df_for_model,
  df_minus_3,
  how = 'left',
  left_on = 'invoicedate',
  right_on = 'invoicedate_minus_3',
  suffixes = ('','_3months')
  )

#remove blank values
df_for_model = df_for_model[pd.notna(
  df_for_model['invoicedate_minus_3']
```

```
)]

#create day of week feature
df_for_model['day_of_week'] = pd.to_datetime(
  df_for_model['invoicedate']
  ).dt.weekday

#create day of the year feature
df_for_model['day_of_year'] = pd.to_datetime(
  df_for_model['invoicedate']
  ).dt.dayofyear

#create week of the year feature
df_for_model['week_of_year'] = pd.to_datetime(
  df_for_model['invoicedate']
  ).dt.isocalendar().week

#rename price column
df_for_model['Price_3months'] = df_for_model['Price']
```

Listing 7.3 Join Future Dates with Lagged Variables

Lastly, to make a prediction, we perform the same steps on the data that we did to train the model (select columns, create features, dummy code, etc.), as shown in Listing 7.4. This is the same code we've used throughout this use case, most recently in Chapter 6, Section 6.1.

```
#select columns to be predictors for the model
X = df_for_model[[
  'Description',
  'Customer ID',
  'Country',
  'month',
  'since_first_sale',
  'Price_3months',
  'day_of_week',
  'day_of_year',
  'week_of_year'
  ]]

#create holiday season feature
X['holiday_season'] = np.where(X['month'] >= 9, 1, 0)

#dummy code the weekday feature
x_weekday_dummies = pd.get_dummies(
```

```
X['day_of_week'],
prefix='weekday'
)

#add the dummy coded weekday feature to the dataset
X = pd.concat(
 [X, x_weekday_dummies],
 axis = 1
 )

#drop the weekday_5 column
X = X.drop('weekday_5', axis=1)
```

Listing 7.4 Adjust to Required Data Format

Now that we have the necessary data to create our predictions, let's load in our model, as shown in Listing 7.5. This is the same code for loading any pickled model.

```
#load the pickle library
import pickle

#load the model
with open('model.pkl', 'rb') as file:
    model = pickle.load(file)
```

Listing 7.5 Loading Pickled Model

Now we can use the model to predict the future! It's almost comical how simple this is — you just need to use the predict function, as shown in Listing 7.6.

```
#read in library
from sklearn.ensemble import GradientBoostingRegressor

#generate predictions
predictions = model.predict(X)
```

Listing 7.6 Create Predictions

This will generate an array of our predictions, just like it did with our test dataset. As we have been, it's best to visualize this data in a graph to see if our predictions pass the sniff test. In Listing 7.7, we're plotting the future predictions and actual data using plt.plot. This results in the output shown in Figure 7.1.

```
#import plotting library
import matplotlib.pyplot as plt
```

```
#set the figure size
plt.figure(figsize=(9, 5))

#plot the future values
plt.plot(
    df_for_model['invoicedate'],
    predictions,
    label = "future predictions"
    )

#plot the historical values
plt.plot(
    df_data_grouped['invoicedate'],
    df_data_grouped['Price'],
    label = "actuals"
    )

#display the legend
plt.legend()
```

Listing 7.7 Visualizing Predictions

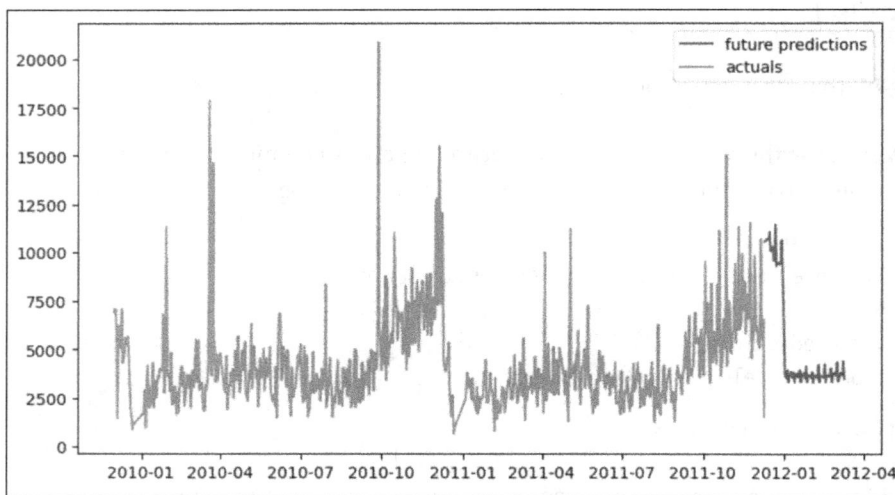

Figure 7.1 Visualizing Predictions

Take some time to review the predictions before you read on.

Overall, these predictions look good! They clearly capture the trends of the end-of-the-year peak and the sharp decline at the start of the year. The last component to consider is that we need to sum these predictions to answer Chris's question of how much sales will be in the next 3 months. When we run the following code, we get a little over $400,000 in sales:

```
sum(predictions)
```

Some of the takeaways from this use case as it relates to generating predictions are:

- You need to follow the same ETL process for preparing your data for a time series use case.
- The majority of the work is prepping the data (shocker).
- It's a good idea to visualize your predictions to ensure they pass the sniff test, especially when generating predictions for the first time.

Use Case 3 Predictions

For this use case, we'll be predicting the next week of crime from the end of the dataset we have. To do this, we'll need to load in the model, but we'll also need to perform the same data transformations we did when training the model to the last 7 days of our dataset. This is the portion of data that was removed when we lagged our target variable by 7 days, so the target variable didn't exist for these values.

Modeling in Practice

Ideally, when building a model like this in practice, you'll have created a set of functions that you can apply to both the training data and the predicting data. This makes the learning process harder, since you're not seeing the actual code both times. However, the goal is to avoid needing to copy and paste the same code you used for training the model to generate predictions.

As the first step, we'll read in our data and apply the target variable lag (see Listing 7.8, and refer to Chapter 6, Section 6.3 for more information).

```
#load in packages
import pandas as pd
import numpy as np

#read the data
df = pd.read_csv("df_summarized")

#sort the values to ensure it's in ascending order
df = df.sort_values(by=['DATE', 'AREA NAME'])

#shift 7 days for each area name independently
df['crime_count_shifted'] = df.groupby('AREA NAME')['crime_count'].shift(-7)
```

Listing 7.8 Load in Data and Lag Target Variable

Now we need to apply all the same transformations we did when building the model in Chapter 6, Section 6.7.3 (see Listing 7.9). This part is satisfying because you can see all the transformations in one spot. This makes it look like we did quite a bit to the model!

```
#we found some bad date values we need to remove
df = df[df['DATE'] != '0']

#convert the DATE column to pandas date for ease of use
df['DATE'] = pd.to_datetime(df['DATE'])

#create numeric day of week column
df['day_of_week'] = df['DATE'].dt.dayofweek

#create numeric day of month column
df['day_of_month'] = df['DATE'].dt.day

#create numeric week of year column
df['week_of_year'] = df['DATE'].dt.isocalendar().week

#create numeric month of year column
df['month_of_year'] = df['DATE'].dt.month

#create numeric month of year column
df['year'] = df['DATE'].dt.year

#rolling average of crime from the last 7 days
df['last_7_days'] = df.groupby('AREA NAME')['crime_count']
  .rolling(window=7, min_periods=1)
  .mean()
  .reset_index(level=0, drop=True)

#rolling average of crime for the last 30 days
df['last_30_days'] = df.groupby('AREA NAME')['crime_count']
  .rolling(window=30, min_periods=1)
  .mean()
  .reset_index(level=0, drop=True)

#standard deviation for the last 30 days
df['last_30_days_std'] = df.groupby('AREA NAME')['crime_count']
  .rolling(window=30, min_periods=1)
  .std()
  .reset_index(level=0, drop=True)
```

Listing 7.9 Apply the Same Data Transformations as When Building the Model

Next, we want to filter for rows where the target variable is missing, as these represent the last 7 days of the dataset. To do this, we filter the data with no shifted values in df_ historicals and then create the dataset in df to predict on, as shown in Listing 7.10. The order of operations is important here because we want to ensure our variables referencing the history are already calculated before we filter down to what we're trying to predict. For example, if we filter out the last 30 days of data, the average won't be able to calculate the last 30 days' worth of data. We're also keeping a separate copy of our data that we'll use to visualize the data after we've created the predictions.

```
#create historical version of data for visualizing data at the end
df_historicals = df[df['crime_count_shifted'].notnull()]

#filter to only null values in the target variable
df = df[df['crime_count_shifted'].isnull()]

#validate the row count
print(df.shape)
```

Listing 7.10 Filter to Only the Values We Want to Predict

Next, we load in our model that we saved in Chapter 6, Section 6.7.3, as shown in Listing 7.11.

```
import pickle

#input file name and open file
file_name = 'model.pkl'
with open(file_name, 'rb') as file:
  #read in the model
  model = pickle.load(file)
```

Listing 7.11 Load Pickled Model

We're now ready to generate predictions! Similar to our previous use case, we'll use the model.predict(x) function, as shown in Listing 7.12.

```
#drop the columns not in the model as inputs
X = df.drop(
  columns=['DATE', 'AREA NAME', 'crime_count', 'crime_count_shifted'])

#predict the results
y_pred = model.predict(X)
```

Listing 7.12 Create Predictions

To gut-check our predictions, let's visualize them. In order to do this, we need to summarize the data at the date level. We'll use the groupby function to summarize the data for both df_historicals as well as df, which becomes df_predictions after the summarization, as shown in Listing 7.13.

```
#group by date for historical data
df_historicals = df_historicals.groupby('DATE')['crime_count'].sum().reset_in-
dex()

#add the predictions onto df
df['predictions'] = y_pred

#group by date for the predictions
df_predictions = df.groupby('DATE')['predictions'].sum().reset_index()
```

Listing 7.13 Prepare the Data for Visualization

Now we can put this into a line graph, using plt.plot to plot both the actuals and historical lines, as shown in Listing 7.14. What's your initial impression of Figure 7.2?

```
import matplotlib.pyplot as plt

#set the size of the plot
plt.figure(figsize=(9, 5))

#plot the predictions
plt.plot(
  df_predictions['DATE'],
  df_predictions['predictions'],
  label = "future predictions"
  )

#plot the actual values
plt.plot(
  df_historicals['DATE'],
  df_historicals['crime_count'],
  label = "actuals"
  )

#show a legend on the plot
plt.legend()
```

Listing 7.14 Create First Plot Predictions

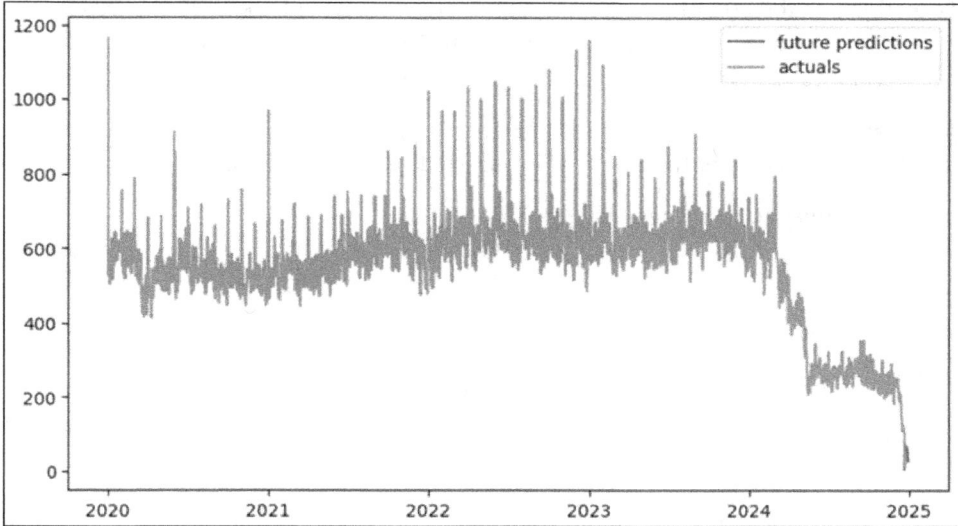

Figure 7.2 First Plot Predictions

If your first impression was anything like mine, you'll notice how hard it is to read! As a first step, let's filter down the historical data to 2024 (see Listing 7.15). Then, we'll rerun the plot. What's your initial impression of Figure 7.3?

```
#filter down df_historicals to just 2024
df_historicals = df_historicals[df_historicals['DATE'] >= '2024-01-01']
```

Listing 7.15 Filter Historical Data to the Current Year

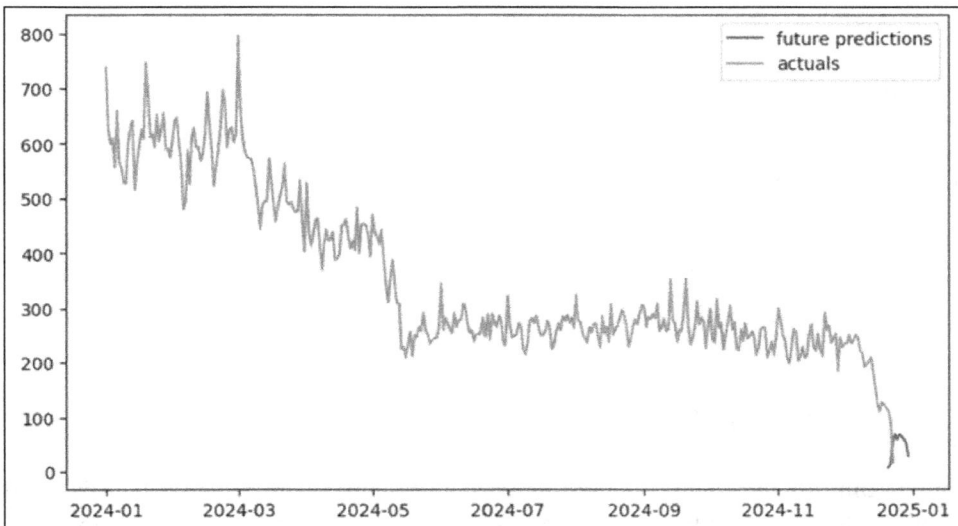

Figure 7.3 Filtered to Just 2024 Data

Something at the end of 2024 seems to be off. Let's take a look at where exactly this occurs in December (see Listing 7.16 and Figure 7.4). Since our data is sorted, we can see the December rows by selecting the first 25 rows with head.

```
#sort and view the last 25 records of the dataset
df_historicals.sort_values(by='DATE', ascending = False).head(25)
```

Listing 7.16 Check Underlying Data

DATE	crime_count
2024-12-23	17.0
2024-12-22	94.0
2024-12-21	113.0
2024-12-20	117.0
2024-12-19	123.0
2024-12-18	128.0
2024-12-17	111.0
2024-12-16	123.0
2024-12-15	154.0
2024-12-14	182.0
2024-12-13	210.0
2024-12-12	203.0
2024-12-11	197.0
2024-12-10	193.0
2024-12-09	218.0
2024-12-08	223.0
2024-12-07	247.0
2024-12-06	252.0
2024-12-05	240.0
2024-12-04	237.0
2024-12-03	252.0
2024-12-02	236.0
2024-12-01	236.0
2024-11-30	229.0
2024-11-29	247.0

Figure 7.4 Last 25 Days of 2024

At a minimum, the crime counts for the last 4 days are suspiciously low. This is where you should go back to Jen and lean on her as the expert on the data. When you bring this up to her, she mentions that there was an issue with the database and anything after December 12, 2024, should be removed. (If you think this is just me playing up the use case, I can assure you I've gotten this exact answer before when working on a project.) While this is unfortunate, we can accommodate it. We'll rerun our process and filter our data right away to only include data after December 12, 2024.

Listing 7.17 shows the entirety of the pipeline we've run to generate the graph. Figure 7.5 shows the updated line graph based on the new predictions.

```
df = pd.read_csv("df_summarized")

#sort the values to ensure it's in ascending order
df = df.sort_values(by=['DATE', 'AREA NAME'])

#filter out dates after 12/12
df = df[df['DATE'] <= '2024-12-12']

#shift 7 days for each area name independently
df['crime_count_shifted'] = df.groupby('AREA NAME')['crime_count'].shift(-7)

#we found some bad date values we need to remove
df = df[df['DATE'] != '0']

#convert the DATE column to pandas date for ease of use
df['DATE'] = pd.to_datetime(df['DATE'])

#create numeric day of week column
df['day_of_week'] = df['DATE'].dt.dayofweek

#create numeric day of month column
df['day_of_month'] = df['DATE'].dt.day

#create numeric week of year column
df['week_of_year'] = df['DATE'].dt.isocalendar().week

#create numeric month of year column
df['month_of_year'] = df['DATE'].dt.month

#create numeric month of year column
df['year'] = df['DATE'].dt.year

#rolling average of crime from the last 7 days
df['last_7_days'] = df.groupby('AREA NAME')['crime_count']
  .rolling(window=7, min_periods=1)
  .mean()
  .reset_index(level=0, drop=True)
```

```
#rolling average of crime for the last 30 days
df['last_30_days'] = df.groupby('AREA NAME')['crime_count']
  .rolling(window=30, min_periods=1)
  .mean()
  .reset_index(level=0, drop=True)

#standard deviation for the last 30 days
df['last_30_days_std'] = df.groupby('AREA NAME')['crime_count']
  .rolling(window=30, min_periods=1)
  .std()
  .reset_index(level=0, drop=True)

#keep historical data for visualization
df_historicals = df[df['crime_count_shifted'].notnull()]

#filter to only null values in the target variable
df = df[df['crime_count_shifted'].isnull()]

#drop the columns not in the model as inputs
X = df.drop(
  columns=['DATE', 'AREA NAME', 'crime_count', 'crime_count_shifted'])

#predict the results
y_pred = model.predict(X)

#group historical data by date
df_historicals = df_historicals.groupby('DATE')['crime_count'].sum().reset_index()

#add predictions to dataset
df['predictions'] = y_pred

#group predictions by date
df_predictions = df.groupby('DATE')['predictions'].sum().reset_index()

#filter to just 2024 data
df_historicals = df_historicals[df_historicals['DATE'] >= '2024-01-01']

#set plot size
plt.figure(figsize=(9, 5))

#plot the predictions
plt.plot(
  df_predictions['DATE'],
  df_predictions['predictions'],
```

```
  label = "future predictions"
  )

#plot the actuals
plt.plot(
  df_historicals['DATE'],
  df_historicals['crime_count'],
  label = "actuals"
  )

#display legend
plt.legend()
```

Listing 7.17 Rerun Entire Process to Generate Line Graph

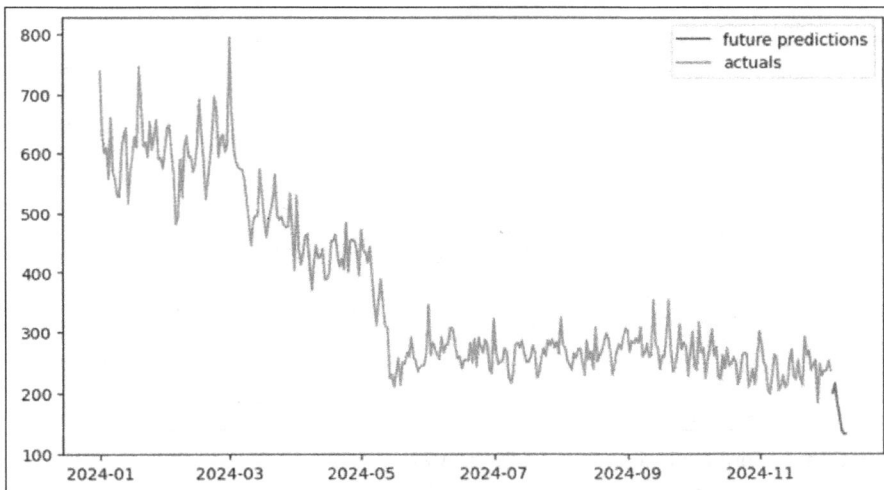

Figure 7.5 Line Graph with Updated Predictions

Generally, these predictions pass the sniff test. The overall trend in 2024 is decreased crime counts. The decline does appear to accelerate at the end of the year. Given that this is the holiday season, it wouldn't be surprising to expect such a decline. To double check, let's rerun the data for 2023 to see if it also has a trend of decreasing crime at the end of the year (see Figure 7.6 and Figure 7.7). It looks like the answer is yes.

In Figure 7.6, we can see that while there are many variations in 2023, the end of the year does appear to have a downward trend. Then, in Figure 7.7, we see the same declining trend up close for November and December.

Having verified this, we can be comfortable in our model and the predictions it's generating!

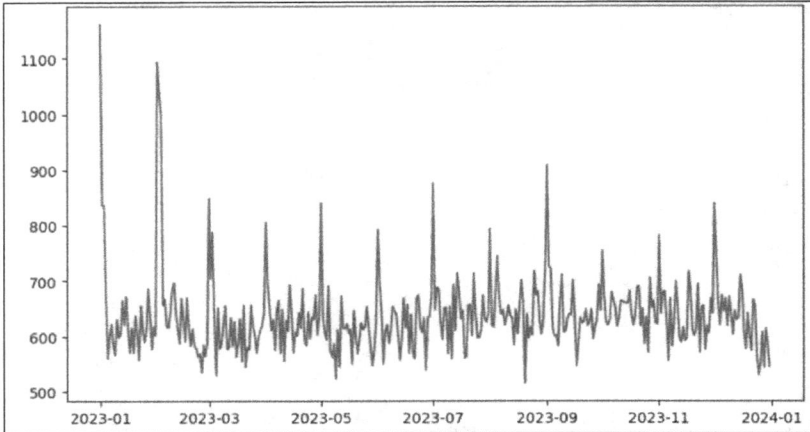

Figure 7.6 Crime Counts in 2023

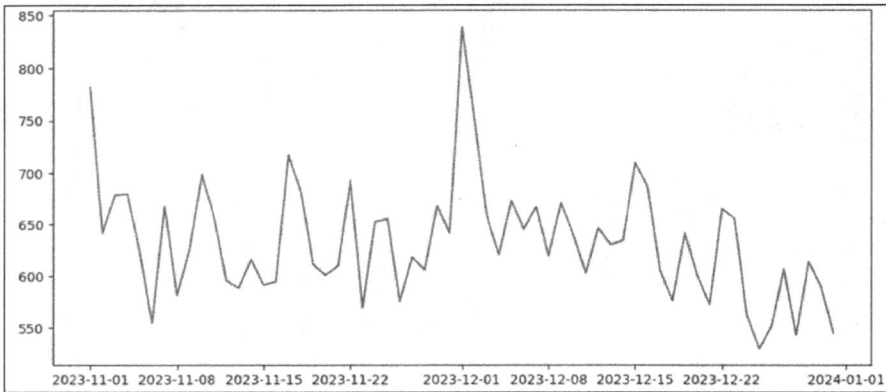

Figure 7.7 Crime Counts in November and December 2023

7.1.3   Saving Your Predictions

Regardless of your use case and tech stack, you should save your predictions to a centralized location. That could be a simple file or it could be a table in a database. It's best to set this up so you're able to see the history of your predictions to reference. A file format could look something like what is shown in Table 7.1.

Column	Purpose	Example Value
Date of prediction	This tells you when the prediction was created.	1/1/2026
Version of model	If you're iterating on your model, this can make it easier to identify which version of the model the prediction came from.	V1

Table 7.1 Example Prediction Data Structure

Column	Purpose	Example Value
Unique ID for use case	This tells you what this prediction is for (for example, a customer ID).	C-4536
Prediction	This is the value of the prediction.	.75

Table 7.1 Example Prediction Data Structure (Cont.)

This type of data structure allows you to *append* (add the news row to the existing data) the predictions on top of each other to reference that history when necessary. If you overwrite your predictions each time the process runs, you can create a challenging situation for yourself. What if your stakeholders ask a reasonable question like "Can I see what the predictions were yesterday?"

If you don't have a database you can write to, you'll need to create a file that you read and write to with each prediction you run. Don't create a unique file for each prediction—you could have hundreds of files to comb through if you ever need to look across all your predictions.

The best approach is to write to a table in a database if it's available to you. It's the most secure and accessible option for sharing the data. It's easy for other processes and tools like Power BI to reference in parallel. The recommendation to append the new data applies the same to the database as it does a file.

Automation presents another looming risk. If you take a hands-off approach and your predictions are continuously being made on their own, then you're less likely to see any oddities in the results. You need to create a process to monitor your model to ensure it continues to behave as expected. We'll get to this in Section 7.2.

7.1.4   Practical Approaches to Consider

Now that you've seen how to generate predictions, a new question arises: How does this process translate into a practical approach you can do on a recurring basis? The starting point goes back to your use case's requirements. You might need to generate predictions on demand or on a daily, weekly, or monthly schedule. Each situation warrants a different recommendation.

Let's start with lower-frequency weekly and monthly schedules. If you don't have a robust tech stack available to you to schedule your code, generating the predictions by manually running your code is a realistic option to consider. A manual approach for a monthly process only occurs twelve times a year. The cost of implementing a machine learning operations (MLOps) platform with scheduling capabilities likely isn't worth it.

At the other extreme, what about the use case of on-demand predictions? This requires you to have script scheduling capabilities already in place. Azure, Amazon Web Services, and Google Cloud Platform all have scheduling capabilities if you're running your code there. Platforms like Domino also enable this type of functionality. It's necessary to run the code automatically to effectively implement and manage a daily prediction process.

In addition to the specifics of your use case, your technology will also play a significant role in the strategy you choose. In general, it's easiest if you can leverage a platform that allows you to write, execute, and schedule code. While it's unlikely you have the ability make the decision on which technologies to purchase, you can find ways to influence your leadership to invest in a solution that expedites your predictive modeling projects. This makes models easier to develop and simplifies their management and maintenance.

7.2   Model Monitoring

The answer to the risk identified in Section 7.1.3 is model monitoring. Throughout this chapter, you've seen how use case-dependent each of these topics are in practice, and model monitoring is no exception! The variance in approaches for model monitoring are significant. If you work at a company with minimal predictive modeling capabilities, you may have no guidelines for this. If you work at a large company, the technology and legal teams may have a very long list of requirements you must meet to ensure your model is properly monitored.

A core piece of this section is the concept of *model drift*. This occurs as more time passes without retraining your model, meaning the predictions have drifted and the accuracy has degraded. As you're using the model's predictions to make decisions, this means you could be making decisions with incorrect information.

While you don't want to retrain your models too often, you do need to find a balance—and model monitoring helps with this. Retraining a model daily can lead to the many challenges we've discussed. However, you also don't want to run a model for years without retraining it, because the underlying assumptions and trends may have changed since the model was originally created.

In the following sections, we'll dive into the key aspects of model monitoring, including why it's important, what you should monitor, what you should consider when monitoring, and your options for retraining your model.

7.2.1   Importance of Model Monitoring

A valuable model is used to make decisions. If you're making decisions with an outdated model, you're introducing additional risk. The important factors of modeling monitoring are a combination of legal considerations, public perception, business value, and simply doing the right thing:

- **Legal considerations**

  Let's say you have a model that predicts whether a consumer should be given a loan. You've done all the work to train the model and it performs well. You set up an automated process that scores each new application, which the loan officers can incorporate into their decision-making when determining whether to approve a loan. However, after a full year, your loan officers tell you they're dealing with upset customers who were denied loans. You revisit the model and find that, over the past year, there have been shifts in the population submitting loan applications that your model hasn't seen before.

  Had you been consistently monitoring the model, you would have started to see the drift in the model's results and could have intervened before the complaints started. Even though there was no ill intent, there are customers who were rejected from receiving a loan, which could result in legal action against your company.

- **Public perception**

  To continue with the loan example, even if legal action isn't taken against your company, an outdated model can tarnish its reputation. The public may now see your company as unfair, potentially influencing consumers to apply for loans elsewhere. If you break consumers' trust, it can be difficult win it back!

  While it's not a predictive modeling example, consider Wells Fargo and their scandal involving fake accounts being opened on behalf of customers without their knowledge (http://s-prs.co/v617002). Wells Fargo took significant steps to correct the situation, but the negative public perception has continued to affect its business.

- **Business value**

  "I want to make decisions with bad information." Hopefully no one agrees with this statement! If your model is being used to make decisions, you want those decisions to be based on up-to-date and accurate information. That is the root of the business value component of model monitoring.

  Say you built a model to predict customer purchasing behavior for a restaurant in late 2019. Imagine how inaccurate that model would be at predicting customer behavior in April 2020 when the full scope of COVID-19 shutdowns were in effect. While this is an extreme example, you'll most likely encounter similar problems on a smaller scale. Monitoring your model allows you to proactively identify shifts and retrain to avoid model drift.

7.2.2   What to Monitor

Just saying that you should monitor your model is quite vague, so what specifically should you be monitoring? You can use a combination of both operational approaches (is the process working?) and analytical approaches (is the model still working well?). We'll break these down into three steps that we'll discuss in the following sections:

1. Confirm the predictions are being generated.
2. Compare predictions to actual outcomes.
3. Evaluate important distributions where applicable.

Confirm the Predictions Are Being Generated

This is purely operational in nature. The goal of this step is to ensure you're on top of any backend processes that break. It's much better to alert your stakeholders of an issue with a process you own than to have them alert you. When you're able to raise these questions, your stakeholder sees you as more competent. The inverse is also true when they're the ones alerting you of an issue with your process.

How you'll implement this step depends on the technology you're using. In general, the goal is to run a parallel validation process that ensures predictions are generated in the correct place. This should be a parallel process rather than one you integrate directly with your prediction workflow. This will ensure that if prediction generation fails, a separate process is in place to identify the missing predictions, which is especially important when you're using automation. In practice, this means two separate scripts that are scheduled: your code for the model that generates the predictions and code that checks whether the predictions have been generated.

While this process may not be the most glamorous, it does have the best return on investment (ROI) for your time. As mentioned, there's a big difference between a stakeholder informing you of an issue and you informing them. While the probability of the process failing may be quite low, the impact of it occurring is significant. It's also relatively easy to establish this type of process. In its simplest form, you're checking whether the current day contains predictions and scheduling the process. This means the investment is quite small compared to the potential return of getting ahead of any issues.

Compare Predictions to Actual Outcomes

This is where you evaluate and monitor model drift. From a mental model perspective, think about this as an ongoing test that's similar to the way you train models. When the actual outcome is available, you can combine it with your model's predictions to track how well your model performs. Plotting predictions and actual outcomes in a line graph is where the term model drift comes from. As time progresses and the model's training data becomes outdated, you'll see the actual outcomes and the predictions drift away from each other. Figure 7.8 shows an example of what this can look like over time. The drifting starts to occur in March and then becomes severe in June. By monitoring your model, you'll see these results as they occur and be able retrain the model to account for new underlying behaviors.

The ability to execute this in practice requires you to join your predictions and your actual data. Having a unique identifier associated with your predictions enables you to do this. In Figure 7.8, we included the date from the predictions to ensure we could create such a graph. If you're predicting something about customers, including the customer ID would enable you to compare the predicted and actual outcomes.

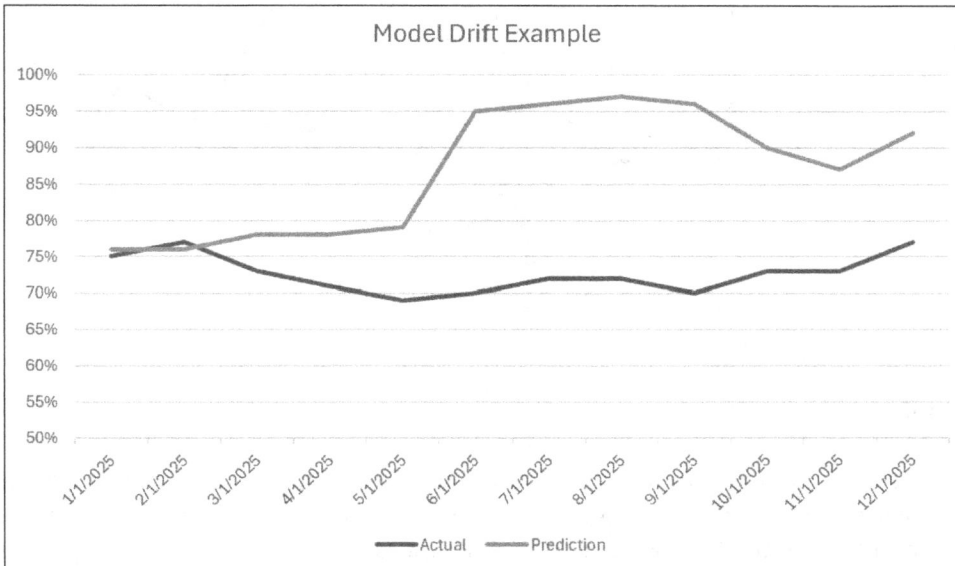

Figure 7.8 Model Drift Example

Displaying this data in a time series manner allows you to see whether the predictions are changing over time. If you only look at a fixed period of time, you lack context and run the risk of over- or underreacting. If you only look at the results in aggregate, it will likely take a while to identify any model drift.

It's helpful to set up a dashboard or recurring report that will push this information to you. Each model is different, but aim to keep an eye on the results each week, or at minimum, a few times a month. That way, you can you avoid a situation where a model should have been retrained a while ago but business leaders are still making decisions from it.

Evaluate Important Distributions Where Applicable

This is a subset of your comparison of actual data and predictions, adding a layer to evaluate how prediction performance varies across different population groups. Let's use the example of a resume scoring model to illustrate this concept.

An important measurement for a resume scoring model is how the model handles demographic information like ethnicity, gender, and age. You don't want the model to create or perpetuate any bias from the data it was trained on. You can monitor this by comparing the actual results to the predictions for each subset of the population, as shown in Figure 7.9.

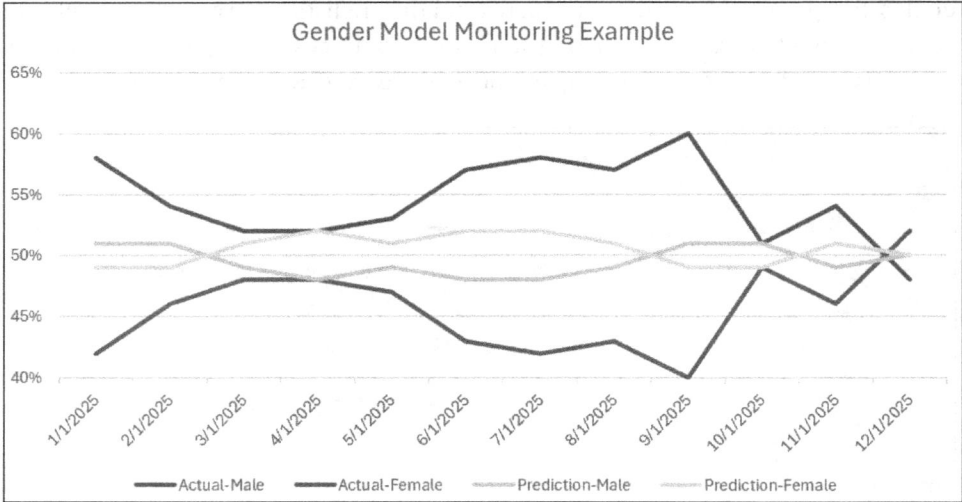

Figure 7.9 Model Monitoring Gender Example

By looking at the data this way, you're able to see how the actual results compare to the model. In this example, the model actually creates a more stable balance between the scores of males and females. This type of finding is an insight you can bring to your stakeholders. This helps you demonstrate the model's value to them, while highlighting that you're actively reviewing this type of information. A key component to modeling monitoring is building trust with your stakeholders. Proactively showing that you're looking into your data this way helps build that trust.

Sometimes the cardinality, or number of options within a column, is too large to make a time series view valuable. In a case like this, you can visualize the data in aggregation, as shown in Figure 7.10.

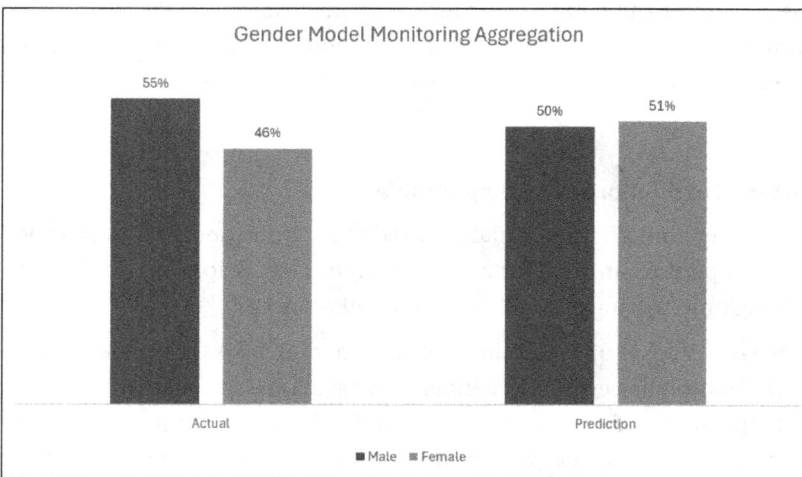

Figure 7.10 Model Monitoring Example in Aggregation

This makes it easier to review the data across multiple values. Even with a binary male–female view, this visualization more concisely shows how the model's predictions produce a more stable and equal distribution compared to the actual outcomes.

7.2.3   Considerations for Model Monitoring

How you implement model monitoring will vary based on the technology you have available. The platform you're using might have built-in functionality for model monitoring, or you might need to build most of your monitoring process from scratch. Either approach works!

It's important to keep the implementation simple and easy to understand. If you create a system that you don't understand, you risk missing important warning signs. Prioritization is key, as addressing the highest-risk items first will help you avoid going down the rabbit hole of chasing any possible edge case.

Another important dynamic to keep in mind as your model matures is the potential for a *circular reference issue*. The goal of building a model is to help your company make better decisions. This means the data you're using to train your model can become a circular reference. As shown in Figure 7.11, as you retrain your model, it will start to reference data that's been influenced by its outputs.

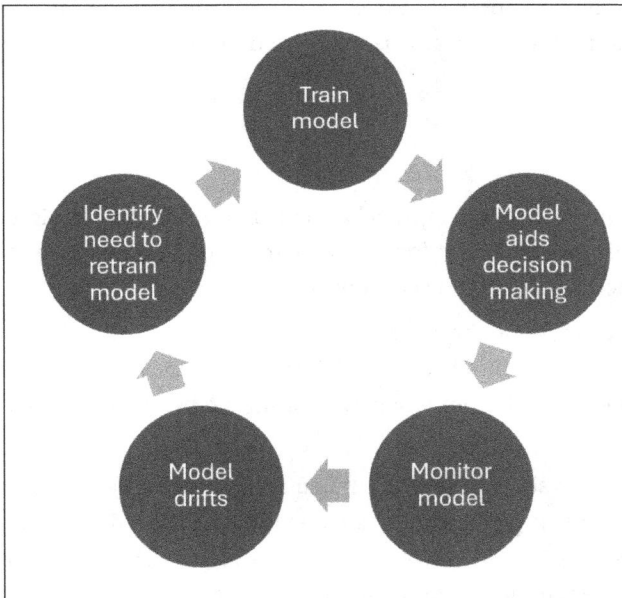

Figure 7.11 Model Monitoring Circular Reference

As your model is more widely adopted, the risk of this circular reference increases. Decisionmakers start using the model and it gets retrained with information that was based on their decisions. That's when the circular reference problem can turn into a risk.

This dynamic is perhaps the most important reason that humans still need to be part of the decision-making process in the vast majority of situations. It's part of what fuels fears about doomsday scenarios where AI runs everything and we're taken along for the ride. Let's return to the loan approval example from the previous section. If the system uses the model's results completely autonomously, who will flag flaws when a loan was approved or rejected incorrectly?

The majority of companies are against a completely autonomous AI or model-based decision-making system. Most companies use models and AI to do the heavy lifting, with humans performing the final review. Personally, I believe this is the way it should be.

The extent of your model's impact on the end decision is an important input to consider as part of your model monitoring process. If you have influence over how it's used, you should augment the decision-making process with a final review by a human decisionmaker. Putting aside the doomsday AI scenario, this will create the best outcome and yield the most useful data for future model training. This improves your future training data because a human corrects the model's decision when it's wrong and adjusts the final output accordingly. The model can then learn from this when it's retrained.

Keep this in mind as you're monitoring your model. It isn't necessarily a problem you'll have during the initial development, or potentially ever, depending on how your model is used. The point is to make sure you monitor this as a potential risk as your model matures.

7.2.4   Retraining the Model

We can split model retraining into two broad approaches. The first is the most straightforward. The second is when you need account for the circular reference concerns that arise when you retrain your model on data it has influenced.

If you simply need to retrain your model because it's drifted, you'll follow the same steps from Chapter 5 and Chapter 6 to rebuild your model. This can also be a good time to revisit your ETL process to identify whether any new data should be included in the model.

But what happens when you need to account for the circular reference problem? This is where you need to get creative. It's important to be a consultant and partner to your stakeholders in these cases, because you must work with them to create processes that enable cleaner training data. Here are a few approaches to consider:

- **Hold out random subset of situations**
  This can get complicated, but it entails not providing a prediction for a random subset of situations. Using a resume scoring model as an example, one approach could be that every 20th resume isn't scored and requires a review from a recruiter. By

doing this and tracking which resumes are scored by recruiters, you're creating a clean set of data for your model to be trained on. In practice, this likely requires technology support, regardless of your application. You could then combine it with the larger dataset if you don't have a sufficient sample size to train your model.

- **Focus on situations where the model differs from the actual result**
  This would also be an ensembled approach, but in this case, you'd retrain your model to focus on the situations where it was wrong with the goal of capturing where it's still missing the mark. This can come with a number of challenges, but it doesn't require the same level of intervention from your stakeholder group as withholding a random subset.

7.3   Measuring the Impact of Your Model

You've built your model and you're making sure to monitor it for continued success. Your job is done, right? Unfortunately not! In many situations you'll be asked to further prove the impact your model has on the business. In some cases, this may be incredibly obvious, while in others you may need to explore more sophisticated methods to demonstrate its value.

In either case, it's wise to think about how you can quantify the value of your model. Even when the value is straightforward, having quantifiable metrics will provide significant returns for you. This is especially true in cases where a new leadership group comes in and doesn't know the benefits of the model. Knowing how to quantify a model's impact also helps you speak to its value in job interviews.

In the following sections, we'll explore key techniques for demonstrating your model's impact: the business sniff test and a variety of code experiments.

7.3.1   Business Sniff Test

The methods categorized under the *business sniff test* umbrella can be somewhat subjective. They're low-cost and potentially high-value approaches. You can use them as a starting point for understanding how much value your model has. You can't quantify these approaches, but they aren't necessarily the marketing tagline you'd be highlighting in an interview anyway.

We'll look at some specific tactics that fit into the sniff test category in the following sections.

Questions

If you're receiving questions from stakeholders and users about the model, this is a positive sign of engagement. If the users don't have questions, it's much more likely that they're simply not using the model than that the model is perfect and there's no need for additional clarification. This is especially true for a new model.

You should look a bit deeper to understand the sentiment of the questions you receive. Ideally, you want to get a mix of constructive feedback and questions that seek understanding or clarification.

Experience Survey Questions

While it's not related to predictive modeling, one of the larger projects I've worked on is implementing an enterprise experience survey. The path to getting the questions and the backend technical components of this survey was long, so I was anxiously awaiting the results of how leaders were using the survey.

Right away, I received a ton of questions about the survey and the specifics of the process. In the moment, I felt overwhelmed, and maybe even a bit attacked. After getting through the questions and reflecting on them, I changed my perspective. Roughly 75% of the questions were about specifics that the users wanted to understand for themselves. This included questions like who receives the survey, when they receive the survey, who may not receive survey, etc. The other 25% of questions were constructive feedback about how to improve the process or enhancements to consider.

This level of engagement quickly showed how the survey process filled a need and added value. It's important to remember that data products are often a small component of your stakeholder's job and responsibilities. If something doesn't add value to them, they're not going to use it.

Adoption

Tracking adoption is similar to evaluating the questions you receive, but it has a more formal approach to measurement. When you're building a model, your stakeholders and users are almost always going to be interacting with some type of tool. Rarely would they be going directly into your code to get the predictions. Measuring the number of individuals accessing the tool can be another strong indicator of value. As mentioned in the previous section, data products are often only a small piece of the user's job. If you're able to see that someone is using your tool, you're likely providing something that is valuable to them.

Microsoft's Power BI is currently the most prominent data visualization tool that tracks the usage of reports and dashboards. This can be a low-effort way to track the adoption of your model. Most data visualization tools will have some type of usage reports available out of the box.

An added benefit of the adoption approach is that you're able to quantify the level of adoption *and* identify who in particular is adopting the model. You often already know who your power users are because they're attending your trainings, asking you questions, and finding other ways to connect with the project. User-level adoption reports can help you identify individuals who were expected to be power users but aren't engaging with the model. You can then gather feedback from these users and understand what adjustments would make the tool or the model more valuable for them.

Prioritization

Tracking questions and model adoption is often a good proxy for value among individual contributors and managers. This approach focuses on direct users of the tool. However, what about leaders who aren't part of the day-to-day operations?

Leadership's prioritization of your model offers insight into how they assess its value. There's rarely a shortage of workflows that you could improve, so leadership continuing to prioritize an initiative means they see value and/or potential in the work. This can also be a circular reference, though. To make decisions on prioritization, leaders often want quantifiable metrics to determine the value of a project.

Early on in a project, you may get more grace in using subjective measures of value. However, over time you'll likely want to consider more quantitative approaches to ensure that your leadership continues to prioritize your work.

7.3.2 Experiments

Experiments are a fascinating phenomenon in the business world. On one hand, it's normal for tech companies to experiment with their progress, with massive systems designed to monitor and measure those experiments. On the other hand, so much of the decision-making in departments like finance and human resources (HR) is rooted in the loudest voice in the room.

When building a model, you can set up an experiment to validate that using its results in the decision-making processes leads to better outcomes. For example, if one group uses the model to make a decision but another doesn't, you can compare outcomes between the groups as a simple experiment.

Experimentation Can Save Time

I've found the lack of experimentation in the areas I've worked in incredibly confusing. It would take weeks or months for a group to debate the right decision, while running an experiment to determine the same thing would have taken a fraction of that time.

In the following sections, we'll explore a new use case that demonstrates the experimentation process and walk through a few examples of experiment types.

Experiment Overview

If you don't already have experience with experimentation in either a corporate or academic setting, it's a good idea to read up on it. Experimentation isn't the main focus of this book, but we'll provide some insight into how experiments work at a high level. It may be hard to understand how experiments can help you prove your model's value if you don't have any foundational knowledge.

> **Reading Recommendation**
>
> If you wish to dig into experiments and how you can leverage them in the business world, I'd recommend reading *Experimentation Works: The Surprising Power of Business Experiments* by Stefan Thomke (Harvard Business Review Press, 2020). I found this book insightful, and it really shaped my mindset around how experiments can improve decision-making.

The goal of an experiment is to measure a cause-and-effect relationship, rather than relying on correlation. As you've likely heard, correlation doesn't equal causation. Experiments are designed to mitigate the risk of *confounding variables*—alternative reasons why what you're observing is going up or down.

Regardless of the experiment's design, there is almost always a *treatment* and *control* group. The treatment group consists of individuals who receive the intervention, or the factor whose effect you're looking to measure. The control group doesn't receive the intervention. They're controlling for other factors that may lead to a change in what you're measuring.

Where the various experimental designs differ is primarily in how the treatment and control groups are selected. That is our focus in the sections that follow, where we discuss natural experiments, quasi-experiments, and randomized control experiments. Figure 7.12 provides a generalized point of view on the frequency of experiments compared to the rigor of experimental design in the business world.

There are three main takeaways from this visual:

- The least rigorous approach is also the most frequently used. This leads to suboptimal decisions being made more often than they should.
- The A/B test is somewhat of an anomaly, given its frequency relative to its rigor. This is primarily driven by the large-scale products that heavily rely upon A/B testing to compare two options.
- With the exception of the A/B test, the most rigorous approaches are the least used.

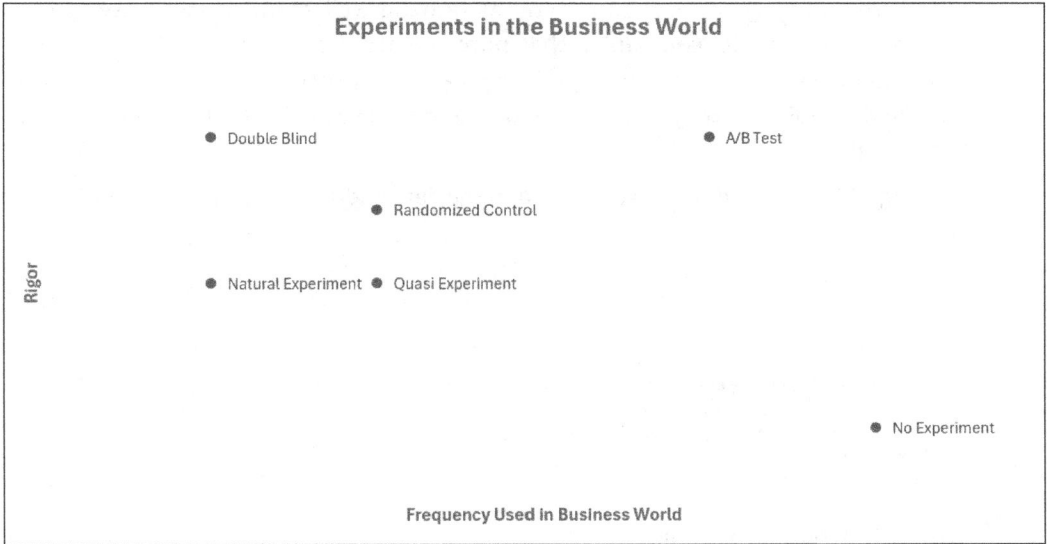

Experiments in the Business World

- Double Blind
- A/B Test
- Randomized Control
- Natural Experiment • Quasi Experiment
- No Experiment

Rigor

Frequency Used in Business World

Figure 7.12 Experiments in the Business World

On the third point, there are some viable explanations for why these aren't used as frequently:

- *Natural experiments* are largely out of your control, since they occur naturally. The one thing you can control is your ability to identify opportunities to treat a situation as a natural experiment.

- *Double-blind* and *randomized control* experiments can be very difficult to execute in a business setting due to legal and ethical considerations. Another complication is the need to consider the design from the start of your project. This is often deprioritized in predictive modeling projects, and it's quite difficult to retrofit the model into one of these designs.

- *Quasi-experiments* are one of the most underutilized analytical techniques in the business world, despite the fact that they can be used in almost any situation. Their goal is to make your control group as similar as possible to your treatment group to reduce as many confounds as possible. Pay special attention to the section on quasi-experiments. You'll see how they can help prove the value of your model and how you can use them for other analysis work.

The power of an experiment is significant, so it's important to know how to leverage it to demonstrate your model's value. The type of experiment you pick often comes down to your use case. A key decision point is often the sensitivity of the situation and potential issues around fairness. For example, if you want to perform an experiment around employee benefits, you can't give two employees with the same qualifications different sets of benefits in the name of experimentation.

Our three existing use cases don't demonstrate this well, so instead we'll use a simplified example of 100 companies that purchase from a business-to-business (B2B) e-commerce site. The company operating the site is worried about the impact of tariffs on their sales, so they created a model to adjust the product offerings for customers accessing the site.

Listing 7.18 shows the code used to read in the data and Table 7.2 shows a sample of the data.

```
import pandas as pd

#read in data
df = pd.read_csv("Experiment Data.csv")

#preview data
df.head()
```

Listing 7.18 Read In Experiment Data

Company	Previous Month Purchases	Company Size	Tariff Exposure	Model Adjusting Offering	Purchases Next Month
1	49	17,072	1	1	62
2	71	13,065	1	1	70
3	13	7,424	0	0	23
4	17	5,292	0	1	52
5	46	23,844	1	0	63

Table 7.2 Sample Data for Experiment Explanation

This data shows all 100 companies have the potential to adjust their offerings, with the model deciding who should or shouldn't. The concern with this approach is that the model may not be improving the outcome. The difference between the previous month and the next may have happened without the model's influence. This is a common situation that you'll likely encounter in the real world, so let's review the natural experiment and quasi-experimental approaches first. Then, we'll revisit the data in a scenario that simulates a randomized experiment.

Natural Experiments

The concept of a natural experiment is simple: something naturally occurs that allows us to define a treatment group and a control group. This can be very helpful in situations where the topic is sensitive and/or the cost of experimenting is high.

Remote Work Natural Experiment

Working in the people analytics space during the COVID-19 chaos was a fast-paced but interesting place to be. Many questions about the employees of the company flooded in from various leaders and many new norms were set. None were more impactful than the shift to remote work.

After the height of the COVID-19 pandemic, many companies needed to wrestle with—and, at the time of writing, are still fumbling through—how or whether employees should return to the office. We now have more research about this topic, but in late 2020, not much was available.

The leaders at the company I worked for wanted to understand the impact of remote work. One of the ways my team was able to answer this question was due to an office closing a few years earlier, which required a majority of associates to work from home. This was the definition of a natural experiment! We tracked key metrics like performance and engagement before and after the office closure. We concluded that the remote working arrangement had a positive impact.

While potential confounding variables may have influenced the result, this type of natural occurrence offered a more effective way to measure the impact of remote work than many other methods available at that time.

To mimic the natural experiment with our sample data, we'll use the `Tariff Exposure` column to split our data into a treatment and control group. We'll also split the data out by samples that the model adjusted and samples the model didn't adjust (see Listing 7.19).

```
#treatment group
df_natural_treatment_model = df[
  (df['Tariff Exposure']==1) &
  (df['Model Adjusting Offering']==1)
  ]

#control group
df_natural_control_model = df[
  (df['Tariff Exposure']==0) &
  (df['Model Adjusting Offering']==1)
  ]

#print the rows and columns of the data
print(df_natural_treatment_model.shape)
print(df_natural_control_model.shape)
```

Listing 7.19 Split Data into Treatment and Control for Natural Experiment with Model Adjusted

We see there are 35 companies with tariff exposure that also had their offering adjusted by the model, which is our treatment group. Our control group is the 5 companies that don't have tariff exposure but did have the model adjust their offering. Now we need to do the same thing for the companies who didn't have their offering adjusted by the model (see Listing 7.20).

```
#treatement group
df_natural_treatment_nomodel = df[
  (df['Tariff Exposure']==1) &
  (df['Model Adjusting Offering']==0)
  ]

#control group
df_natural_control_nomodel = df[
  (df['Tariff Exposure']==0) &
  (df['Model Adjusting Offering']==0)
  ]

#see number of rows and columns
print(df_natural_treatment_nomodel.shape)
print(df_natural_control_nomodel.shape)
```

Listing 7.20 Split Data into Treatment and Control for Natural Experiment with No Model Adjusted

This results in 30 for each group. Now we're ready to run the statistics—although we won't go too deep, as statistics isn't the core focus of this book. First, let's take the average of the purchases in the next month across each scenario (see Listing 7.21).

```
#calculate the mean for each dataset
print(df_natural_treatment_model['Purchases in Next Month'].mean())
print(df_natural_control_model['Purchases in Next Month'].mean())
print(df_natural_treatment_nomodel['Purchases in Next Month'].mean())
print(df_natural_control_nomodel['Purchases in Next Month'].mean())
```

Listing 7.21 Average of Purchases Next Month

Table 7.3 shows the results.

Tariff Exposure?	Received Model Adjustment	No Model Adjustment
Tariff exposure	68	57
No tariff exposure	52	32

Table 7.3 Average of Natural Experiment Groups

When reviewing the results, you'll notice some interesting findings:

- When controlling for tariff exposure, the companies that received the model adjustment had more sales.
- When controlling for the model adjustments, those with tariff exposure also had more sales.

Now let's run the various scenarios through t-tests to confirm what we're seeing in the averages (Listing 7.22). If you're unfamiliar with t-tests, they're tests to measure the independence of two groups. To execute the t-test, we use the `ttest_ind` function from the stats library. We run the t-test on both scenarios—one with the model adjustment and one without—and then print out the p-value to evaluate the results.

```
#import stats library
from scipy import stats


#run t test on data
t_statistic, p_value = stats.ttest_ind(
  df_natural_treatment_model['Purchases in Next Month'],
  df_natural_control_model['Purchases in Next Month']
  )


#print the p-value
print("P-value:", p_value)


#run t test on data
t_statistic, p_value = stats.ttest_ind(
  df_natural_treatment_nomodel['Purchases in Next Month'],
  df_natural_control_nomodel['Purchases in Next Month']
  )


#print the p-value
print("P-value:", p_value)
```

Listing 7.22 Natural Experiment T-Test

The p-values from these tests are well below the 0.05 threshold. This is a strong indicator that the model adjusting the product offerings has a statistically significant and positive impact on sales.

Quasi-Experiments

As mentioned, this technique is one of the least used techniques in the analytics space. The ideal of randomized treatment and control groups simply isn't always possible. Quasi-experimental approaches allow you to create similar treatment and control groups when randomization isn't possible.

Before we dive into our example dataset, let's start by talking about this theoretically.

Let's say you're part of the team that owns the internal large language model (LLM) for your company, and you're tasked with proving the value of the LLM. The metrics your team wants to measure include productivity, engagement, and performance. The simplest approach would be to compare these metrics for users who have used the LLM at least once compared to those who haven't. However, think about all of the potential confounds:

- **Type of role**
  Are certain types of roles more or less likely to use an LLM in their day-to-day job compared to others? Are those types of roles also then correlated with the measure we're evaluating? The answer is likely yes to both of these questions. Someone who works in a tech-focused job probably has time to test a new tool like the LLM, while someone who works in a call center role might not have the same opportunity.

- **Age**
  While it's not *always* true, it's generally correct that younger individuals are more likely to try new tools. This could then influence who is or isn't using the LLM, which may be the cause for the difference, rather than the LLM usage itself.

The solution to this is to control for these confounds. One approach is called *matching*, where the treatment group is matched to individuals who weren't part of the treatment. This creates a control group that more closely resembles the treatment group than the whole population of treated and untreated individuals.

This matching is done numerically and is similar to clustering (we'll go through an example of matching using our sample data later in this section). You feed an algorithm the numeric data for each individual, and it finds the most similar match.

This can be a very powerful approach to use, so it can seem a bit like magic. Don't be fooled: It does have its limitations, the primary one being you're only able to control for observable attributes. Using the employee example, you likely can't control for an individual having a partner or children at home who also use an LLM, which could then influence the individual's adoption of the tool.

The other limitation in practice can be the computational power required to perform on large datasets. In this example, if you have 30,000 employees and 10,000 have used the LLM, you have to find a match for those 10,000 individuals. You want the algorithm to be able to find the optimal match for each member of the treatment group, which requires it to make a very high number of calculations on the backend. Doing this over a few hundred records isn't a concern, but doing it over thousands can quickly become a problem.

Now, let's go through the algorithm. One important note is that because we're using Anaconda, we don't have the ability to download new packages. The preferred package for this would be `causalinference`; however, it's not available in our computing environment, so we'll have to be more creative with our solution. If you're writing your code locally or have the ability to edit your environment, you can download this package. Using `causalinference` would be the most practical approach, but from a learning perspective, writing the code from scratch will help you better understand what's happening behind the scenes.

For our approach, we'll leverage the `pairwise_distances` function to accomplish the matching. This is a relative of the `kmeans()` clustering algorithm we discussed in Chapter 5, Section 5.6. This does involve some programming-specific knowledge, but we'll explain each step so you can follow along—even if you don't have that experience yet!

First, we need to split the data into treatment and control groups. In this case, we've already applied to the model to adjust the offerings to the users of the website. Then, we'll select only the columns we want the algorithm to match the data on (in our case, `Previous Month Purchases`, `Company Size`, and `Tariff Exposure`), as shown in Listing 7.23.

```
#filter to only the treatment group
df_treatment_full = df[df['Model Adjusting Offering'] == 1]

#select only specific columns
df_treatment = df_treatment_full[[
  'Previous Month Purchases',
  'Company Size',
  'Tariff Exposure'
  ]]

#filter to only the control group
df_control_full = df[df['Model Adjusting Offering'] == 0]

#select only specific columns
df_control = df_control_full[[
  'Previous Month Purchases',
  'Company Size',
  'Tariff Exposure'
  ]]
```

Listing 7.23 Splitting into Treatment and Control for Matching Algorithm

This next step is incredibly important. When working on a process like this, you must normalize the data first. If you're not familiar with *normalization*, it's the act of ensuring all columns are on the same scale. In this example, we have 3 columns on very different scales. The `Previous Month Purchases` column is generally between 0 and 100, the `Company Size` column is in the thousands, and the `Tariff Exposure` column is either 0 or 1.

Without normalizing the columns, we'd run into the issue of the Company Size column having an outsized impact on the end result. The matching process calculates the distance between each of these rows, meaning the Company Size with the largest scale would end up dominating the calculation.

To account for this, we'll ask a generative AI tool to create a function for us. You can prompt Copilot to create a function that normalizes the data on a scale of 0 to 1, which it does quite well, providing a useful solution right away.

When to Use Generative AI

Generative AI can be incredibly helpful to speed up your programming process for these types of functions. Personally, I've never been great with lambda functions and have always found the workflow confusing. With generative AI, I'm able to avoid the trial and error that would have likely ensued if I built this function from scratch myself.

It's still wise to test that the solution works. Copilot even provides the standard documentation for creating Python functions without you needing to ask for it!

For this use case, we want everything to be normalized to 1, which is what the Tariff Exposure indicator represents (see Listing 7.24).

```
#create function to normalize the data
def normalize_dataframe(df):
  """Normalizes each column of a Pandas DataFrame to the range [0, 1].

  Args:
   df: The input Pandas DataFrame.

  Returns:
   A new DataFrame with normalized columns.
  """
  return df.apply(lambda x: (x - x.min()) / (x.max() - x.min()))
```

Listing 7.24 Normalize to 1 Function

If you haven't used your own functions before in Python, there's often a mental gap that you'll need to overcome. The act of writing the function, as shown in Listing 7.24, doesn't apply this to your data and perform the calculation. You must also run the function, as shown in Listing 7.25. This will then apply whatever code your function is intended to run to your data.

```
#apply normalization function to each treatment and control
df_treatment = normalize_dataframe(df_treatment)
df_control = normalize_dataframe(df_control)
```

Listing 7.25 Apply Normalization Function to Matching Treatment and Control Data

Figure 7.13 is a sample of what the data looks like after being normalized compared to its original state. The left represents the original data and the right represents the normalized data. Notice how the Tariff Exposure hasn't changed since we specified the scale to be from 0 to 1; however, the Previous Month Purchases and Company Size columns have significantly changed.

Previous Month Purchases	Company Size	Tariff Exposure	Previous Month Purchases	Company Size	Tariff Exposure
49	17072	1	0.526316	0.489444	1.0
71	13065	1	0.815789	0.333657	1.0
17	5292	0	0.105263	0.031453	0.0
24	29732	1	0.197368	0.981649	1.0
80	12124	1	0.934211	0.297072	1.0
14	23522	1	0.065789	0.740212	1.0
73	26570	1	0.842105	0.858715	1.0
19	29437	1	0.131579	0.970180	1.0
47	16307	1	0.500000	0.459702	1.0
76	13700	1	0.881579	0.358345	1.0

Figure 7.13 Example of Normalized Data

Now that we've sufficiently prepared our data, let's walk through how to implement a matching algorithm yourself. We won't explore the libraries that could be used for this, as they unfortunately aren't part of the Anaconda environment at the time of writing. If you haven't used for loops before, don't worry—we'll provide an explanation in sufficient detail for you to follow what's going on.

For Loops

Let's pause for a quick explanation of *for loops*. Think about them as a way to perform the same operation without having to copy and paste it each time. A for loop takes a list and iterates over it. For example, let's take this list of programming languages:

```
programming_languages = ['Python', 'R', 'SQL']
```

You could simply print out each manually to display them:

```
Print('Python')
Print('R')
Print('SQL')
```

However, this is time consuming, especially as the list you're working with grows. For loops are the solution. You can give the for loop the list of programming languages and have it print out each one, as follows:

```
For language in programming_languages:
  Print(language)
```

The output is the same, but you didn't have to repeat the same print function each time.

If you're familiar with writing functions in Python, you can refer to Listing 7.26, which shows all of the code to write the function in one spot. However, starting with Listing 7.27, we'll break down the code to provide a more detailed explanation of what it's doing.

Overall, this function compares each row in the treatment dataset to each row in the control dataset using pairwise distance. These results are then appended to a single DataFrame, which we'll use to create the one-to-one matches. The resulting DataFrame is a long list of distances for each unique combination of treatment and control roles.

```python
#import pairwise_distances function
from sklearn.metrics.pairwise import pairwise_distances


#create blank dataframe to add matches to
all_matches = pd.DataFrame()


#loop through treatment list to find matches
for treatment_row in list(range(0, len(df_treatment))):
  treatment = df_treatment.iloc[treatment_row].to_numpy().reshape(1, -1)
  treatment_company = df_treatment_full.iloc[treatment_row, 0]
  distances_list = list()
  for control_row in list(range(0, len(df_control))):
    control = df_control.iloc[control_row].to_numpy().reshape(1,-1)
    distance = pairwise_distances(treatment, control, metric = 'euclidean')[0][0]
    distances_list.append(distance)
  data = {
    'Treatment Company': treatment_company * len(distances_list),
    'Control Company': df_control_full['Company'],
    'Distance': distances_list
    }
  temp_df = pd.DataFrame(data)
  all_matches = pd.concat([all_matches, temp_df], ignore_index = True)
```

Listing 7.26 Full Matching Code

An important first step is to create a blank DataFrame (see Listing 7.27). We'll use this to aggregate the results as we loop through the data. When performing an operation like this, having a blank list or DataFrame to append to is one of the most common approaches.

```python
from sklearn.metrics.pairwise import pairwise_distances


all_matches = pd.DataFrame()
```

Listing 7.27 Create Blank DataFrame

Now it's time for our for loop. Overall, there are two for loops being used. The first allows us to identify each row of the treatment dataset, as shown in Listing 7.28. How does it do this? We're creating a list of numbers with the `list()` function and `range()` function together. This results in a list of numbers from 0 to the size of our treatment dataset, which is what we want. We'll then use this to filter our treatment dataset to a single row.

The `treatment_row` part of the for loop is where beginners get tripped up the most. This is just like naming your DataFrames; you can name it anything you want, but it helps to choose something that makes sense when others (or you) have to revisit the code. As the for loop iterates over the list, `treatment_row` becomes the number of the row we're trying to filter the treatment data to.

```
for treatment_row in list(range(0, len(df_treatment))):
```

Listing 7.28 First Matching For Loop

Now we're inside the first for loop, as shown in Listing 7.29. This stage of the for loop is specific to the treatment data, and we're using `treatment_row` to filter the treatment data to a single row. We do this for both the full normalized dataset and a smaller list of columns for matching. It's easy to forget, but this is all a way to compare the outcomes of each group against each other. We need to ensure we bring along the full scope of the data!

The first line filters to a single row in the treatment data and then adjusts it into the format required by the distance algorithm. The third line creates a blank list, to which we'll append each of the distance calculations for this specific row of treatment data. This list will ultimately become a column that we append to the main dataset. The location of `distances_list` within this portion of the code is important. It will run with each new row of treatment data we focus on, meaning it starts from scratch each time we calculate the distances for a new row.

```
treatment = df_treatment.iloc[treatment_row].to_numpy().reshape(1, -1)
treatment_company = df_treatment_full.iloc[treatment_row, 0]
distances_list = list()
```

Listing 7.29 Filtering to Single Treatment Row

Listing 7.30 shows our second for loop. Here, we're using something called *nested for loops*, which allow us to compare multiple lists against each other (which is our use case). This is the same code structure as the treatment data, which allows us to filter to each of the rows in the control data.

```
for control_row in list(range(0, len(df_control))):
```

Listing 7.30 Second Matching For Loop

Listing 7.31 now shows the first operation within the second for loop:

1. The first line filters the control data in the same way as the treatment data.
2. The second line calculates the distance between each of the rows. The 0s ensure that only the number is extracted as the result from the calculation is an array.
3. The third line appends the distance to the `distances_list` we created.

```
control = df_control.iloc[control_row].to_numpy().reshape(1,-1)
distance = pairwise_distances(treatment, control, metric = 'euclidean')[0][0]
distances_list.append(distance)
```

Listing 7.31 Calculate Distance and Append to List

Listing 7.32 is where everything comes together. We're creating a dictionary with all the necessary data inputs (`Treatment Company`, `Control Company`, and `Distance`). We put this into its own temporary DataFrame that's then appended to the main `all_matches` DataFrame that we created at the very beginning of the code.

After all the control rows have been compared to the first row of the treatment data, the loop starts over again with the second row of the treatment data and compares each row to the control data again. This will continue to occur until all treatment rows have been compared against all control rows of the data.

```
data = {
  'Treatment Company': treatment_company * len(distances_list),
  'Control Company': df_control_full['Company'],
  'Distance': distances_list
  }
temp_df = pd.DataFrame(data)
all_matches = pd.concat([all_matches, temp_df], ignore_index = True)
```

Listing 7.32 Create Temporary DataFrame and Append to Main DataFrame

We now have a very long list of distances between our treatment and control data (see a sample in Table 7.4). The next step is to write the code to find the best 1-to-1 matches.

Treatment Company	Control Company	Distance
60	3	1.07
60	5	.40
60	6	1.03
60	7	1.15
60	8	.15

Table 7.4 Example of Matching Distance Output

While this book isn't about writing code like a programmer, thought exercises like these come up often. They require you to think about how to build a function that performs the operation you want without relying on copying and pasting code. Most of my predictive modeling projects have involved moments where I've had to think through problems like this one—most often in the data preparation step. My thought process for this specific problem is that we need to introduce some degree of randomization into the algorithm. When creating these matches, we want to assign a single match for each treatment group and to use each control company only once. This means we'll be in a situation where the order of execution matters!

To introduce the randomization, we'll use a library called random. This allows us to randomize the order of the treatment companies after we take their unique value in Listing 7.33.

```
#load random library
import random

#create list of the treament companies
treatment_companies = all_matches['Treatment Company'].unique().tolist()

#randomize the order of the treatment companies
random.shuffle(treatment_companies)
```

Listing 7.33 Randomize Order of Treatment Company List

In these situations, it's best to start small and build from there. First, we'll adjust the list we iterate over to only include two values for testing purposes, and we'll print out the results so that they are displayed. Listing 7.34 shows this process.

```
#test with subset of the companies
for treatment_company in treatment_companies[0:2]:
  temp_df = all_matches_for_filtering[
    all_matches_for_filtering['Treatment Company']==treatment_company
    ]
  min_distance_row = temp_df[temp_df['Distance'] == temp_df['Distance'].min()]
  print(min_distance_row)
  print(min_distance_row['Control Company'].iloc[0])
  print(min_distance_row['Distance'].iloc[0])
```

Listing 7.34 Initial Test of Identifying Best Matches

If you run this code, you'll notice that it's possible for the same control group to be matched to the treatment group. To account for this, we need to create a DataFrame of the matches that we can filter after each treatment group (see Listing 7.35).

```
#rename data frame
all_matches_for_filtering = all_matches

#run test with subset of companies
for treatment_company in treatment_companies[0:2]:
  temp_df = all_matches_for_filtering[
    all_matches_for_filtering['Treatment Company']==treatment_company
    ]
  min_distance_row = temp_df[temp_df['Distance'] == temp_df['Distance'].min()]
  print(min_distance_row)
  print(min_distance_row['Control Company'].iloc[0])
  print(min_distance_row['Distance'].iloc[0])
    all_matches_for_filtering = all_matches_for_filtering[
    all_matches_for_filtering['Control Company'] != min_distance_row['Control
    Company'].iloc[0]
    ]
```

Listing 7.35 Update to Filter Out the Control Group That Is Used After Each Iteration

Now we've removed the risk of the same control company matching to multiple treatment companies. As a next step, we'll want to ensure we can put our results into a DataFrame. To do this, we'll add the values to lists and append them as shown in Listing 7.36. This takes the code and puts it into a for loop. To ensure these values are evaluated, we'll create treatment_company_vals, control_company_vals, and distance_vals. This allows us to append each iteration's values into these lists, which will ultimately be put into a DataFrame for easy evaluation.

```
#create new df
all_matches_for_filtering = all_matches

#create blank lists
treatment_company_vals = []
control_company_vals = []
distance_vals = []

#test on subset of companies
for treatment_company in treatment_companies[0:2]:
  temp_df = all_matches_for_filtering[
    all_matches_for_filtering['Treatment Company']==treatment_company
    ]
  min_distance_row = temp_df[temp_df['Distance'] == temp_df['Distance'].min()]
  print(min_distance_row)
  print(min_distance_row['Control Company'].iloc[0])
  print(min_distance_row['Distance'].iloc[0])
```

```
treatment_company_vals.append(treatment_company)
control_company_vals.append(min_distance_row['Control Company'].iloc[0])
distance_vals.append(min_distance_row['Distance'].iloc[0])

all_matches_for_filtering = all_matches_for_filtering[
  all_matches_for_filtering['Control Company'] != min_distance_row['Control
  Company'].iloc[0]
  ]
```

Listing 7.36 Appending the Values to the Lists So They Can Be Appended Into a DataFrame

It's better to append these values to a list rather than create them as a row and then append them to a DataFrame because appending is much faster than adding rows to a DataFrame. To add the data into a DataFrame, we apply the code shown in Listing 7.37 outside of the for loop.

```
#add lists into dictionary
data = {
  'Treatment Company': treatment_company_vals,
  'Control Company': control_company_vals,
  'Distance': distance_vals
  }

#create a data frame from the dictionary
final_matched_df = pd.DataFrame(data)
```

Listing 7.37 Create DataFrame from 1-to-1 Matched Data

Now we can remove the limitation of the first 2 values of the list and run our process! You can remove the print statements in this step to keep your notebook from getting too long. Think for a moment whether there are any potential flaws or challenges in our approach. Is there anything that we could improve?

One specific component we could improve is the randomization. How do we know we're getting the absolute best matches if we're going in a specific order? To test this out, let's run the full code (starting with the randomization code) and add up the final distance. Table 7.5 shows the results after 10 runs.

Run Number	Sum of Distance
1	16.4
2	15.9
3	16.5

Table 7.5 Results of Running the Full 1-to-1 Matching Process

Run Number	Sum of Distance
4	16.3
5	16.7
6	15.9
7	16.3
8	15.5
9	15.9
10	15.2

Table 7.5 Results of Running the Full 1-to-1 Matching Process (Cont.)

While the results are all generally around 16, they vary quite a bit! To address this, we can add a final piece to our code. We will run this process 10 times and take the results with the lowest overall sum of distance. This will hedge against the risk of an unusually bad match if we only run the process once. To do this, let's take a brief moment for an introduction to writing functions, commonly referred to as *user-defined functions (UDFs)*.

User-Defined Functions

Writing your own functions is a major topic in Python programming books, often covered in a large section or even an entire chapter. However, for the purposes of this book, we'll attempt to explain them in a single note box.

The syntax for writing your own functions is simple: It begins with def, the keyword to identify you're writing a function, followed by the name of the function you want to create and parentheses that specify the information that can be passed into the function. Here's an example:

```
def name_of_function(parameter1, parameter2, parameter3):
    code to run
```

The names of the parameters you create are the *only* objects or variables the code within your function will understand. For example, if you define df earlier in your notebook and also name one of your parameters df, the function will only use the DataFrame that was passed in, even if the name is the same. If want to research this further, it's the concept of global versus local variables.

In the example shown in Listing 7.38, other_df will be printed because it's passed to the print_df function being created. This is because print_df takes a parameter called df, which is then printed per the function's code. Whichever DataFrame is provided in the parentheses when print_df is ultimately run is what gets printed (in this case, other_df).

```
df = pd.read_csv("file_location/file_name.csv")
other_df = pd.read_csv("file_location/other_file_name.csv")
def print_df(df):
  print(df)
print_df(other_df)
```

Listing 7.38 Read In and Print CSV Files

UDFs help you avoid code duplication and simplify your code. As in our next example, breaking up large chunks of code will make the code easier to read and edit in the future.

If you want to improve your code and write like a programmer, UDFs are a great topic to read up on and become comfortable leveraging in your code.

Luckily, with this example, the code is relatively easy to convert into a function. We only have one input, which is the long DataFrame of the distances between the treatment and control companies. This means the only change we need to make is generalizing how all_matches is defined by replacing it with df (see Listing 7.39).

```
#create function to match treatment and control together
def create_one_to_one_match(df):
  import random

  treatment_companies = df['Treatment Company'].unique().tolist()
  random.shuffle(treatment_companies)

  all_matches_for_filtering = df
  treatment_company_vals = []
  control_company_vals = []
  distance_vals = []

  for treatment_company in treatment_companies:
    temp_df = all_matches_for_filtering[
      all_matches_for_filtering['Treatment Company']==treatment_company
      ]
    min_distance_row = temp_df[temp_df['Distance'] == temp_df['Distance']
      .min()]
    #print(min_distance_row)
    #print(min_distance_row['Control Company'].iloc[0])
    #print(min_distance_row['Distance'].iloc[0])

    treatment_company_vals.append(treatment_company)
    control_company_vals.append(min_distance_row['Control Company'].iloc[0])
    distance_vals.append(min_distance_row['Distance'].iloc[0])
```

```
    all_matches_for_filtering = all_matches_for_filtering[
        all_matches_for_filtering['Control Company'] !=
          min_distance_row['Control Company'].iloc[0]
        ]

  data = {
    'Treatment Company': treatment_company_vals,
    'Control Company': control_company_vals,
    'Distance': distance_vals
    }

  final_matched_df = pd.DataFrame(data)

  return final_matched_df
```

Listing 7.39 Convert Code to UDF

Now that we have our function, we need to put it into additional code that executes 10 times, comparing the total distance across the dataset to retain the iteration with the lowest distance matched. By executing 10 times, we create 10 different scenarios through sampling, and we can pick the best match. If we only did this once, we risk producing a result that isn't a good representation of the data. The approach shown in Listing 7.40 is the complete approach; we'll then break it down to explain each component.

```
#create list of values from 1 to 10
iterations = list(range(1, 11, 1))

#create large number
winning_distance = 90000*90000

#create loop to run the matching function 10 times to find best matching result
for iter in iterations:
  match_iteration = create_one_to_one_match(all_matches)
  match_distance = sum(match_iteration['Distance'])
  if match_distance < winning_distance:
    winning_df = match_iteration
    print(match_distance, winning_distance)
    print("New best match score")
    winning_distance = match_distance
  else:
    print("Best match score remains")
```

Listing 7.40 Test for Best Match Over 10 Iterations

The first line creates what our for loop will iterate over. Python's range function isn't inclusive of the last number, meaning to get numbers 1 through 10, we need to provide 11 to the function, as follows:

```
iterations = list(range(1, 11, 1))
```

Our approach here is to pick a very high number. This will be the variable used to compare the total distance. Any time a new match is identified, the variable will be overwritten.

```
winning_distance = 90000*90000
```

Within the for loop, we can put our function into action. We're identifying the DataFrame from the function as match_iteration, which is then used to sum the total distance so we can compare DataFrames on this metric.

```
match_iteration = create_one_to_one_match(all_matches)
match_distance = sum(match_iteration['Distance'])
```

Lastly, we use an if statement to compare the scores against each other. The first matching DataFrame will always win by design, because we picked such a high number. As it goes through the 10 different matching iterations, it will execute any time a lower distance score is created.

```
winning_df = match_iteration
print(match_distance, winning_distance)
print("New best match score")
winning_distance = match_distance
```

The first component creates a single DataFrame called winning_df that we can reference when the code finishes running. The print statements help us follow what happens with the loop as it's executing. Finally, we need to overwrite the winning_distance number before the next iteration of the loop.

The last part of the code is the else, which in this case is just a print statement that lets us know that the code ran, but that iteration didn't have a lower distance score.

With all of this coding, it can be easy to lose track of *why* you're doing what you're doing. All of this is a means to determine whether the adjustment to the company's website experience led to a statistically significant difference in sales.

In order to measure for a statistically significant difference, the first step is to filter the full control DataFrame to include only the companies matched to the control group, as shown in Listing 7.41.

```
#filter to matched control data
df_control_matched = df_control_full[
  df_control_full['Company'].isin(winning_df['Control Company'].unique())
  ]
```

Listing 7.41 Filter Control to Only Matched Data

Before the t-test, we'll check the averages for both the treatment and control groups using the .mean() method, as shown in Listing 7.42.

```
#print mean of the treatment and control groups
print(df_control_matched['Purchases in Next Month'].mean())
print(df_treatment_full['Purchases in Next Month'].mean())
```

Listing 7.42 Treatment and Control Group Averages

The results show the control group has an average of 51 and the treatment group has an average of 66. This is a pretty large difference, and we can expect the difference to be significant when we run the t-test.

If you have a statistics background, you could make the argument for a paired t-test. For those without a statistics background, a paired t-test measures two related groups, where each group has a pair that is compared against the other. However, that approach could exacerbate any error and variance introduced by the matching process with this type of data. Having two additional similar populations was the goal of the matching process, so it's better to perform an independent t-test. The code is shown in Listing 7.43.

```
#load stats library
from scipy import stats

#run t test on data
t_statistic, p_value = stats.ttest_ind(
  df_control_matched['Purchases in Next Month'],
  df_treatment_full['Purchases in Next Month'])

#print the p-value
print("P-value:", p_value)
```

Listing 7.43 T-Test on Matched Control Data

The resulting p-value is essentially 0, indicating the difference is statistically significant.

While the matching process is not directly related to predictive modeling, it can help test the impact of your model, which is a significant benefit. It's also an incredibly useful tool to leverage when conducting other analysis and analytics work. Lastly, this may have given you a glimpse of how valuable libraries can be for reducing the amount of code you need to write. Much of the matching code we created likely exists in libraries that aren't available in the Anaconda environment we're using. Having to write your code in this way helps you gain an appreciation for how much time and energy libraries can save you!

Randomized Control

The randomized control experiment is considered the ideal approach for testing the effectiveness and impact of a change—but why?

Think back to the matching example in the previous section. What was the goal of the matching process? Using observed information about each company, the goal of the process was to create two groups that were as similar as possible. The limitation was that there are only so many data points you can gather with the matching approach.

When using the randomized control group approach, you don't have to rely on what you do or don't know when creating the groups. By randomly selecting the groups, as the size of the groups grow, the random nature of their selection makes them more similar. That's when you can compare the two groups. The irony is that, from a coding perspective, the randomized control approach is much simpler to execute.

As discussed earlier, the randomized control approach is less commonly used in the business setting than you'd expect. This is often due to the proactive effort required and concerns about perceived fairness. When you're building models on the job, timelines are often rushed, and people are hoping to just get a model in place. One of the first steps they skip is measuring whether the model is effective.

As the one building the model, it's easy to get trapped in the mindset of "the model is accurate, I've done my job." Resist this mindset! The data scientists who get the most adoption of their models are the ones who translate the model back into the business value it provides. A low area under the curve (AUC) or mean absolute error (MAE) doesn't prove business value on its own. Your stakeholders and leaders will appreciate you taking it a step further to show how the model provides value.

Another ironic point is that stakeholders and leaders are often the reason the randomized control approach isn't used. It's unfortunate because you both want the same thing: a way to measure the business value of the model, which translates into the impact on the business. Why is it the business leaders—rather than technical experts— who prevent a more robust measurement approach?

In my experience, much of this is an education gap and desire for speed. You need to implement the randomized control approach to measurement as soon as the project starts. This small window can be tricky to navigate, given that at the beginning of the project, trust and rapport have yet to be built. This is where the soft skills required in a data analytics role become critically important.

Sometimes, it's not worth pushing back if your stakeholders aren't willing to use this approach. This is why knowing how to use matching is so important, because it's a reliable secondary option once your model is built and being used. Hopefully, through the process of building your model, you'll develop a good working relationship with your stakeholders. After the project reaches a steady state, it's best to hold a retrospective meeting so you and your stakeholders can discuss what went well and what could be improved, all with the goal of learning from the project for the future. This would be an appropriate setting to bring up the use of a randomized control measurement. Team members and leaders are usually more open to ideas in these types of meetings. Plus, you'll have an entire project worth of examples to show them where the randomized control approach could have further solidified confidence in the measurement results.

7.4   Summary

This chapter tackled the final—but often overlooked—stage of the machine learning lifecycle: deployment, monitoring, and measurement. You saw that building a model is only half the battle. Getting it into production, keeping it stable, and ensuring it continues to deliver value are what make it truly useful. Here's a recap of what you've learned:

- You learned how to load a pickle file and generate a prediction. This is the easy part. The hard part is ensuring the prediction data undergoes the same steps as the data you used to train the model.

- You learned about the model monitoring process. On the surface, this is an intuitive concept. It makes sense to keep tabs on your model and what it's doing. However, you now have more practical considerations on what specifically you should monitor.

- Finally, you learned about ways to measure the impact of your model. This was split into the business sniff test (quick and easy) and experiments (longer and move involved). Both have their time and place, but there's often too little focus on creating experiments to measure the impact of models in the real world.

This brings us to the conclusion of the book in the next chapter. We'll look back on our journey and summarize the key lessons you can take with you.

Chapter 8
Closing Thoughts

This chapter summarizes the themes and topics covered throughout the book. If nothing else, I hope this serves as a reminder of how much you've learned. The content is wide ranging and complex, so kudos to you for working through it!

Now that you've completed the practical chapters of this book, it's worth stepping back to review everything you've learned. This final chapter wraps up the book with summaries of key concepts and questions to guide your continued learning.

8.1 Learning How to Learn with Generative AI

Generative AI is currently the key differentiator in combining learning, proficiency, and efficiency when building machine learning models. The pace at which generative AI is evolving and improving is incredible. Even throughout the course of writing this book, the generative AI landscape has continued to change, as has my opinion on it. If you want to be an effective machine learning model practitioner, it's *essential* that you use generative AI to help problem solve your code. Learning how to use generative AI will be critical when you're stuck on a problem. Rather than spinning your wheels for a day, you can use generative AI tools to work through it in a fraction of the time.

> **Real-World Example**
> I was bringing in National Oceanic and Atmospheric Administration (NOAA) hurricane data. The way NOAA stored the data was very unique and not something I had worked with before. I tried thinking through the problem myself for a few minutes but then brought it to Copilot. After about five minutes of prompting, I was able to get a fully functioning solution. I suspect getting the same result on my own would have taken me around two hours. The time savings are undeniable.

A word of caution, though: Take full advantage of the explanation of the code that your generative AI tool provides. Don't become a mindless zombie who just copies and pastes code from a generative AI tool and immediately runs it, as mistakes will show up quickly. Use the provided solutions as an opportunity to learn about the approaches and frameworks of how the code operates. Generative AI tools can almost function as a coding mentor when you're using them this way.

Here are some key takeaways to think about:

- How can you integrate generative AI into your coding practices?
- Consider setting a time limit for how long you'll think about a problem before bringing it to generative AI. For example, if you can't figure out a coding problem after 15 minutes, you'll input it into a generative AI tool to avoid it becoming a multiple-day problem.

8.2   Learning How to Learn with Use Cases

This is the nontechnical component. You must learn how to understand the business you're working with. While this book is primarily technical, it's critically important that you also know how to learn about use cases. While there is no magical answer to this (if there was, management consultants probably wouldn't get paid as much as they do), the common thread I've experienced both first- and secondhand is the need to be curious. If you're actually interested in the business and the use case, you'll find the quality of your questions increase.

Don't feel like the questions you ask need to have a direct tie to your use case. Getting a holistic understanding of your business mean you'll have more information to draw connections to your use case. When you're able to connect the dots on behalf of your stakeholder, your value to them significantly increases. While it's simple advice, curiosity is the closest thing you can get to a magic solution when it comes to understanding a use case.

Here is the key takeaway to think about: What questions can you ask of your stakeholders to better understand their business?

8.3   Explore and Visualize Your Data

It can be tempting to jump into the modeling part of the process; however, before you start cleaning the data, take the time to be an analyst and explore it. This book showed you some techniques for this, though it's by no means a comprehensive list of options. Your objective should be to get ahead of any gremlins so you can clean the data and understand potential patterns before building the models.

You don't need to be a statistician, but the more you think with a mindset of distributions and statistical patterns, the easier it will be to make downstream decisions in the modeling process, such as which error metric to select.

Another key aspect is visualizing your data. Humans are inherently visual beings—that's just how our brains work. Learning how to visualize your data as part of both the exploration and the explanation stages of your model will have high returns. If nothing else, from a purely practical standpoint, not having to save your data out to a comma-separated values (CSV) file to put into a PivotTable or chart will save you a lot of time.

Here is a key takeaway to think about: Is there a consistent approach you can develop to execute on any new dataset you use? This can help you be grounded in the data.

Further Resources

If you're looking to upskill in the statistics space, consider Khan Academy as a free resource: *http://s-prs.co/v617003*.

If you're interested in learning more about data visualization, check out IBM's Coursera course that covers a breadth of these topics in 20 hours: *http://s-prs.co/v617004*.

8.4   Cleaning Your Data and Dummy Coding

You cannot successfully build models if you don't know how to effectively clean data. Exploring and visualizing the data is a big step toward cleaning it, but you also need to know how to execute it. As you gain experience, you'll develop more tools to clean your data. You'll begin to see patterns and trends in the data that needs cleaning, which allows you to reuse techniques you've used before to save time.

While most improvement in data cleaning comes with experience, the other essential component is getting better at Python. The more you learn in Python, the more tools you'll have for new situations you come across.

A particular point within data cleaning to highlight is dummy coding. So much of the world's data is in text form. As you learned, the actual dummy coding itself isn't hard, but it's critically important to consider the cardinality of a column to avoid creating high dimensionality when dummy coding. This goes back to understanding your business case as well as the exploration of your data.

Here is a key takeaway to think about: Where you're working today, seek out opportunities to build models to gain more experience. Most organizations have data, so find a way to position this as a development opportunity with your boss and organization.

Further Resources

If you're looking to get more Python learning and experience, DataCamp (*https://www.datacamp.com/*) is my favorite platform. They combine text, video, and hands-on learning into their modules.

8.5   Machine Learning Models

You learned about a handful of different models and how to use them. Ultimately, you saw the best model was the Gradient Boosting Machine (GBM). I've found this to be the case in my experience as well. While this book focused on the packages within scikit-learn (sklearn), there are other machine learning packages that carry different benefits. From a learning perspective, staying within sklearn enables you to focus on foundational concepts; in practice, however, other models are also available.

Here is a key takeaway to think about: There are many different machine learning models. Knowing how to learn them is nearly as important as understanding each model in depth.

Further Resources

XGBoost (*http://s-prs.co/v617005*) is one of the most popular GBM models in practice, so learning how it works will benefit you.

8.6   Hyperparameters and Grid Search

Learning how to edit and adjust your model's hyperparameters means understanding how your model works at a base level. Because each model is different, the hyperparameters for models differ as well. While it'd be awesome if there was a process that tells you the optimal answer every time, it's not realistic, given the computational expense of running an infinite number of unique combinations. The next best thing is using grid search, which allows you to select a handful of hyperparameter options for the model to test.

Here is a key takeaway to think about: While understanding all details of how the model operates isn't necessary, further enhancing your knowledge and experience with each of the hyperparameters helps you develop a sense of where to start and what to expect when you start building a model.

Further Resources

DataCamp has a 4-hour course on hyperparameter tuning in Python, which is a good resource to learn about techniques that go beyond the scope of this book. You can find at the following URL: *http://s-prs.co/v617006*.

8.7   Variable Lagging

Machine learning models don't have the ability to see historical context; each row is independent from another. The only way to give a model true historical context is through variable lagging. This is a critical component to learn, especially if you're using a machine learning model for time series predictions. Luckily, the basics of this can be done using either `shift` or `rolling`.

Here is a key takeaway to think about: Use variable lagging in all your models. The more you use this technique, the more comfortable you'll be with it. Plus, it almost always makes you model better because you're giving it access to data it didn't previously have.

8

8.8   The End

I want to thank you for reading this book, and I hope you found it valuable in your learning journey. I've had a number of individuals help me in my own journey, so my goal in sharing my experiences in this space is to pass along both my knowledge and my approach to learning a technical topic like building machine learning models.

Feel free to connect with me on LinkedIn (*https://www.linkedin.com/in/jason-hodson-21a19ab7/*) as well as interact with the book's GitHub page (*https://github.com/jason-hodson/applied_machine_learning*). I would love to hear from you about your experience with this book. I welcome any feedback about what worked well and what didn't!

8.9   Acknowledgments

No book is possible without the help of others. Many helped, but I want to specifically discuss a handful of individuals who made a notable impact on the output.

First, I want to thank Maysoon Shafi. You brought the opportunity for me to work on this project at exactly the right time. Having just been told I likely won't have a job in the next few months, I found that writing this book gave me the outlet I needed to distract myself from the stress that comes with interviewing for a new job. While it may have seemed like an inconsequential gesture for you, it had an outsized impact on me.

Next, Reba Koenen. You singlehandedly increased the quality of this book by multiple levels. You spent hours reviewing an initial draft, combing through the sections and giving your feedback to ensure the content was relevant, engaging, and clear. I cannot thank you enough for taking the time to do this review. By far my favorite running joke with our reviews was my consistent incorrect use of "it's" and "its."

Lastly, I want to thank Maureen Kalas, Greg Russell, and Alex Kalluf. You all contributed to the book in a meaningful way—sometimes it was giving me a story, sometimes it was brainstorming a topic with me, and sometimes it was simply mentioning a topic that became an entire section of the book.

The Author

Jason Hodson has worked in data-centric roles for nearly a decade. He currently works as an HR analytics manager, and he has prior experience in a forecasting role using the full range of applied machine learning. In a previous role, Jason wrote the end-to-end code for an enterprise hiring manager and candidate experience process, collaborating with recruiting leaders to understand and leverage data from a company-wide survey. He's built large data models and dashboards and taught nontechnical users how to adopt and use them. Jason has been a technical mentor in all his roles, helping others develop their analytics and programming skill set. The common thread across Jason's career is his ability to be a translator for stakeholders, peers, and junior team members. His learning journey also gives him a unique perspective: Before earning a master's degree in business analytics, he was entirely self-taught. This has made his approach to teaching more practical, allowing concepts to translate better (and faster) into the business world.

Index